ALSO BY STEPHEN BUDIANSKY

Biography

Mad Music: Charles Ives, the Nostalgic Rebel

Her Majesty's Spymaster

History

Blackett's War

Perilous Fight

The Bloody Shirt

Air Power

Battle of Wits

Natural History

The Character of Cats

The Truth About Dogs

The Nature of Horses

If a Lion Could Talk

Nature's Keepers

The Covenant of the Wild

Fiction

Murder, by the Book

For Children

The World According to Horses

CODE WARRIORS

NSA's Codebreakers and the Secret Intelligence War Against the Soviet Union

Stephen Budiansky

Alfred A. Knopf · New York · 2017

THIS IS A BORZOI BOOK PUBLISHED BY ALFRED A. KNOPF

www.aaknopf.com

Knopf, Borzoi Books, and the colophon are registered trademarks of Penguin Random House LLC.

Library of Congress Cataloging-in-Publication Data
Names: Budiansky, Stephen, author.
Title: Code warriors : NSA's codebreakers and the secret intelligence war against the Soviet Union / by Stephen Budiansky.
Description: New York : Alfred A. Knopf, 2016. | Includes bibliographical references and index.
Identifiers: LCCN 2015045330
| ISBN 978-0-385-35266-6 (hardcover) | ISBN 978-0-385-35267-3 (ebook)
Subjects: LCSH: United States. National Security Agency—History. | Cryptography—United States—History | United States—Foreign relations—Soviet Union. | Soviet Union—Foreign relations—United States.
Classification: LCC UB256.U6 B83 2016 | DDC 327.7304709/045—dc23 LC record available at http://lccn.loc.gov/2015045330.

Jacket image: Russian M-125 Fialka Cipher Machine (detail). Courtesy of Glenn Henry, Glenn's Computer Museum.
Jacket design by Chip Kidd
Maps by Dave Merrill

Manufactured in the United States of America
Published June 6, 2016
Second Printing, January 2017

TO DAVID KAHN,

who led the way

Contents

Abbreviations

AEC	Atomic Energy Commission
AFSA	Armed Forces Security Agency
ASA	Army Security Agency
ASAPAC	Army Security Agency, Pacific
COMINT	communications intelligence
CSAW	Communications Supplementary Activities, Washington (U.S. Navy)
DF	direction finding
ELINT	electronic intelligence
FISC	Foreign Intelligence Surveillance Court
GC&CS	Government Code and Cypher School (UK)
GCHQ	Government Communications Headquarters (UK)
GRU	Main Intelligence Directorate (Soviet military intelligence)
HF	high frequency
HUAC	House Un-American Activities Committee
KGB	Soviet Committee of State Security (1954–91)
MGB	Soviet Ministry of State Security (1946–54; predecessor to KGB)
MI5	UK counterintelligence
MVD	Soviet Ministry of Internal Affairs (1946–91)
NKGB	People's Commissariat for State Security (Soviet foreign security service, 1943–46; predecessor to MGB and KGB)
NKVD	People's Commissariat for Internal Affairs (Soviet internal security service, predecessor to MVD; incorporated State Security, 1934–43)
NSC	National Security Council
NSG	Naval Security Group
ONI	Office of Naval Intelligence
Op-20-G	Office of Naval Communications cryptanalytic section
OSS	Office of Strategic Services
PFIAB	President's Foreign Intelligence Advisory Board

RAM	rapid analytical machinery
SAC	Strategic Air Command
SCAMP	Special Cryptologic Advisory Math Panel
SIGINT	signals intelligence
SIS	Secret Intelligence Service (UK)
TICOM	Target Intelligence Committee (Allied project to capture German cryptologists and cryptologic material at end of World War II)
USAFSS	U.S. Air Force Security Service
USCIB	U.S. Communications Intelligence Board
USIB	U.S. Intelligence Board

Author's Note

In May 2013, a twenty-nine-year-old computer security expert who had worked for three months as a $200,000-a-year contractor for the National Security Agency in Hawaii told his employer he needed to take a leave of absence for "a couple of weeks" to receive treatment for the epileptic condition he had recently been diagnosed with. On May 20, Edward J. Snowden boarded a flight to Hong Kong, carrying with him computer drives to which he had surreptitiously copied thousands of classified intelligence documents. Their contents, revealing copious details about NSA's domestic surveillance of telephone and e-mail communications, would begin appearing two weeks later in a series of sensational articles in the *Guardian* and the *Washington Post.*

It was a move he had been secretly preparing for some time, having secured the job with the specific aim of gaining access to classified NSA material. (He was ultimately able to do so only by duping more than twenty coworkers into giving him their computer passwords, which he said he needed for his duties as a systems administrator; most of the colleagues whom he betrayed were subsequently fired.) Snowden would later explain that he chose Hong Kong as his place of intended sanctuary because "they have a spirited commitment to free speech and the right of political dissent"—an assertion that would have come as a surprise to members of the city's pro-democracy movement, whose peaceful mass protests the following year would be efficiently crushed by the Hong Kong authorities at the behest of their Chinese Communist Party masters in Beijing.[1]

A month later, his U.S. passport canceled and under indictment for theft of government property and violations of the Espionage Act, Snowden fled to Moscow. There the government of President Vladimir Putin, a former lieutenant colonel of the Soviet KGB whose increasingly dictatorial control of the media, ruthless suppression of political opposition, and chest-thumping nationalist bellicosity was reviving the worst memories of the Cold War superpower confrontation, soon granted Snowden asylum, then temporary residency in Russia.

Snowden's political naïveté was honestly come by: a self-taught computer whiz who never finished high school, a supporter of the quixotic campaign of the libertarian presidential candidate Ron Paul, he was given to sweeping, conspiratorial pronouncements about his duty to expose "the federation of secret law, unequal pardon, and irresistible executive power that rule the world."[2] Whatever his motives, there was no denying the impact of his revelations concerning NSA's surveillance programs, particularly those involving the unauthorized monitoring of American citizens. No single incident in NSA's sixty-one-year history came close to bringing so many of its most secret activities into the harsh glare of public scrutiny or so shook public confidence in the agency's mission.

Three of the programs in particular seemed to epitomize a secret agency out of control, venturing well beyond the bounds of legitimate foreign intelligence gathering. The architects of the post–World War II permanent intelligence establishment, recognizing the fundamental incompatibility of deeply ingrained American beliefs in open government, liberty, and privacy with the tools of the shadowy intelligence trade—there had been much indignant talk about an "American Gestapo" when reports surfaced toward the end of the war that FDR was considering a plan to preserve the Office of Strategic Services (OSS), the forerunner of the Central Intelligence Agency, into the postwar period—sought to draw a sharp line that would resolve the dilemma. CIA and NSA would be strictly limited to foreign intelligence targets. Abroad, anything might go; it was after all a dangerous world, and the United States, having been wrenched from its long dream of isolationism, was determined never to be caught by another Pearl Harbor. But at home, the rule of law and American values would be maintained as always; a man's home would still be his castle, and a warrant issued by a court would be required to search his personal effects or spy on his conversations.

Yet inevitably there were gray areas, and now it seemed that in the aftermath of the September 11, 2001, terrorist attacks on New York and Washington by the Islamic fundamentalist group al-Qaeda, the gray areas had spread across the entire horizon. One of the NSA programs

Snowden revealed, the Bulk Telephony Metadata Program, employed secret orders issued to American telephone companies to obtain records of the duration and number dialed of *every* call made in the United States, and stored this information—billions of records, amassing five years' worth of calling data—in vast data warehouses where they could be searched by NSA analysts.

The second, Prism, was an even more comprehensive dragnet; it collected from major Internet servers the contents of the billions of e-mails, Web videos, voice-over-Internet phone calls, and other data that passed through the United States to other countries. (A related "upstream collection" program directly tapped undersea fiber-optic cables to intercept the same kinds of Internet traffic.)[3]

A third program, the SIGINT Enabling Project—SIGINT stood for signals intelligence, NSA's core mission—provoked outrage not only from civil libertarians but also from the high-tech computer and cybersecurity sector, which had enjoyed a close working relationship with NSA dating back to the very beginnings of the computer age in the postwar years. It was an industry upon which the agency depended more than ever as a source of expertise in the age of the Internet. Yet it turned out that NSA was at the same time undermining the industry's commercial products by devising ways to insert hidden vulnerabilities that rendered otherwise unbreakable public encryption systems "exploitable." According to a comprehensive description in one document leaked by Snowden, the methods NSA used included covert hacking of devices and networks as well as "investing in corporate partnerships" to ensure that cryptologic weaknesses, known only to NSA, were built into the products.

For years there had been conspiratorial whispers about the agency's secret sabotaging of advanced digital encryption schemes via "back doors"; now it seemed that the conspiracy theorists had if anything underestimated the reality. Many in Silicon Valley took it as a personal betrayal: they felt they had been duped, and, more to the point, it was a terrible commercial blow to American businesses that offered computer security products now known to have been deliberately compromised.

Worse, it was a reckless and dangerous policy; as one computer expert told a White House review group, anything that makes it easier for NSA to spy on computer systems and the Internet "also inevitably makes it easier for criminals, terrorists, and foreign powers to infiltrate these systems for their own purposes."[4] (Even a system designed to allow law enforcement and intelligence agencies access to secure communications under a legally regulated framework would be the equivalent of leaving a key under the doormat, warned another high-level group of academic cryptologic experts; any such stored master key would itself become a target for hacking by Chinese government agencies, Russian organized crime syndicates, and others who had repeatedly demonstrated their skill in raiding U.S. government computer systems.)[5]

Six months later a federal district court found the bulk collection programs an unconstitutional violation of the Fourth Amendment's privacy protections, calling particular attention to their "Orwellian" sweep. NSA was exploiting a vast loophole in Section 215 of the post–9/11 Patriot Act—which permitted the government to demand business records that were "relevant" to an investigation—to indiscriminately sweep up vast amounts of U.S. citizens' data without a warrant, mining it for evidence of contact with foreign targets, then declaring it "relevant" after the fact when it did find such evidence. Such a procedure stood completely on its head a hallowed prohibition, enshrined in centuries of English common law and the U.S. Constitution, against exactly this sort of dragnet-like "general warrant."

Earlier secret court orders, subsequently declassified by President Obama's administration in an effort to respond to the Snowden revelations with a demonstration of commitment to "transparency," revealed that the special federal court charged with overseeing NSA's foreign-intelligence-gathering activities, the Foreign Intelligence Surveillance Court—itself the product of reforms enacted in the wake of revelations of NSA's surveillance of American citizens during the 1960s and 1970s—had repeatedly chastised NSA for "substantial misrepresentations" of its bulk collection practices and for overstepping court-mandated rules to "minimize" the collection of U.S. citizens' data intercepted in the course of monitoring a legitimate foreign target.[6]

These initial disclosures undeniably raised important questions about domestic surveillance policy, legality, and morality. The FISC itself acknowledged that Snowden's "unauthorized disclosure" of one of the court's rulings had "engendered considerable public interest and debate," and agreed that authorized declassification of additional rulings would similarly "contribute to an informed debate." In May 2015, the U.S. House of Representatives voted 338–88 to end NSA's bulk metadata collection program, and subsequently passed a Senate bill explicitly phasing out any such programs under Section 215, an outcome that would have been hard to imagine absent the public debate Snowden set off.[7]

But the classified information that continued to dribble out at well-timed intervals from Snowden soon crossed the line from defensible whistle-blowing to reckless exposure of ongoing foreign intelligence operations. Snowden—and even more so his chief journalistic collaborator, Glenn Greenwald of the British newspaper the *Guardian*—saw the world in simplistic terms: one was either a tool of the "establishment . . . elite," sycophantically "venerating" and meekly obedient to "institutional authority," or one daringly engaged in "radical dissent from it," Greenwald asserted. Among the laundry list of disclosures Greenwald reported were NSA's monitoring of radio transmissions of armed Taliban militants in northwest Pakistan, the bugging of twenty-four embassies (all identified by name), and technical specifics of listening devices used to intercept fax messages of foreign diplomats. Snowden and Greenwald presented these all as equally shocking evidence of the rise of a "menacing surveillance state" that was threatening to bring about the end of "privacy," "internet freedom," and "intellectual exploration and creativity" throughout the world.[8]

There was an astounding historical and moral blindness in lumping all these together. No one familiar with the diplomatic, military, and intelligence history of the preceding half century or more would have seen anything even worthy of surprise, much less anything illegal or immoral, in U.S. efforts to intercept and decipher the communications of foreign governments and military organizations using any means possible. Nor could they have so recklessly doubted the essential importance of foreign signals intelligence in safeguarding national security during that fraught chapter of world conflict.

By the same token, there was an element of remarkable historical obtuseness in the public position that NSA and its supporters took in response to the disclosures: U.S. intelligence officials denounced Snowden as a mere "traitor," insisted that criticisms of the agency were based only on "gross mischaracterizations" of its activities, asserted that grave damage had been done by *all* of Snowden's disclosures, and flagrantly exaggerated the effectiveness of the dragnet collection programs.[9] After first insisting that NSA's post–9/11 bulk surveillance efforts had foiled fifty-four terrorist plots, NSA officials were forced to revise that claim to thirteen, then "one or two"—then zero.

The agency's reputation suffered even more when it became clear that the indignant denials made by NSA officials to Congress about the extent of its monitoring of Internet and telephone communications employed tortuous language to imply the exact *opposite* of the truth. NSA's director, General Keith Alexander, repeatedly insisted, for example, that the agency "does not collect information under section 215 of the Patriot Act" on the location of cell phone calls placed in the United States; he omitted to mention that NSA *was* collecting such information, on a massive scale, but relying upon another legal theory for its authority to do so.

NSA's director and deputy director likewise repeatedly asserted that NSA was not "targeting" the communications of U.S. nationals "anywhere on earth" without a warrant; it was, however, "incidentally" sweeping up vast amounts of such traffic in its warrantless bulk collection programs and storing it in huge data warehouses for later analysis as needed.[10]

The roots of the crisis NSA faced as a result of Snowden's disclosures reached deep into its own history. The bureaucratic imperatives, habits of mind, and institutional culture that drove the agency to engage in such a breathtakingly comprehensive technological intrusion into private communications, despite manifest questions about its legality and even practical value, and then to instinctively and ineptly attempt to cover it up when details became public, characterized NSA from its very beginnings. Going back even to NSA's predecessor agencies in the 1930s and 1940s, there had *always* been a sense that the mission was to collect literally everything possible in the way of signals intelligence, about friends and foes alike, even as the resulting flood of incoming data routinely overwhelmed any ability to analyze the haul. There had *always* been an obsessive pursuit of technical proficiency that pushed to the side any sober weighing of actual intelligence requirements, which often resulted in vast efforts expended on marginally important sources at the cost of huge human and diplomatic risks. And there had *always* been an impulse to push to the very limits claims of legal authority under national security necessity and presidential prerogative. NSA inherited those same institutional values from its predecessors (along with many of their top officials) when it was established in 1952.

No habit of mind was more deeply ingrained among signals intelligence professionals than the absolute prohibition on even hinting at the government's capabilities at intercepting signals and breaking codes; given that even the most trivial disclosure might alert a target to take protective countermeasures, undoing years of painstaking effort and resulting in the loss of a vital intelligence source, NSA for years tried even to keep its very existence a secret. The principle behind the agency's cult of silence was valid enough as a general proposition but was also self-serving in the extreme in the face of public questions now being raised about the fundamental legality and propriety of NSA's activities—as well as simply unrealistic at a time when everyone knew not only what NSA was but what it did, and cryptography itself had become a public commodity, no longer the purview of great governments alone. It also ironically deprived the agency of just what it needed most at the moment of Snowden's disclosures: to be able to explain to the American public the value and success of its legitimate mission.

Nothing had so solidified NSA's reflexive secrecy and institutional tendencies to keep repeating the same mistakes as the bizarre looking-glass world of the Cold War, the unprecedented four-decade-long peacetime confrontation that began with the swift crumbling of the Grand Alliance of World War II after the briefest period of hope of a safer and saner world following the defeat of Nazi Germany in 1945, and ended with the astonishing fall of the Berlin Wall in 1989. The Cold War imperatives of secrecy and "deniability" and its never-ending technological arms race, which encompassed the world of intelligence just as it did the battlefield, only reinforced the sense of impunity and the conviction on the part of NSA officials that no one but themselves could be trusted to know, much less understand or judge, their activities; in a manner all too familiar among political zealots, religious sects, academic departments, and other cults and closed societies, they became convinced of their own virtue.

More deeply, the Cold War clouded moral choices and removed traditional democratic checks on government. Nothing had really prepared the United States to embrace as a permanent necessity the kinds of morally dubious actions that had been accepted hitherto only as the exigencies of war, emergency measures in a kill-or-be-killed fight that—it seemed obvious to nearly all Americans—would be promptly dismantled as soon as the victory was won and peace and normalcy returned. Democracies never could and never would really figure out how to reconcile their democratic values with the employment of methods that were, after all, the hallmarks of totalitarian and police states. Stalin's Soviet Union used surveillance, deception, secrecy, betrayal, official lies, and the permanent militarization of society as its tools of power; to Americans those were the very evils they were fighting against in the struggle to contain the USSR and the expansion of world Communism. The inherent contradictions and almost insane logical disconnects of the Cold War nuclear standoff, in which a constant war footing became the only way to prevent a war that would have meant the end of human civilization were it actually to occur, ineluctably shaped the institutions that were at the forefront of this shadowy fight, none more than NSA.

NSA continues to this day to be extremely chary of revealing any details of its successes against Soviet cryptology during the Cold War, even though nearly all of the systems it targeted then have long been rendered obsolete by the digital age. Yet throughout the Cold War, signals intelligence would be the primary—often the sole—source of intelligence about Soviet intentions and capabilities; its military, economy, and industry; and the myriad technical specifications of its bomber and missile force and air defenses that the Dr. Strangelovean strategists of Armageddon needed to know in order to constantly recalibrate the credibility of America's nuclear deterrent and maintain a twenty-four-hour-a-day vigil for signs that the Soviets were about to strike first.

The history-changing significance of "Ultra" and "Magic" intelligence in World War II, the Allies' breaking of the Nazi Enigma cipher and Japanese high-level diplomatic and military code systems, is now common knowledge. "SIGINT successes during the Cold War," one former NSA director has acknowledged, "were no less significant in terms of gravity and magnitude for changing world history and for protecting the interests of the United States, our allies, and democracy."[11] Those successes were likewise the product of an intellectual and technical triumph of mathematics, linguists, and engineers whose stories of inspiration, struggle, and insight equal those of their famed World War II predecessors.

Despite NSA's best efforts to the contrary, it is possible to piece together significant strands of the story of its work in the Cold War from an array of declassified, if highly fragmentary, sources. (I am particularly grateful to Rene Stein of the National Cryptologic Museum Library for doing her utmost, under often impossible circumstances, to help me locate useful materials.) The capabilities and institutional traditions forged during this long chapter of modern intelligence and military history explain much that is both admirable and dysfunctional about NSA today. At its best, the agency galvanized innovation in computing and higher mathematics, delivered vital intelligence on foreign threats available in no other ways, and deftly and brilliantly targeted key sources with technical and espionage wizardry. At its worst, it obsessively pursued the unattainably grandiose scheme of collecting literally every signal on earth, undermined communications and Internet security for everyone, evaded legal oversight, and became the victim of its own secrecy with an unchecked culture of impunity, obfuscation, and byzantine bureaucratic politics. Reconciling the inherently clandestine and often dirty business of intelligence with the principles and ideals of an open democratic society will never be completely possible. But understanding how these things came to be is one place to start.

The human and political story of NSA in this era is inseparable from the technical story of codes and codebreaking, an undeniably fraught subject for the layman. It is impossible to understand the intellectual challenges and triumphs of the men and women who carried out this work, or even the most basic decisions they made, without an appreciation of the scientific problems and goals that drove them. I have tried to capture the essential contours of the technical story and give a sense of what the codebreakers were up against without assuming any specialized knowledge of cryptology or mathematics on the part of the reader; for those who are interested, more detailed technical explanations can be found in the appendixes.

CODE WARRIORS

Prologue

"A Catalogue of Disasters"

In the late afternoon of October 31, 1949, two young Latvian émigrés came aboard a fast motorboat in the West German harbor of Kiel and quickly disappeared belowdecks as the boat headed quietly out into the darkening waters of the Baltic Sea. At the helm was a former naval officer of the Third Reich, Kapitänleutnant Hans-Helmut Klose. The vessel under his command, though it now belonged to the British Baltic Fishery Protection Service and flew the Royal Navy's White Ensign, was one he knew thoroughly. The *S208* was a captured German E-boat, the type of fast torpedo boat that Klose had captained while serving in Hitler's navy. Overhauled in Portsmouth after the war, stripped of its torpedo tubes and armament and fitted with silent underwater exhausts, the *S208* was capable of a blazing forty-five knots.

Klose also knew the waters of his destination intimately, for one of his assignments during the war had been to drop Nazi agents behind Russian lines in the Baltic territories. Now he was doing exactly the same job for his new masters, the British Secret Intelligence Service (SIS).[1]

Neither the British spymasters nor their counterparts in the newly formed American CIA were very particular about the war records of the men they recruited to try to penetrate Stalin's Russia in the opening years of the Cold War. Vitolds Berkis, the younger of Klose's two passengers, was the thirty-one-year-old son of a former Latvian diplomat. He had been a more-than-willing Nazi collaborator during the war, joining an SS intelligence unit in Riga before fleeing the country in 1944 as the Russian armies rolled west. Western intelligence officials, milking émigré organizations and combing displaced persons camps for zealous—or desperate—anti-Communist nationalists who could be induced to return to their Soviet-occupied homelands, plucked Berkis out of an internment camp in Belgium. His fellow recruit had an even more unsavory past. Andrei Galdins had served the entire war in an SS execution squad of Latvian collaborators that was responsible for murdering half of Latvia's Jews.

Brought to London, the men had received six months' intensive training in tradecraft while living in a comfortable four-story Victorian house, complete with a cook and housekeeper, in the exclusive area of Chelsea, and provided £5 a week pocket money. They were drilled in Morse code and radio and cipher procedures, the use of invisible ink, arranging letter drops, shaking surveillance, resisting interrogation. On excursions to Portsmouth harbor, and later into the wilds of Dartmoor and the Scottish Highlands, they practiced firing small arms and machine guns, hand-to-hand combat, rowing and swimming silently through the ocean at night, surviving off the land, circling past villages without being detected.

As they boarded the *S208,* each carried a large brown suitcase. In one was a radio set; in the other two pistols, two submachine guns, ammunition, 2,000 rubles, codebooks, false passports; around their waists they wore money belts crammed with rolls of gold coins.

Landing that night by rowboat on an isolated beach west of Ventspils, about one hundred miles from the Latvian capital of Riga, the men got ashore unseen and made their way to the home of a priest who, according to SIS contacts in the émigré community, was a trusted link to partisans already organizing against the Soviet occupiers. The men explained that they were the first of a wave of British-trained agents on the way, and by the next day they were ensconced in a safe house for the winter with promises that they would be able to make contact with partisans in the forest when the weather improved and it was safer.[2]

Encouraged by the reports Berkis and Galdins sent back to London, their SIS handlers continued to dispatch new agents at roughly six-month intervals. Some of the earliest agents to make their way into the Baltic states, who had landed from Sweden at the end of the war, were smuggled out on return trips of *S208.* Brought to London for debriefing and further training, they added their confirmation of the success of the operation. Eager to get into the game, flush with cash, and chafing at occasional hints of British condescension toward them as inexperienced newcomers, CIA quickly organized its own penetration route, in 1949 and 1950 dropping agents by parachute into Ukraine and Lithuania and laying ambitious plans for six or more drops a year. An unmarked C-47 transport plane, piloted by two Czech flyers who had served in the RAF in the war, would take off from the U.S. Air Force base in Wiesbaden, Germany, skimming the treetops at two hundred feet to evade radar as it crossed the Soviet frontier, then at the last minute popping up to five hundred feet—the minimum safe altitude for a parachute jump—just over the drop site.[3]

In their training, the émigrés had been told to focus their efforts on obtaining information on Soviet military installations and airfields, above all any evidence of developments relating to rocketry and atomic weapons. Yet the actual reports coming back were oddly thin, little more than what was available in newspaper articles and other published sources. When occasionally pressed by their handlers to provide better information, the men brushed it off, indignantly insisting that they were not mere spies but "freedom fighters" dedicated to liberating their homelands; their mission was to "overthrow the Communists," not gather crumbs of intelligence for the British and Americans. And so the flow of agents, radios, weapons, and money continued.[4]

Skeptics in Washington and London were just as dismissively brushed off. Determined to prove that the wartime derring-do of SIS and the American OSS was alive and well and relevant in the postwar world, the heads of operations of the British and American intelligence services refused to see the obvious. Stewart Menzies, head of SIS since 1939, was a throwback to the romantic days of cloak-and-dagger espionage, a spymaster from central casting. He was charming and aristocratic and slightly mysterious, had been a star athlete as a boy on the playing fields of Eton, belonged to all the right London clubs, rode to hounds with the Duke of Beaufort's fox hunt, never missed the Ascot races, and was the principal source of the untrue but widely believed rumor that he was the illegitimate son of King Edward VII. He had acceded to the post of "C" (the traditional designation of the head of the British secret services; it was the actual initial of SIS's first chief, George Mansfield Smith-Cumming, who signed his memorandums that way in green ink) by beating out two rival candidates; the clinching factor was his producing a sealed envelope containing a letter purportedly written by his predecessor, Admiral Sir Hugh "Quex" Sinclair, who had died in office, endorsing him for the job. As Harold Adrian Russell "Kim" Philby, the master Russian mole who duped Menzies and a generation of SIS colleagues, would later sarcastically observe of his chief, "His intellectual equipment was unimpressive, his knowledge of the world, and views about it, were just what one would expect from a fairly cloistered man of the upper levels of the British establishment." Ironically, Menzies owed most of his internal prestige to having presided over Bletchley Park's brilliant codebreakers and their astonishing intelligence coups during the war—a new breed of technocratic spies who represented the antithesis of the old world spycraft he embodied.[5]

On the American side, Allen Dulles, appointed in 1950 to take charge of the new CIA's operations as deputy director for plans, was the former OSS station chief in Switzerland during the war and something of a romantic himself, enthralled by the image of parachuting in an army of spies to bring down the Communist enemy. By April 1951 it was becoming unmistakably evident that the entire operation had been blown, that the émigré groups were thoroughly penetrated by Soviet intelligence and had been for years, that the agents sent in one after another had in fact all been either killed or turned immediately upon their arrival. CIA director Walter Bedell Smith, an Army general who had served as General Dwight D. Eisenhower's chief of staff in Europe, and then as ambassador to the Soviet Union after the war, dispatched his old Army comrade General Lucian Truscott to assess the agency's covert operations. "I'm going over to Germany to see what those weirdos are up to," Truscott remarked, and it did not take long for him to confirm that the whole operation was a colossal mistake, wasting lives "just to see what happens when men put their feet in the water," without producing a scrap of useful intelligence.

"The only thing you're proving" by continuing to drop agents into Soviet-controlled territory, acidly remarked Truscott's aide at one confrontational meeting with the local CIA chief, "is the law of gravity."[6]

But still CIA officials tried to defend the effort. "Even if they don't send back good intelligence, we're causing the Russians a lot of headaches," insisted one. Dulles, confronted the following year by an aide who told him bluntly, "Our operations have failed and there is no alternative to offer," shot back, "You can't say that." Later, when the failures became truly undeniable, Dulles, who would become CIA director in 1952, still tried to justify it all. "At least we're getting the kind of experience we need for the next war," he insisted.[7]

In fact, as it would later be discovered, every single one of the 100 agents dispatched by E-boat to the Baltics by the SIS from 1944 to 1954 was under Moscow's control, wittingly or unwittingly, from the very start. So too were the 150 agents parachuted into Ukraine from 1949 to 1954—at least those not immediately captured and shot after having been lured into the web of KGB deception. So too was an entire Polish "resistance" movement fabricated by the Soviets in 1949 that duped the CIA into sending radios, weapons, supplies, agents, and a million dollars in gold—all humiliatingly exposed in a mocking two-hour diatribe broadcast on Polish radio in December 1952. It was not until 1956 that SIS finally abandoned the last of these fiascoes, an operation that trained agents in Turkey to be parachuted into the Soviet Caucasus. Dozens went in. The failure rate was 100 percent.[8]

Most horrific of all was the fate of scores of Albanian émigrés recruited by SIS and CIA from 1947 to 1952 to gather intelligence and lay the groundwork for an uprising against the Communist government of Enver Hoxha. "Operation Valuable," it was optimistically named. All were betrayed by Kim Philby, who in many cases knew in advance the exact time and location of their landing by sea or air, even the names of individual men, and passed the information to his Soviet controllers, who in turn passed it to the Albanians, who were ready and waiting to ambush the infiltrators. Some were killed in a hail of bullets the moment they touched the ground or came ashore; others were hunted down and butchered by the police and militia—burned to death in a barricaded building, dragged to a bloody pulp tied behind a jeep—or captured, tortured, and put on display in show trials in which they "confessed" their guilt in zombie-like voices.[9]

"There was not one successful operation," said a U.S. Army intelligence officer in Germany, looking over the whole sordid history of the effort to slip spies behind the Iron Curtain. Anthony Cavendish, who ran many of the SIS parachute infiltrations, admitted afterward that they "must seem now nothing more than a catalogue of disasters."[10] A 1955 review ordered by President Eisenhower of intelligence sources that might be able to provide strategic warning of a Soviet nuclear attack on

1

The Russian Problem

The men and women who aimed to supplant Mata Hari with the steady, efficient, reliable, and ever so much safer methods of science and technology began their first tentative foray into what they would always call, with a certain clinical detachment, "the Russian problem" on February 1, 1943. Almost all of the codebreakers who would work over the next several years to achieve the first breaks into the labyrinth of Russian communications secrets were newcomers to the business—which made them no different from the other thirteen thousand new recruits, military and civilian, who would by the war's end swell the explosively growing ranks of the U.S. Army and Navy signals intelligence headquarters in Washington.[1] The variety of their backgrounds was extraordinary: career officers and new draftees, young women math majors just out of Smith or Vassar, partners of white-shoe New York law firms, electrical engineers from MIT, the entire ship's band from the battleship *California* after it was torpedoed by the Japanese in the attack on Pearl Harbor, winners of puzzle competitions, radio hobbyists, farm boys from Wisconsin, world-traveling ex-missionaries, and one of the world's foremost experts on the cuneiform tablets of ancient Assyria.

In June 1943, Cecil Phillips was an eighteen-year-old chemistry student at the University of North Carolina who had just been rejected for the draft because of flat feet ("to my great pleasure and surprise," he later admitted); with no plans for the summer, he wandered into the U.S. Employment Service office in his hometown of Asheville, in the Blue Ridge Mountains, to see if he could get a job. The person there told him there was a lieutenant from the Army Signal Corps over at the town post office who had a large quota of clerk positions to fill.

"How would you like to go to Washington and be a cryptographer?" the lieutenant asked him.

"That sounds interesting," Phillips answered.

The lieutenant, clearly surprised that someone actually knew what he was talking about, blurted out, "You mean you know what that means?"

Phillips did, having once owned a Little Orphan Annie decoder ring, though that was about as far as his knowledge went. But it was good enough for the lieutenant, who administered him a general aptitude test on the spot, signed him up as a $1,440-a-year GS-2 junior clerk, and told him to report in a week to an address in Arlington, Virginia.[2]

Brought in under nearly identical circumstances was a young home economics instructor, Gene Grabeel, who was teaching high school near Lynchburg in central Virginia and dissatisfied with her job when she met a young Army officer in the post office who was looking for college graduates to go to work at an undisclosed location near Washington, to do a job he could not offer any details about. (The officer, an infantry lieutenant who just days earlier had been posted with the First Army at Governors Island, New York, did not know himself what the work involved. Driven largely by the need to process volumes of Japanese army traffic that had suddenly become readable due to breakthroughs in several systems, the Army would hire four thousand new civilian employees for its signals intelligence operation in 1943 alone. In the rush to meet such burgeoning manpower requirements, the recruiters were as green as everyone else. The lieutenant had been ordered to report to Arlington Hall on Monday the week of Thanksgiving in November 1942; he spent the next day filling out administrative paperwork; Wednesday he was given a crash course on recruiting procedures; and on Thanksgiving morning he found himself at the post office in Lynchburg trying to collect warm bodies, without even having had a chance to find out what his new outfit did.) Grabeel had been thinking about trying to get a job with the federal government and asked her father what he thought of the idea. He told her she might as well "go to Washington for six months and shuffle papers." She was off to the capital as soon as she found a replacement teacher to take over for her.[3]

Arlington Hall Junior College had been a finishing school for girls before the Army abruptly seized it under emergency war powers in June 1942. The site was convenient to the Pentagon, the colossal new

Army and Navy headquarters, soon to be the largest office building in the world, which was arising on a marshy flat along the Potomac River less than two miles away, just across Arlington National Cemetery. The school's one hundred acres also offered security and ample room for expansion. One of the Arlington Hall recruiters found a prewar postcard of the school, depicting a stately residence hall, manicured lawns, tennis courts, and indoor and outdoor horseback riding arenas and stables, and shamelessly deployed it in his recruiting pitch. What the new arrivals found instead were two huge, dreary, hastily erected warehouselike office blocks built in the traditional U.S. government style with long corridors and a series of perpendicular wings, surrounded by barbed wire and guardhouses. Workers sat side by side at long tables arrayed in rows. Air-conditioning remained a dream for the future; the buildings were sweltering in the humid Washington summers and overrun year-round with legions of rodents. There were no horses in the deserted stables, but there was a drill field and barracks for the enlisted men. The lieutenant who had paraded the prewar postcard avoided eye contact with his recruits when he later encountered them at Arlington Hall.[4]

More than 70 percent of the staff at Arlington Hall were civilians, and by the war's end more than 90 percent of those were women. A similar balance of the sexes quickly took hold at the Navy's signals intelligence headquarters, across the Potomac River. The Navy had a deep tradition of never permitting a situation to arise where an officer might have to take orders from a civilian, and insisted on putting all of its new hires in uniform. But with its establishment in summer 1942 of the WAVES—Women Accepted for Voluntary Emergency Service, which allowed women to serve in the Navy as officers and enlisted personnel—the service was also able to freely recruit women for codebreaking duty, and some 80 percent of its cryptanalysts by the war's end were female.[5]

The joke making the rounds in Washington the first year of the war was that if the Army and Navy could capture enemy territory as fast as they were seizing it in the nation's capital the fighting would be over in a couple of weeks. Not long after the Army claimed Arlington Hall, the Navy took possession of its own girls' school, Mount Vernon Academy on Nebraska Avenue in northwest Washington. In February 1943 the Naval Communications Intelligence Section (known as Op-20-G in the arcane numbering system the Navy's bureaucratic administrators had devised to designate its myriad branches and offices) moved into its new home at what was now officially called the Naval Communications Annex; construction crews went to work at once ripping out the school's graceful colonnaded walkways to make room for functional buildings; the headmistress's residence, with its elegant polished hardwood floors, was converted into the post exchange, selling Cokes and cigarettes; and a double line of wire fences went up around the perimeter, guarded by marines patrolling with submachine guns.[6]

That summer there began arriving at Nebraska Avenue's Building 4 the first of what would soon be a phalanx of one hundred massive electromechanical calculating machines. Built by the National Cash Register company at its factory in Dayton, Ohio, at a staggering total cost of $6 million, the "bombes," as they were called, weighed two and a half tons apiece and housed sixty-four motor-driven wheels whose electrical contacts spun at speeds of up to 1725 rpm; when they were all running together they drew a quarter of a megawatt of electricity, enough to power a thousand homes.[7]

Their internal electrical logic was the genius of the eccentric British mathematician Alan Turing. Building on prewar work by a small team of Polish cryptanalysts who shared their results with the British just weeks before the Nazi invasion, Turing in the fall of 1939 developed a comprehensive mathematical solution to the Nazis' legendary Enigma cipher machine. By guessing a few probable words contained in an enciphered Enigma message, he showed how to eliminate in one mathematical-logical leap about 10^{14} of the permutations that the Enigma's scrambling machinery relied upon to baffle any would-be codebreaker. That left only a few hundred thousand possibilities to test to recover the unique daily setting of the scrambling rotors used on each of the high-level radio networks that employed the Enigma, the key to unlocking thousands of extremely secret signals a day: orders to U-boats prowling the Atlantic, reports on the dispositions and plans of Rommel's troops in the North African desert or the state of the Nazis' Western Wall defenses along the coast of France. The whirring wheels of the U.S. Navy's bombes, re-creating the internal wiring of the rotors of the actual Enigma machines, could in twenty minutes try every possible starting setting of the Enigma (each of the three or four rotors of the Enigma could be set at twenty-six different positions, which meant there were 26 × 26 × 26 × 26 = 456,976 possibilities in the case of the four-rotor version used by the German U-boats). When the bombe reached a position consistent with a chain of letters built up from a short sequence of matching plaintext and cipher text, a circuit was completed through a series of interconnected cables used to program the bombe for each test, causing an electrical relay to trip and triggering a clutch and brake that would bring the device jolting to a halt, revealing the daily jackpot.[8]

At the height of the Battle of the Atlantic nearly half of the staff at Nebraska Avenue was working on the Enigma problem; most of the rest were trying to keep up with the huge volume of Japanese navy traffic and the ever-changing complexities of its code systems, most of which used codebooks rather than cipher machines like the Enigma.[9*] IBM punch card machines turned out to be well suited to the exhaustive cataloging and searching required to break into these Japanese codes. The process involved combing through intercepted messages to hunt for repetitions of their four- or five-digit numerical cipher groups—a possible sign that two different messages had been prepared with the same string of obscuring key. Compared to a modern computer, punch card machines were undeniably primitive, but they could carry out massive data searches that would have overwhelmed a human being. Cards punched with the code groups of tens or even hundreds of thousands of messages could be automatically placed in numerical order by an IBM card sorting machine, then printed out in massive catalogs by a printing tabulator to be scanned by eye for any repetitions. The Japanese army codes worked much the same way as the Japanese navy codes, and by 1943 Arlington Hall and Nebraska Avenue were operating hundreds of IBM machines and paying the company three-quarters of a million dollars a year in rental fees, while burning through hundreds of thousands of punch cards a month.[10]

What had begun as little more than tiny back-office research groups (at the outbreak of World War II in September 1939 the Army's Washing-

---

*A code, strictly speaking, is a system in which an entire word is represented by an individual symbol or numeral; in a cipher, each plaintext letter is replaced by another letter (or other symbol).

ton signals intelligence staff was nineteen people, the Navy's thirty-six) had become veritable decryption factories, working round-the-clock shifts and tied to a sprawling global network of outstations that fed an uninterrupted stream of intercepted communications. Besides the thirteen thousand workers in Washington there were thousands more in the field manning a dozen principal intercept posts from Winter Harbor, Maine, to Bainbridge Island, Washington, and to points around the world as far-flung as the Aleutian Islands, the Canal Zone, Guam, and Recife, Brazil, each bristling with antennas and equipped with multiple shortwave receiver sets.[11] Teams of enlisted men took down Morse code messages by hand and then retransmitted the copied traffic via encrypted landline or radio teleprinter links back to Washington or other cryptanalytic processing centers in Hawaii and Australia.*

By October 1943 Arlington Hall had up and running a semiautomated decryption processing line for Japanese army traffic that punched incoming teleprinter messages onto paper tape, converted the paper tape to IBM cards, matched the resulting decks of punch cards with other sets of cards punched with the corresponding sequence of cipher key, subtracted one from the other to reveal the underlying code groups and punched those on a third set of cards, and then used a library of cards containing code groups whose dictionary meanings had been recovered to print out the complete decoded message. In some cases Arlington Hall was reading a message before its intended Japanese recipient, who had to manually flip through his codebooks and work out the message on paper and pencil, could do so.[12]

The size and extent of the American wartime cryptanalytic empire

---

*Morse code, which dated to the earliest days of telegraphy and radio, dominated the military, maritime, and commercial communications of the era. Morse code was not a "code" in the sense of a security measure, but a system for representing the letters of the alphabet; a radio operator manually pressed a key to pulse a radio transmitter on and off to produce the pattern of long and short beeps that stood for each letter. A Morse code signal took up much less radio bandwidth than a voice signal, was far less susceptible to jamming, interference, or misinterpretation, and required only the simplest radio gear to send and receive. (The last commercial Morse message was sent in 1999.) By the 1930s a growing portion of commercial and government radio traffic was being sent using radio teleprinter, which linked two automatic typewriter terminals over the airwaves, eliminating the need for manual keying and copying but requiring bulky and expensive specialized equipment.

reflected the global reach of the conflict, but it also reflected the global nature of communications, and thus of intelligence opportunities ripe to be exploited. One of the most valuable sources of information on German preparations for the Allied D-Day landings would prove to be the reports of Japan's ambassador in Berlin, in cables radioed six thousand miles back to his government in Tokyo using the Japanese diplomatic system known to the American codebreakers as Purple, a cipher machine they had cracked in 1940, sight unseen, in one of their most stunning achievements of pure mathematical cryptanalysis.

Some of the size of the wartime enterprise, to be sure, also reflected what the British liaison officer sent to Arlington Hall, Captain Geoffrey Stevens, diagnosed with ill-concealed irritation as the American genius for constructing "hopelessly overorganized" operations.[13] The work involved a huge amount of painstaking drudgery, and the American solution was to parcel out tasks exactly like a factory assembly line. When Cecil Phillips arrived at Arlington Hall on June 22, 1943, his first job was to stamp the date on incoming messages. Having demonstrated his competence at that task, he was promoted to stapling. But his first boss saw something in the high school graduate that perhaps Phillips himself did not, and started setting aside an hour or two a day to teach him the rudiments of cryptanalysis. On May 1, 1944, Phillips was led to the back of one of the wings on the second floor of B Building, where a fifty-by-fifty-foot area had been partitioned off with plywood screens from the rest of the open wing. A small opening between the screens, just large enough for one person to squeeze through, led past a desk where an Army captain sat with his back to the entrance, keeping a sharp eye on the several dozen people at work at long tables.[14]

Among them were Gene Grabeel, who had been assigned to this mysterious unit during her fourth week on the job; Richard Hallock, the Assyrian linguist and archaeologist from the University of Chicago, now an Army lieutenant; another newly commissioned lieutenant, Ferdinand Coudert, a partner in his family's venerable New York international law firm, Coudert Brothers (among its clients had been the czar of Russia and the French and British governments), who held an MA degree in Slavic studies from Harvard and knew Russian, French, German, Serbo-Croatian, Bulgarian, and Japanese; and Frank Lewis and Genevieve Grotjan Feinstein, two of the comparatively veteran cryptanalysts of the

Signal Intelligence Service, both of whom had been hired in a modest expansion of the service made possible with funds appropriated in response to the "limited national emergency" declared by FDR upon the outbreak of the war in Europe. Feinstein, a quiet mathematician, had made the crucial break into Purple in 1940; Lewis, a voluble polymath and musician, contributed to the solution of the Japanese army codes and a seemingly impenetrable German diplomatic one-time-pad system. (Lewis was to gain a devoted following in the outside world for the notoriously difficult cryptic crosswords he composed every week for six decades for the left-wing intellectual journal the *Nation* starting in 1947.) The captain in charge of the section was William Smith, a classmate of Coudert's at Harvard who had recently been working as an editor for the famous one-volume *Columbia Encyclopedia* at Columbia University Press.[15]

This, Phillips was told, was the Russian problem, and it was where he would be working from now on, and he was not to mention it even to anyone else at Arlington Hall. Phillips, Grabeel, and Lewis would likely have been astonished had they been told they would spend much of the next four decades of their lives on the problem.

The decision in 1943 to begin studying Russian traffic was odd on several counts, not least that Arlington Hall was already overwhelmed with the volume of enemy coded messages it was trying to read following America's entry into the war.

Even with the influx of thousands of new hires, Arlington Hall was struggling to keep up with an avalanche of operationally urgent traffic for which an hour's delay could mean the difference between a successful battlefield coup de main or an opportunity forever lost. The solution of the Japanese army codes meant processing upwards of three hundred thousand messages a month. Reflecting the importance of that priority, a reorganization of the production branch that summer reallocated cryptanalytic work between two sections. The larger, B-II, now focused exclusively on Japanese army messages. The other, B-III, now called the General Cryptanalytic Branch, got literally everything else—all of Japan's non-army traffic (diplomatic, military attaché); all of Nazi

Germany's traffic, military and diplomatic; and a whole kitchen sink of diplomatic and commercial codes and ciphers of some thirty other countries, enemies and neutrals alike, among them Sweden, Finland, Turkey, Bulgaria, Spain, Portugal, Vichy France, China, Mexico, Chile, Brazil, and the Vatican.[16]

William F. Friedman, who for a decade after the end of World War I had been the sole cryptologist employed by the Army Signal Corps, and who then in the 1930s had patiently and painstakingly trained a cadre of young mathematicians to resurrect an Army codebreaking bureau, was one who thought there was no choice now but simply to abandon work on *all* diplomatic codes, much less take on any new problems, and focus exclusively on enemy military traffic. Friedman was the opposite of a defeatist. Although he had only a limited mathematical background himself—he had taken a single freshman course in mathematics as a student at Michigan Agricultural College early in the century—he had early on discovered an aptitude for codes, and a conviction that the main requirement for successfully breaking one was the confidence that it was breakable. He was the first to work out a solution to the new generation of cipher machines that used rotating wired wheels that advanced to a new position as each letter was typed in on a keyboard, generating a completely different substitution alphabet for each successive letter of the message.* He astonished the inventor of the Hebern electric cipher machine in 1923 by breaking ten enciphered test messages in six weeks of work, four of which he spent developing the theory for how to do it. Friedman's precise habits and systematic approach were abundantly

---

*Rotor machines, of which the German Enigma (invented in 1923) is the most famous example, remained a mainstay of cryptography into the 1970s. Each wheel was electrically connected to its adjacent wheels via a series of contacts, one for each letter of the alphabet, that ran around its opposite faces. A maze of wires within each wheel interconnected the two faces, scrambling the identity of the letters in random order. As the wheels rotated, a new scrambling pattern continuously emerged. Thus the cipher letter A might stand for the plaintext letter G at one position, Q at the next, foiling the simple method of cracking a substitution cipher by counting the frequency of each letter in the cipher text and assuming that the most common one stands for a high-frequency letter such as E. A cipher that employs an unvarying substitution alphabet for the entire message is known as monoalphabetic; the continually varying substitutions produced by a rotor machine constitute a much more secure "polyalphabetic" cipher.

on display in the classic foundation textbooks he prepared in the 1920s and 1930s (*Elements of Cryptanalysis*, expanded into the four-volume *Military Cryptanalysis*), which would train several generations of cryptanalysts, bringing logic and flair to what previously had been unsystematized chaos, and in his always fastidious personal appearance with his trademark Hollywood-idol dapper mustache, perfect bow ties, and two-toned shoes. Characteristically, he played an intensely disciplined game of golf and prided himself on being an expert ballroom dancer. William Friedman, said a colleague who worked with him closely in later years, simply was never "scared of the magnitude of the problems facing him."[17]

By 1943 the organization had long outgrown Friedman's talents as an administrator. After a hospitalization for a nervous collapse in January 1941, he was eased out of responsibilities and, at age fifty, had become something of an elder statesman and adviser. But he remained a legendary and revered figure among the rank-and-file cryptologists of the service, and the men who were now in charge of its most important technical branches were all his protégés. The three young math teachers Friedman had hired in 1930 in the first expansion of the service, and to whose training and mentoring he personally devoted the next decade, now ran virtually the entire technical side. Frank Rowlett, whose outward demeanor as a soft-spoken, courtly Virginia gentleman concealed an often prickly and acerbic contempt for incompetence, was head of B-III; Solomon Kullback, a perpetually rumpled verbal bulldozer of a New Yorker, was in charge of B-II; and Kullback's fellow New Yorker Abraham Sinkov—he and Kullback had been classmates at City College and were both teaching in the New York City public schools when Sinkov spotted a notice for an examination for a government job as a mathematician—ran the cryptanalysis center the Army had set up in Brisbane, Australia, to bring additional manpower to bear on the Japanese army problem.

In a detailed proposal to the agency's chief on July 29, 1943, Friedman argued that given the duplication of effort between the Americans and the British in intercepting and processing diplomatic traffic, it made sense just to let the British have it all so that Arlington Hall could focus, without further distractions, on the much more vital Japanese army material.[18]

In any case, the intelligence value of much of the diplomatic codebreaking effort was minor at best. The U.S. Army codebreakers had originally taken on the job, in fact, mainly because they had nothing else to do. Eager to give his cryptanalytic trainees some real-world experience, Friedman had found precious little else on the airwaves for them to work on during the pre–Pearl Harbor years. German and Japanese military traffic, much of it short-range transmissions sent using low-power radios, was almost impossible to intercept from monitoring stations in the United States. Diplomatic signals, by contrast, went over transoceanic commercial networks operating high-power shortwave transmitters. The decision to copy and decipher the diplomatic traffic of dozens of neutral countries had always had more to do with its ready availability than with its intelligence significance.

There was also the fact that as sensitive as were the issues raised by spying on neutral nations, spying on an ally and now comrade in arms like the Soviet Union was something else entirely. Winston Churchill, for all his long-standing enmity toward and distrust of the Soviets, had promptly ordered a halt to the monitoring of Soviet communications by British intelligence upon Stalin's joining the fight against Hitler in June 1941. Although no similarly explicit order seems to have been given to the U.S. Army and Navy codebreakers following the United States' entry into the war in December 1941, both had shelved their prewar work on Russian systems given the obviously much more pressing priorities of fighting a global war across two oceans.[19]

And finally there was the profoundly discouraging consideration that that earlier work on Russian codes had made precious little progress. Soviet diplomatic systems, in fact, were widely viewed as utterly unbreakable. In the 1920s, allowing discretion to be trumped by political pressures, the British government on several occasions had released the verbatim texts of intercepted and decoded Soviet cables exposing anti-British espionage, propaganda, and subversion by Soviet officials operating under diplomatic cover—and also alerting the Soviets to the weaknesses of conventional code systems. The final break in British-Soviet relations in 1927 was accompanied by a wholesale change in Soviet codes; Soviet trade missions and diplomats subsequently began using unbreakable one-time pads to encipher nearly all of their cable messages.[20]

As cogent and logical as Friedman's arguments were, they were no match for formidable political and bureaucratic imperatives whose force did not depend strictly on logic—a consideration that someone with Friedman's orderly mind had a hard time appreciating. As early as August 1942, Captain Stevens, the British liaison officer at Arlington Hall, picked up hints that the Americans were eager to take another crack at the Russian problem, allies or no allies, and other priorities notwithstanding. Stevens, who was extremely diligent at keeping his ear to the ground, reported back to London that the Americans had never stopped collecting and filing Russian diplomatic traffic for possible future study. "Sooner or later they will inevitably try to break this," he wrote, "since they do not trust the Russians further than they could throw a steam-roller."[21]

Worries that Stalin might be secretly negotiating with the Japanese were part of that mistrust.[22] But the decision by the Army in 1943 not only to retain its far-flung net to capture worldwide diplomatic communications, but to extend it to encompass the Soviet Union, had little to do with any articulable concerns about the Soviets, or even a mistrust of them in general. In the aftermath of Japan's surprise attack on Pearl Harbor, the War Department had ordered a sweeping review of how the Army handled signals intelligence, and its major conclusion was that the way to make sure nothing ever again fell through the cracks was to miss nothing in the first place. Before the war, the Army had tended to regard signals intelligence as a technical and arcane subspecialty. It did not even come under military intelligence but had been shunted aside under the direction of the radio experts of the Signal Corps. The Navy's codebreakers similarly came under the Naval Communications department. Ad hoc arrangements for passing on to the State Department and White House any items of importance gleaned from decoded messages had been made by the Army and Navy cryptanalytic units on their own initiative, but neither had any defined intelligence responsibilities beyond the narrow confines of service requirements. Even within the military services they had struggled for recognition and resources. Many regular officers viewed the entire business of codebreaking as a colossal waste of time. Captain Thomas H. Dyer, one of the Navy's few trained prewar cryptanalysts, would recall with undiminished bitterness in 1945 the "general attitude" he encountered in those years, which "ranged from one of apathy to one of ridicule or open hostility. . . . The vast majority of those whose support and cooperation were essential viewed it as the visionary project of impractical dreamers."[23]

Alfred McCormack, a formidable New York lawyer who had come to Washington right after the Pearl Harbor attack to ask his former law partner John J. McCloy, now assistant secretary of war, for "the toughest assignment he had," was given the job of coming up with a new blueprint for signals intelligence that reflected the realities of America's sudden thrust onto the world stage. McCormack swiftly concluded that the mission had been construed far too parochially. In an age of global conflict and global responsibilities, it was not enough just to monitor the radioed orders of enemy commanders on the battlefield. Simply, the United States "must know as much as possible about the objectives, the psychology and the methods of our enemies and potential enemies (and of our Allies as well) in order to make the right decisions." *Every* country was a legitimate target, friend and foe alike; intelligence on everything from its industry to agriculture, from its politics to internal social forces, might provide the one crucial detail that would make the difference between triumph and catastrophe for American policy.[24]

The wartime influx of men and money was the chance to make such an all-encompassing intelligence-gathering system a reality. It didn't hurt that in fulfilling this greatly enhanced signals intelligence mission the War Department would also greatly enhance its prestige in the corridors of Washington political power, and its influence in postwar planning. The man brought in on McCormack's recommendation to implement the vast expansion of the mission and scope of the Signal Intelligence Service, Colonel Carter W. Clarke—McCormack then became his deputy for the rest of the war—was in early 1942 already looking to the end of the war, and beyond, in enthusiastically endorsing McCormack's vision. Clarke wrote in May 1942:

> Our primary task is to paint for our superiors as completely a realistic picture as possible of the activities "behind the arras" of all those associated with and against us with the end purpose of enabling an American peace delegation to confront problems

of the peace table with the fullest intimate knowledge it is possible to secure of the purposes and attitudes, covert and overt, of those who will sit opposite them.[25]

In a nation that would see eleven million men mobilized, a seventy-acre factory arise almost overnight on a patch of scrubland west of Detroit to begin disgorging a finished B-24 bomber every hour, fleets of cargo ships by the thousands constructed in as little as two days apiece start to finish, and the elemental forces of the universe transformed on a remote New Mexico mountaintop into a weapon of unimaginable destructive power, the idea of monitoring, intercepting, and deciphering every coded message transmitted anywhere in the world seemed neither far-fetched nor unworthy of the attempt. It was, Frank Rowlett later said, a foregone conclusion in the post–Pearl Harbor climate that the interception and deciphering of diplomatic signals would continue, that Friedman's prudent objections would be gently brushed aside, and that Russia, ally or not, would be added to the ever-expanding list of new targets.[26]

"Get everything" would henceforth be a virtual article of faith for the U.S. signals intelligence community. "Whether or not it actually can be done in practice," insisted Rear Admiral Joseph R. Redman, the head of U.S. Naval Communications, in a memo he wrote at the end of the war, it was the fundamental mission and responsibility of the U.S. communications intelligence organizations to come "as closely as is humanly possible" to reading "*every* enemy and clandestine transmission."[27]

There were some new practical reasons for encouragement, too, in tackling "the Russian problem." Finland had a small but well-regarded cryptanalytic bureau, and messages from Japan's military attachés in Helsinki and Berlin to Tokyo, deciphered and read by Arlington Hall in early 1943, revealed that the Finns had made some progress on the "unbreakable" Soviet one-time-pad diplomatic ciphers and were sharing their results with their Japanese counterparts; a series of lengthy cables offered a wealth of basic technical details about the Russian systems and how they worked.

A considerable amount of raw Russian traffic was also now becoming available to the American codebreakers. In January 1940 the Army's adjutant general had sent a letter to the president of RCA, David Sarnoff, asking if a Lieutenant Earle F. Cook might be assigned to the company for six months "for a course of study." As Cook later explained, "All of this nonsense was a cover. Looking over the traffic was what I was there for." RCA, one of the major carriers of commercial cablegrams, supplied Cook a private room and photographic equipment in its Washington offices on Connecticut Avenue; each morning he would arrive and make copies of all of the messages that had been handed in the previous day for transmission on the company's international circuits. (Frank Rowlett would recall seeing Cook's thumbs in many of the photos.) The clandestine arrangement—almost certainly illegal—set a precedent that would come back to haunt NSA three decades later. But no one at the time seemed overly concerned, and in any case America's entry into World War II in December 1941 made the question of legality moot for the duration when official wartime censorship began, requiring all cables to be turned over to the government for inspection.[28]

The Army was able to intercept additional Russian traffic using its monitoring stations at Fort Sam Houston, in Texas, and Fort Hunt, just south of Washington, D.C., to copy telegrams transmitted, in Morse code, over the international radio networks of other commercial operators. In June 1944, the U.S. Army negotiated an agreement with the Soviet government to establish a direct radio teleprinter link from the Pentagon to Moscow via an American radio station in Algiers; the two governments shared use of the channel and operated their own teleprinter terminals at each end. The major purpose was to improve on the unreliable and interference-prone over-the-North-Pole radio link the cable companies operated between the capitals. What the Russians did not know was that a teleprinter installed in A Building at Arlington Hall automatically copied everything passing over the circuit. For several years it would prove the most important source of enciphered Russian traffic available to the American codebreakers.[29]

The Navy's codebreakers at Op-20-G meanwhile found that they could pick up some Russian naval signals from the Far East at their monitoring stations on the West Coast. The thirty-two-year-old Navy lieutenant (j.g.) in charge of Station S at Bainbridge Island, Washington, was Louis W. Tordella, a math professor who had been teaching at

Loyola University when the war broke out. Tordella had tried to offer his services to the Army first, visiting the headquarters of the Fifth Army in Chicago and explaining to a bored and contemptuous major that he had a PhD in mathematics, was an amateur radio operator, and had done work on codes as a hobby. The major replied, "When we want you we'll draft you." The Navy proved more amenable, sending Tordella a correspondence course in cryptology; shortly after passing that he received a commission and orders to report to Op-20-G in Washington. After a few months assigned to the Enigma bombe project he was sent to Bainbridge as the second in command.[30]

In July 1943, Tordella received an order from Washington to put his four best intercept operators on the job of transcribing Soviet Far East Morse code traffic. It took a month for him to get hold of a single Russian typewriter. He had to send the copied traffic back to Washington on paper forms, by airmail. Most of the coded messages proved to be relatively simple hand-cipher systems used by the Soviet army, navy, police, railroads, and Communist Party and dealt with the most mundane matters: weather broadcasts, orders to icebreakers, reports on production of dehydrated vegetables.[31]

Still, the progress made by Op-20-G's cryptanalysts back at Nebraska Avenue on the traffic was encouraging on technical grounds if nothing else; their first break into one of the Soviet military systems came just a few months later in October 1943, and every month some two thousand new messages were intercepted and added to the growing trove of Russian traffic available for study, soon to be beefed up by the assignment of additional monitoring stations—Station H at Wahiawa on the island of Oahu in Hawaii, Station W at Winter Harbor on Maine's Schoodic Peninsula, and Station AX at Adak in the western Aleutian Islands—to the job of pulling in these far-off signals from the airwaves.[32]

But it was the diplomatic traffic that clearly offered far richer intelligence pickings. The Japanese military attaché cables—an entire special cipher subsystem, known to Arlington Hall as JAT, had been set up by the Japanese exclusively for the purpose of exchanging cryptologic intelligence—had identified several different Soviet diplomatic systems and confirmed that they were all some form of enciphered code. The basic principle of operation was well understood; nearly all of the Japanese army and navy systems that Arlington Hall and Nebraska Avenue had been attacking for years worked the same way. A codebook assigned words numerical values; in the case of the Soviet codes these were typically four digits long, 0000 to 9999, allowing for ten thousand different words. To each of the code groups in a message to be transmitted, a second set of digits, drawn in sequence from a book or pad containing random numerical groups of "additive" (or "additive key") was then added.* That second step obscured the actual meaning of the message under an additional layer of concealment intended to baffle any would-be codebreaker; it ensured that even when the same word was repeated, it would appear in the enciphered transmission as an entirely different four-digit number each time it occurred in the same or subsequent messages. The recipient of the message reversed the process, subtracting the additive key to obtain the original code groups and then looking up their meanings in the codebook.

The vulnerability of an enciphered code was that in any heavily used system it was inevitable that some messages would eventually be enciphered using overlapping sequences of additive drawn from the same pages of the key book. From such an overlap (a "depth," in cryptanalysts' jargon) it was possible for a codebreaker to begin the long, laborious process of "stripping" the additive from the enciphered messages to reveal their underlying code groups and then start to figure out their individual meanings. There was usually an "indicator" buried within the message that told the intended recipient from what starting point in the key book the sequence of additives used for enciphering that particular message had been drawn; breaking the indicator system was one way to identify overlapping messages directly. The JAT cables offered a few clues about how the Soviet indicator systems worked, but none were enough to be of much help.

The other way to find two overlapping messages was sheer brute force. One indication that two messages were in depth was if they contained some of the same numerical groups, indicating that the same word had been enciphered with the same key in each. A single repetition of a particular numerical group could easily be the product of chance,

*The addition was modulo 10—that is, digit by digit, without carrying—so that the resulting number was always the same number of digits long; for example, the code group 7829 combined with the additive 9234 yielded the enciphered code group 6053.

but a "double hit"—the same *pair* of groups appearing in two different messages in the same relative positions—was much more likely to be the product of the two messages actually being in depth. (See appendix A for a further explanation of the process.) If the Soviet messages really had been enciphered using one-time pads, any such brute-force search for depths would be a guaranteed exercise in futility. In a one-time-pad system, each sequence of additive key is used to encipher a single message, then the sheet is torn off and destroyed and never used again. It imposed huge logistical burdens to produce and distribute the key pads required to sustain such a system, but it undeniably offered unbreakable security.

In October 1943, however, Richard Hallock decided it was worth trying a long shot. He had the first and last five groups of ten thousand of the messages punched onto IBM cards: the opening and closing of any message were the parts most likely to contain stereotyped phrasing, and thus repetitions of their underlying code groups—words such as TO MOSCOW, FROM NEW YORK, REFERENCE YOUR NUMBER, PART 2 OF 2.

The results were unmistakable. Seven pairs of messages contained double hits, meaning they were almost certainly in depth, enciphered using the same sequence of additive key. At least a few of the one-time-pad pages had clearly been used a second time, an astonishing and monumental security blunder.[33]

From that meager start, the Russian section over the following months was able to begin stripping the additives from paired messages in depth and to recover and identify a few of the frequently used code groups in the message openings, particularly the stereotyped beginnings of multipart messages. More massive IBM runs produced a bank of "hypothetical additive": the idea was to subtract, in turn, each of these frequently used code groups from the beginning of every message to calculate the resulting additive key that would have been used to encipher it, then see if any portion of that hypothetical key, when added back to any of the frequent code groups, matched the openings of other messages, indicating possible further one-time-key reuses. The largest set of messages were being sent by Soviet purchasing commissions operating in the United States under Lend-Lease; these "trade" messages, employing the code system Arlington Hall designated ZET, made up about half of all the traffic. The other four systems the American codebreakers identified appeared to be diplomatic traffic from embassies and consulates, and in July 1944, Cecil Phillips was placed in charge of those systems. A few months later, in February 1945, he was looking at one batch of messages from New York to Moscow in the system known as ZDJ. He soon found something odd about the first cipher group of each message. The numbers were not random; when he counted up the frequency of each digit he found that the digit 6 appeared about 20 percent of the time, rather than the 10 percent that would be expected. Phillips took the results to Genevieve Feinstein; she glanced at them, and said at once, "That looks like clear key." She had noticed the same off-kilter bias in the hypothetical additive bank generated from the ZET trade messages. (The uneven distribution was probably the result of a temporary glitch in the machine the Soviets used to generate random numbers for the pads.) Checking the numbers against the hypothetical additives revealed repeated matches between the groups Phillips had spotted and the additive groups located at the very first position of each key page used to encipher the trade traffic.[34]

The nineteen-year-old Phillips had just stumbled on two stunning discoveries. One was that the Soviets were using the first key group of the one-time-pad page as the indicator to tell the recipient of a ZDJ message which page had been used. That offered a huge shortcut to finding a matching message in depth. The other was that some of the one-time-pad pages used in the trade messages had been reused in the four diplomatic systems as well. There was thus a slim but real chance that all could be cracked open.

Even by the standards of the extremely tight security that surrounded everything having to do with codes and codebreaking, the secrecy of the Russian problem was exceptional. The WAVES who reported for duty at Nebraska Avenue would vividly remember for the rest of their lives the introductory lecture they were given their first day. Ushered into what had been the chapel of the girls' school (it was now the Navy Chapel), they were addressed by a deadly serious officer who informed them that if they ever let slip a single word about their work, ever, they would be shot.[35] It was hyperbolic, but in the context of the times none of the young women were prepared to doubt that he meant it. The Russian sections of the Army and Navy codebreaking units were secrets wrapped within that secret. At Arlington Hall the project was referred to only as the "Special Problems Section" or by its designation B-III-b-9 on the organizational chart, which indicated only that it had something to do with Frank Rowlett's General Cryptanalytic Branch, responsible for every code other than Japanese military. The Navy at first called its Russian section Op-20-GZ, then it became the "Foreign Language Research Section," or Op-20-GV; shortly after that the name was changed yet again, to Op-20-3-G-10, and the staff were issued new red badges bearing nothing but the number 10, which meant nothing to anybody.* Less than two weeks after the surrender of Nazi Germany on May 17, 1945, the U.S. Navy commander in chief, Fleet Admiral Ernest J. King, ordered an increased emphasis on the project, but suggested it ought to be moved out of Nebraska Avenue altogether for security reasons.[36]

It was obviously a matter of extreme political sensitivity to be spying on an ally, and at one point in 1944, Carter Clarke apparently decided that even the White House need not be kept apprised of Arlington Hall's efforts.[37] It was likewise too sensitive a matter to share with the British, despite the unprecedented collaboration that had been forged between the two nations' spy agencies since the start of the war. In the summer of 1940, in the dark days of Britain's lonely struggle against Nazi Germany after the collapse of France, and desperate to find some way to budge America off its isolationist neutrality, Prime Minister Winston Churchill had launched a multifront diplomatic and charm offensive with the aim, as he later candidly told the House of Commons, to get the two nations "somewhat mixed up together." A delegation of top U.S. Army and Navy officials, invited to London in August 1940, was showered with British intelligence about Germany and Japan, technical details of weapons systems, and, a few weeks later, an even more astonishing offer

*A welter of code names further concealed the actual target of the projects. Following its standard practice of giving the codes of each country a color designator, the Army initially referred to the Russian systems as Blue. The Navy used the code name Rattan, sequentially designating each system it had identified with the letter R followed by a hyphen and a number. In June 1945 the Navy changed the code name to Bourbon and switched all the R's in the system descriptors to B's.

to exchange cryptographic information with the Americans. William Friedman, instinctively an anglophile, and keenly aware how far the U.S. Army's codebreakers lagged in their efforts against Axis military systems—in fact, Friedman's group had virtually no German, Italian, or Japanese army messages even to work on at that point, having been unable to pick up that traffic from the Army's intercept stations, and it had barely begun to tackle the mathematical conundrums of cryptanalyzing the Enigma—leapt at the offer. Within six months the codebreakers of the U.S. Army and Navy and the British Government Code and Cypher School (GC&CS) were considerably more than "somewhat" mixed up together. The Americans matter-of-factly handed over to their now-astonished new colleagues the complete solution to the Japanese Purple machine, including a cryptanalytically reconstructed copy of the machine they had built, and the British, albeit a bit more warily at first, began to share their Enigma work and results along with more complete exchanges of raw traffic from a great variety of targets.[38]

Although they still hoped to use their considerable head start on the all-important Enigma problem to keep control over that crucial success, the British must have known in their hearts that the Americans would not be content to remain the junior partner indefinitely; by 1942 it was already apparent that Britain needed both American manpower and America's formidable industrial capacity and precision engineering know-how to keep up with the demand for additional Engima-cracking bombes, and the men and women to operate them. A formal agreement in May 1943 between the U.S. Army and GC&CS provided for complete cooperation on all work against Axis military and air force code systems; under the agreement a large contingent of Americans joined the Enigma project at Bletchley Park, and the U.S. Army set up its own intercept station in England, at Bexley in Kent. A separate, less formal understanding reached with the U.S. Navy the previous October established full collaboration on the German U-boat Enigma problem, with the full-scale production of more than one hundred U.S.-made bombes to take over most of the work.[39]

But the BRUSA agreement, as the Army-GC&CS deal was known, said nothing about sharing work on diplomatic ciphers or neutral countries, and each side still held back on cryptologic matters that touched interests too close to home. Following America's entry into the war,

Churchill told FDR that he had ordered GC&CS not to try to decode any American messages, and he briefly considered proposing a more explicit "gentleman's agreement" that each "would refrain from trying to penetrate each other's cyphers." But the chief of the British Secret Intelligence Service, Stewart Menzies, talked him out of it, noting that no such agreement could have much practical force; whatever the leaders might agree, "the temptation to have a peep would be more than some experts could resist." The two countries also adopted a joint, extremely secure cipher system for exchanging secret intelligence between Washington and London. But the British quietly developed their own private cipher machine as well, called Rockex, which employed a one-time paper tape enciphering device and which they used to keep sensitive messages from the prying eyes of their ally: the British had learned that if they wanted to influence joint military plans, the only hope was to get to the lower-level U.S. staff officers who were drawing up the American proposals at their early stages, and many Rockex messages were instructions to the British military mission in Washington on what ideas they ought to try to plant in the minds of their American counterparts.[40]

As the end of the war drew near, the marking "N.B." began appearing with increasing frequency next to certain paragraphs of memoranda and reports from Op-20-G and Arlington Hall. The letters stood not for "nota bene" but "No British," and the restriction was most frequently appended when the subject of reading the coded traffic of neutrals or allies—Russian, Free French, Dutch, Latin American—was discussed, or when details about new developments in U.S. electromechanical codebreaking machines were mentioned, particularly the devices known as "statistical bombes," which might make even the most secure U.S. cipher machines vulnerable to decryption. The British for their part began to be suspicious, as Arlington Hall's liaison man in London reported, that the Americans were "utilizing the war to exploit British cryptographic knowledge" in matters unrelated to actually winning the war, in particular involving countries that fell within the British Empire's traditional sphere of influence in the Near East and elsewhere. (The suspicions were well founded: the informal Army-Navy committee that coordinated U.S. signals intelligence policy urged in February 1945 that "advantage should be taken of the present opportunity to obtain all possible information from the British.")[41] The working relationships between the two countries' signals intelligence operations had grown extraordinary close, on both a professional and personal level: many lifelong friendships and more than one marriage between American and British codebreakers were made at Bletchley Park.[42] But by the spring of 1945 it was far from clear what, if any, collaboration would continue after the war. Significantly, although the British, like the Americans, resumed work on Soviet codes in the summer of 1943, neither had yet revealed the fact to the other.

The greatest doubts came from Op-20-G. The U.S. Navy had never been fully sold on the idea of getting too close to the British, and now with the end of the war in sight many American naval officers began to suggest that continuing the wartime arrangements, especially if that meant working together to break the Soviets' messages, would be a mistake. Some of this reflected the remarkably resilient anglophobia in a service that had never quite been able to forget if not the War of 1812 then at least the infuriating condescension with which the Royal Navy had welcomed its American counterpart when they were fighting as allies in World War I; some was the U.S. Navy's customary view that it did not share its secrets with anyone, starting with the U.S. Army, civilians of any nationality, or politicians up to and including the president of the United States.

But even the cooler heads in the Navy—and there were few cooler than that of Captain Joseph N. Wenger, the chief of Op-20-G and one of the Navy's most experienced cryptanalysts from the prewar era—argued that the interests of Britain and America were bound to diverge in the postwar world. Wenger, a 1923 graduate of the Naval Academy, had been a steady force pushing the Navy to take cryptanalysis and signals intelligence seriously. As radio intelligence officer for the U.S. Asiatic Fleet during Japan's prewar naval exercises in the Pacific, he had intensively studied Japanese call signs and communications procedures and shown that even without reading the contents of messages it was possible to derive considerable information about an enemy's force structure, movements, and intentions, the process that would come to be known as traffic analysis. He subsequently pressed for the establishment of permanent intercept stations around the Pacific, and at times almost single-handedly prodded Op-20-G into the era of modern cryptanalysis, pioneering the use of IBM punch card equipment and leading the push for the United States to build its own bombes to attack the naval Enigma problem.[43]

Wenger praised the professional and personal ties and goodwill that had grown up between the two allies' codebreakers during the war—the British government would award him the CBE in acknowledgment of his wartime contributions and close collaboration with his counterparts at GC&CS—but he was not about to let warm feelings get in the way of cold judgment. "The fact that we are military allies in war does not necessarily mean that we shall be commercial or political allies in peace," Wenger warned in May 1945. Should that not be the case, he said, continuing to work on intimate terms with the British on signals intelligence "might deprive us of a vital advantage we might otherwise enjoy." The British did have some advantages to offer when it came to attacking the Russian codes. Their intercept sites were better located, they could tap British-owned cables that carried some of the traffic, and they were in a better position to "gain physical possession" of codebooks and other materials owing to their "world-wide intelligence organization" that was "unhampered by doubts as to the proprieties of methods used." But, overall, Wenger confidently asserted, the U.S. codebreakers had surpassed their British counterparts in technical proficiency: they no longer needed the help. Wenger proposed that any future exchanges with the British should be on a strict "barter basis."[44]

More troubling, in the view of Admiral Richard S. Edwards, the vice chief of naval operations, was what all of this implied for the postwar order. "I do not think we should 'gang up' with one ally against another ally," he advised:

> In the troubled times that lie ahead we shall have to side with one or another of our friends as differences of interest arise, but for the sake of the peace of the world we should do so openly and frankly. If we secretly join the British in this project, the secret is virtually certain to leak out in the course of time with results disastrous for our relations with USSR. The possible gain is not worth the probable cost.[45]

But the Army held firm to its view that it would be equally foolish simply to pull the plug on so fruitful a collaboration, and in the end what tipped the balance in the Navy was a pessimistic assessment that, based on all past experience of American peacetime parsimony and idealism, budgets were sure to be slashed and intelligence operations reined in, while the more cynical and worldly-wise British would carry on as usual; however far ahead the United States might now be in knowledge and ability, the British were going to forge ahead once again and the United States would need to stick with them not to be left in the dust.

On June 2 the British representative broached the question of working together on the Russian problem, and three days later Admiral King and the Army chief of staff, General George C. Marshall, approved, but with the stipulation that the exchange of information remain informal, that a statement drawn up by the Army-Navy Communications Intelligence Board outlining the agreed-upon terms "be shown but not given" to the British representative, and that their own memorandum to the board giving the go-ahead be burned immediately upon reading.[46]

Although the British had a long history of success in attacking Soviet cipher systems and were well positioned around the world to intercept Russian communications, GC&CS had not only stopped all of its active work on Soviet military and diplomatic signals upon the Soviet Union's entry into the war in June 1941 but had also subsequently discarded "a room full" of accumulated Russian one-time-pad traffic that it considered unsolvable. "We weren't sorting it, couldn't do anything so we just threw the lot away," recalled Brigadier John Tiltman, a legendary GC&CS cryptanalyst who had worked on Russian codes in India in the 1920s and later made the crucial break in the Nazi teleprinter cipher machine. When Kim Philby's spying for the USSR became known in the 1950s, Tiltman briefly came under suspicion that he had destroyed the Soviet material to cover for Philby, and he was interrogated by MI5, the British counterintelligence service; but there was certainly nothing to that idea. Tiltman, who would later serve as the British liaison to the National Security Agency, would forever regret the missed chance to expose the Soviets' most damaging mole inside British intelligence.[47]

Whether America had the money or the stomach for conducting *any* foreign espionage in peacetime was the more salient question than what form its intelligence cooperation with the British might take. It was a deep-seated American belief, as the historian Eric F. Goldman wryly put it, "that foreign policy was something you had, like measles, and got over with as quickly as possible." The total global war that had wrested the United States from its splendid isolation had done little to alter the American feeling that war was something aberrant and exceptional, that once it was over the best thing for the country was to get back to "normal" as quickly as possible, which meant leaving the world once again to its own troubles.

In the months following the Japanese surrender, thousands of GIs dissatisfied with the pace of demobilization staged rowdy mass protests—some were described by local commanders as "near mutiny"—in Paris, Manila, Guam, Yokohama, Honolulu, Vienna, and Frankfurt, chanting, "We want to go home!" A colonel with the American occupation force in Japan dismissed the riots as the work of "a lot of Communists and hotheads," but the slogan caught on like wildfire and newspaper columnists noted that it reflected the widespread feeling of American troops, who were singularly unimpressed by "pleas about our duties in the world at large" or America's newly acquired "international commitments."[48]

Yet even the decidedly internationalist-minded men who had been working since literally the moment the war started to draw a plan for the postwar world, those whose whole premise was that America's past isolationism had contributed to the calamity that had engulfed Europe and Asia in six years of horrific bloodshed, believed that peace this time would be a real peace: new international institutions would guarantee a long era of stability and the growth of global democracy that would make the aggressive realpolitik and crude balance-of-power politics that had governed the relations between states in the previous intervals between wars no longer necessary.

"We have profited by our past mistakes," Franklin D. Roosevelt assured the American people in 1942. "This time we shall know how to make full use of victory." There could be no return to America's prewar isolationism, but FDR and his aides had no notion of a future dominated by perpetual American interventionism or open-ended military commitments. Rather, peace as most Americans understood and expected it could be safeguarded at home precisely by extending America's most deeply held principles to the world at large: openness, democracy, the rule of law, economic justice. The collective security system of the new United Nations and the promotion of self-determination would remove the major causes of war; new international monetary institutions would stabilize the world's economies to lift peoples out of poverty and prevent a repetition of the global depression that Hitler had ridden so effectively on his demagogic path to power.[49]

Even the most patriotic supporters of the war would have been forced to acknowledge that America's four years of total mobilization had deeply compromised many of those American principles. Conscription, censorship, rationing, wage and price controls, regimentation of industry, and other government intrusions into daily life were seen as justifiable wartime emergency measures but had no place in most Americans' conception of a decent place to live in the long run. When word leaked in February 1945 of a proposal that had been drawn up by William J. Donovan to transform his wartime OSS into a single permanent, civilian-run peacetime spy agency, it triggered outraged denunciations in Congress and the press—led by the anti-interventionist, anti–New Deal, anti-FDR, anti-UN *Chicago Tribune*—about an "American gestapo," and the plan's hasty withdrawal from consideration by the Joint Chiefs of Staff.[50]

The leak had probably come from the Joint Chiefs themselves; like the FBI's J. Edgar Hoover, the military services loathed Donovan and wanted to stake their own claim to controlling any postwar foreign intelligence operations. But they had a point. Shortly before his death, Roosevelt asked his military aide, Colonel Richard Park Jr., to conduct an informal investigation of the OSS's activities, and Park's findings, passed on to the new president, confirmed many of the worst fears of Donovan's detractors. Park was scathing about the OSS's recklessness, amateurism, sloppy security, and lavish expenditures with little to show for it. Donovan's organization was "hopelessly compromised" and completely under the thumb of the British intelligence service; it had carried out many "badly conceived" and unauthorized operations that had resulted in grave embarrassment to the State Department and "interference with other secret intelligence agencies of this government." The latter was a pointed reference to a clumsy OSS break-in of the Japanese embassy in Lisbon in 1943 that had infuriated the U.S. Army codebreakers: for a while it looked like the Japanese might react by changing all of their diplomatic code systems, undoing Arlington Hall's crowning successes in breaking Purple and the Japanese military attaché codes.

The credulity that OSS officers had shown in paying dubious sources for even more dubious information almost defied belief. For over a year the OSS station in Rome, including its counterintelligence chief, James Jesus Angleton, had staunchly defended one of its prized agents, a shadowy character who professed to have inside political information from the Vatican; even as his reports became more and more ludicrous (at one point he reported a secret plan to construct an airfield within the Vatican garden), the OSS passed them on to Washington, firmly vouching for their reliability. In fact, the agent had simply made everything up. Park concluded that "if the OSS is permitted to continue with its present organization, it may do further serious harm to the citizens, business interests, and national interests of the United States." Within weeks after V-J Day, one of the first acts of a White House committee charged with liquidation of wartime agencies was to abolish the OSS and transfer to the State Department the one part of the organization that Park thought worth saving, its Research and Analysis Section, which had done "an outstanding job."[51]

And in the summer of 1945 the idea that the United States might be swiftly plunged into a new global confrontation even requiring the cloak-and-dagger skills of an outfit like the OSS seemed remote, even to American foreign policy experts who thoroughly knew the Soviet Union and who had been growing increasingly concerned by Stalin's swift and brutal moves to eliminate democratic opposition in Poland, Romania, and other Eastern European territories under Soviet military occupation. It would have seemed fantastic to most ordinary Americans, for whom the Russians were still gallant military allies who had borne the bloody brunt of the fight to defeat Nazism, and for whom the whole justification for this epic struggle that had sent millions of American boys into battle far from home was to bring about a safer world free from the terrors that had caused such suffering and turmoil. Harry S. Truman was a straightforward man, not a naturally eloquent one, but the brief speech he delivered upon arriving in Berlin in July 1945 for the last meeting of the wartime allies movingly expressed the faith of Americans that this time the terrible sacrifices of war would not be in vain:

> We are here today to raise the flag of victory over the capital of our greatest adversary. . . . We are raising it in the name of the people of the United States, who are looking forward to a better world, a peaceful world, a world in which all the people will have an opportunity to enjoy the good things of life, and not just a few at the top.
>
> Let us not forget that we are fighting for peace, and for the welfare of mankind. We are not fighting for conquest. There is not one piece of territory or one thing of a monetary nature that we want out of this war.
>
> We want peace and prosperity for the world as a whole. . . . If we can put this tremendous machine of ours, which has made victory possible, to work for peace, we can look forward to the greatest age in the history of mankind.[52]

Truman had been thrown into the presidency with no foreign policy experience—Roosevelt had literally never spoken to him about the war or foreign affairs or his plans for the postwar world—but he was a quick study and had spent night after night in the secure Map Room on the ground floor of the White House reading files of correspondence between Roosevelt, Churchill, and Stalin, long cables from the American ambassador in Moscow, Averell Harriman, intelligence assessments of Soviet actions in Poland, enough to leave little doubt that the Soviets' true intentions and interests in Eastern Europe had little in common with America's, or with the promises Stalin had made for free elections. Truman accordingly needed little convincing when Harriman, hastening back from Moscow to confer with the new president, urged a firm line with the Soviets based on a tough and realistic assessment that they simply could not be trusted. But Harriman still believed that the USSR, devastated by a war that had left twenty-seven million of its citizens dead and its economy in ruins, needed the United States and would not risk an outright break in relations, and that enough give-and-take was possible to keep the alliance intact. Truman replied that he understood that "100 percent cooperation" from the Soviets was not possible, but would be happy with 85 percent.[53]

Truman was also a thorough pragmatist who believed in swift decision making and delegating authority. On August 22, 1945, one week after the Japanese surrender, the Joint Chiefs drafted a letter for the secretaries of war and navy to forward to the president urging in the strongest possible terms that the wartime communication intelligence programs—which, they pointed out, had been vital to the defeat of the U-boats, the discovery and countering of German secret weapons, and detailed foreknowledge of Japanese troop movements and plans—be continued, with the same absolute level of secrecy that applied to atomic weapons.[54] Truman responded with a one-sentence order on August 28: there was to be no public release ("except with the special approval of the President in each case") of "information regarding the past or present status, technique or procedures, degree of success attained, or any specific results of any cryptanalytic unit acting under the authority of the U.S. Government or any Department thereof." Two weeks later he equally swiftly approved a proposal from Marshall and King that "in view of the disturbed condition of the world," the "present collaboration" with the British GC&CS be continued "by formal agreement." Truman left it to the secretaries of state, war, and navy to determine whenever "the best interests of the United States" required the arrangement to be extended or discontinued.[55]

Two arguments had carried the day in Truman's mind. One, as Captain Wenger presciently observed in a thoughtful memorandum he prepared on the future of signals intelligence, was that the advent of atomic weapons meant that peacetime intelligence was no longer merely a source of long-term strategic assessments of potential enemies and their military and political developments, but the only protection against an annihilating surprise attack that could come at any time, without any other warning. The example of World War II repeatedly proved that "effective intelligence . . . means, in a large measure, communications intelligence." Even in a world of collective security arrangements and international law, there was ample recent proof, too, that inside information gleaned from deciphering the diplomatic cables of allies and foes alike conferred a priceless advantage that it was hard to give up. Secretary of State James F. Byrnes had sent a worried memorandum to the secretary of war immediately following V-J Day seeking assurance that the Army's codebreakers would continue to supply his department with "the product of its cryptanalytic activities in the diplomatic field," whose value, Byrnes said, "will be equally great, if not greater, as we face postwar problems." One of its most notable recent contributions "in the diplomatic field," as Byrnes and Truman surely were aware, was the precise details of the advance negotiating positions of the fifty nations attending the conference that met in San Francisco in April 1945 to draft the UN Charter. Especially valuable were cables revealing the French delegation's efforts to secure the support of smaller countries, which allowed the United States to skillfully outmaneuver France's attempt to rally opposition to the U.S. insistence that the major powers hold a veto in the Security Council.[56]

The other consideration that made Truman's decision to approve the secret continuation of the Army's and Navy's "special intelligence" activities an easy one was the very fact of its secrecy. All of the experts repeatedly emphasized the grave danger that might be done were any hint of the work to leak out; that secrecy in turn meant that there was simply no need to justify it to the American public or Congress, or to risk the kind of newspaper debates in which the plans for a postwar OSS had become embroiled. That cut two ways, however. As NSA would find time and again, it was a little late to start trying to secure public support only after a scandal had broken that brought its activities to notice, usually in the most unfavorable light.

Even with support from the top, there were powerful forces of bureaucratic inertia, plus some inexorable facts about the difficulty of trying to re-create the wartime successes of cryptanalysis in peacetime, that were threatening to unravel the whole enterprise through the fall of 1945. Upon receiving Admiral King's instructions in May to increase attention to the Russian problem, Op-20-G had ordered a huge increase in personnel assigned to the section, from 106 to 743, but throughout the fall the numbers kept going the other way after peaking at a little under 200. The Navy codebreakers were fighting both the tide of demobilization of officers and enlisted men eager to get out of the service and the basic lack of career opportunities for those who stayed in: specializing in communications intelligence was never the way to get ahead in the U.S. Navy.

"The place is beginning to look like a deserted barn," Wenger glumly observed at the end of the year. At Arlington Hall, where a majority of the workforce was civilian, the same thing was happening; everyone who had been added to the Russian section in the spring, briefly swelling the staff to 91, promptly left after V-J Day. "This falling off can be traced to their feeling of insecurity in the face of a dim and tentative personnel policy on the part of higher authority," Rowlett acerbically noted in his annual report on the General Cryptanalytic Branch that fall.[57]

Yet a huge amount of effort was being expended on trivial targets: that was where bureaucratic inertia came in with a vengeance. In the absence of important intelligence results, the Army and Navy codebreakers just kept filling their reports with unimportant ones. The "Magic Daily Summary" that went to top officials in the State Department and White House, which had once brimmed with decrypts of high-level Japanese and German messages, now devoted page after page to desultory scraps about the future of French schools in Syria, the impending marriage of the Belgian regent, and the Vatican's views on upcoming elections in Colombia. "As a matter of fact, we have been getting disappointingly little of real value from C.I. [communications intelligence] since V-J day," Rear Admiral Thomas B. Inglis, the head of naval intelligence, complained in January 1946. Arlington Hall was reading the diplomatic codes of forty-five governments, including Saudi Arabia, Liberia, Luxembourg, Denmark, Ireland, and Panama, but though it was now intercepting more than fifty-five hundred Russian messages a month it had nothing yet worth reporting from that effort.[58]

Attempts to effect a postwar merger of the Army and Navy signals intelligence units had meanwhile dissolved into an acrimonious bureaucratic struggle that continued through the fall and winter of 1945. The Army strongly favored the idea, and some in the Navy did, too, arguing that a combined organization would be better able to resist outside control, avoid wasteful duplication, bring strength through numbers to bear on the most important problems, and prevent a repetition of the often absurd situation that had arisen during the war when the British, with whom each of the American operations often had better contacts than they did with each other, ended up acting as the go-between, allowing GC&CS to "play one service against the other" for its own purposes, Wenger noted.[59]

But Admiral King hewed firmly to the traditional line that the Navy had to be in full command of its own operations and not allow a situation ever to arise in which some civilian or general could tell them what to do. Op-20-G was in any case insisting that its 1942 agreement to hand over all work on diplomatic codes to the Army had been merely a "wartime expedient" (and was "invalid" in any case), and that the work should once again be shared now that the volume of military and naval traffic was rapidly dwindling. The Army was perfectly amenable to that, but on the larger issue of postwar organization and control King and Marshall were at a complete impasse. Wenger confided to a colleague in February 1946 that "everything has been in a state of confusion here" owing to the unresolved Army-Navy differences and "not a single basic decision affecting our future" had been made while the battle raged.[60]

The deeper anxiety that gnawed at both the British and American codebreakers was how fragile the whole structure was. Without a steady stream of coded material to work on, it would be extremely difficult to maintain either the prestige or the proficiency of the organizations, even if their personnel and budget problems were solved. Successful breaking of a code system required above all what the cryptanalysts called "continuity": if work was dropped even for a short time it often required a huge research effort to recover lost ground. And William F. Clarke, a veteran of the GC&CS naval section whose experience stretched back to the First World War and who still recalled with undiminished bitterness the blockheaded attitudes he had frequently encountered from naval officers over the years and the endless battles that GC&CS had had to fight to "convince the doubters of its necessity," gloomily foresaw more of the same in the coming years, especially when "those at the moment in the highest places" of government who had come to fully understand the value of signals intelligence were replaced by others without wartime experience in its use.[61]

But February 1946 suddenly brought a flurry of developments that profoundly shaped the course of the postwar signals intelligence establishment. The Army and Navy agreed to a cease-fire in their dispute, with an agreement to keep their two establishments separate for now but with a rotating "coordinator" who would allocate tasks that fell within their joint responsibility for diplomatic codes, and with a new interagency board that included a representative of the State Department to set communications intelligence policy. After months of bureaucratic wrangling over the overall government intelligence structure, President

2

Unbreakable Codes

It took Igor Gouzenko, a twenty-six-year-old code clerk in the Soviet embassy in Ottawa, forty hours to find someone to defect to.

It did not help that the short, tubby, ashen-faced man was so petrified, he was barely able to explain himself to the Canadian newspapermen and government officials he approached for help. The larger problem was that to believe what he was saying—that his country had been engaging in a massive act of perfidy against a wartime ally, running a vast espionage network that extended to officials in Canadian government departments and the British High Commission, scientists working on the Canadian-British atomic energy program, even a member of the Canadian Parliament—required standing the world of 1945 on its head. Mackenzie King, Canada's long-serving prime minister, registered in his diary the shock he felt when later confronted with the truth of Gouzenko's claims:

> I think of the Russian Embassy being only a few doors away and of them being a center of intrigue. During this period of war, while Canada has been helping Russia and doing all we can to foment Canadian-Russian friendship, there has been one branch of the Russian service that has been spying on [us]. . . . The amazing thing is how many contacts have been successfully made with people in key positions in government and industrial circles.[1]

Gouzenko's key proof was the texts of 109 cables he had transmitted to Moscow for Colonel Nikolai Zabotin, the resident head of the GRU, Soviet military intelligence, for whom he had been working since his arrival in Ottawa in June 1943. Fearing he was about to be sent back to the Soviet Union, Gouzenko had been making preparations for months, selecting the cables that he felt sure would buy him and his pregnant wife and young son a permanent home in the West. On the night of September 5, 1945, Gouzenko stuffed the documents inside his shirt and slipped out of the embassy—only to find himself brushed off as he tried for the next twenty-four hours to get the editor of the *Ottawa Journal,* the minister of justice, the Crown Attorney's office, and the Royal Canadian Mounted Police to take his story seriously.

The following night, exhausted from a day of fruitless trudging from office to office in the late summer heat, knowing that the embassy by now would be sure to be alarmed by his absence and the missing documents, the Gouzenkos returned to their apartment, and their despair turned to panic when a few minutes later a furious pounding on the door began and a Russian voice—Gouzenko recognized it as Zabotin's driver—called his name. The man eventually gave up and went away. But at around ten o'clock four men from the embassy returned, with an officer of the NKVD, the Soviet state security service, in charge. A neighbor had meanwhile offered to let the Gouzenkos stay with them for the evening, and from across the hall they heard the Russians break down the door of their apartment. The local police at last arrived, and after a brief, tense confrontation in which the Russians claimed diplomatic immunity and tried to order the police off what they declared was Soviet property, the NKVD officer led his men away into the night.[2]

The next day things began to happen. The Mounties took the Gouzenkos into protective custody and moved them to a safe house, a small, isolated lakeside summer cabin ninety miles from Ottawa, where Igor Gouzenko's months-long debriefing began.[3]

His most explosive revelations had to do with Soviet penetration of the atomic bomb program. The cables directly implicated a dozen Canadian scientists who had turned over to Zabotin myriad technical details gleaned from their American and British counterparts. Chief among the scientists was the physicist Alan Nunn May, who was part of a team building a large heavy-water reactor on the Chalk River north of the Canadian capital. Nunn May also served on two Canadian government committees that gave him access to high-level atomic secrets. One of

Gouzenko's cables consisted of a long report written by Nunn May at Zabotin's request that described the entire organization of the Manhattan Project, including the work being done to produce weapons-grade uranium-235 and plutonium at Oak Ridge, Tennessee, and Hanford, Washington, and the names of the scientists who headed each section. Another cable revealed that Nunn May had given the Russians a ten-page, single-spaced typewritten technical report on the bomb project plus a microgram sample of uranium-233, a rarer isotope that the U.S. scientists believed might offer an easier pathway to producing weapons material from reactors.[4]

In February 1946, still fearful of provoking a breach in relations with the Russians at a critical moment but feeling forced to act after the popular American newspaper columnist Drew Pearson, in his radio broadcast on February 3, broke the story of Gouzenko's defection, Prime Minister King authorized the arrest of Nunn May and some twenty others implicated in the cables. Like nearly all of the U.S., British, and Canadian scientists who would soon be revealed to have engaged in espionage for the Russians, Nunn May was an idealistic Communist who was convinced that he was advancing the cause of world peace by sharing U.S. atomic secrets with the Soviets. "The whole affair was extremely painful to me," he would later explain, "and I only embarked on it because I felt this was a contribution I could make to the safety of mankind. I certainly did not do it for gain."[5]

An early visitor to Gouzenko's lakeside cabin was Frank Rowlett.

In response to a prompt entreaty from the British, the Canadians agreed to allow an American expert on "crypto matters" to come and interview the Soviet defector, and Rowlett left Washington with two days' notice on September 25. He was driven out to the secret location accompanied by an RCMP inspector and a Canadian mathematician, Professor Gilbert Robinson, who had worked on signals intelligence during the war and had conducted a preliminary interview of Gouzenko to find out what cryptologic information he might have.

Rowlett returned without any breakthroughs to report but with a significant haul of small details for the team at Arlington Hall working on the Soviet diplomatic codes. Gouzenko confirmed that the diplomatic traffic was enciphered entirely with one-time pads, each containing fifty five-number groups. The GRU's messages, which Gouzenko exclusively handled, employed a different codebook from the ZET trade and ZDJ diplomatic systems that Arlington Hall had made the most progress on to date, but some of the details he was able to supply about the construction of the GRU codebook probably applied to those other systems as well.

The GRU book was what was known as a one-part code: the words in the codebook were placed in alphabetical order and simply assigned their corresponding code numbers in ascending numerical sequence, from the beginning to the end of the alphabet. That permitted a single codebook to be used for both encoding and decoding; by contrast, a two-part code, in which numbers were assigned in random order to the words, required the preparation of two separate books, one arranged in alphabetical order for encoding, the other by numerical order for decoding. A one-part code eased some of the logistical problems for the codemakers, but it offered a great boon to the codebreakers, making it much easier to guess the meaning of an unknown code group by revealing where it fell in the alphabet relative to other already known groups.

Another detail that Gouzenko supplied about the codebook was potentially even greater help. There was a special system, a code table within the code, used to spell out words using Latin or Cyrillic letters. This was needed whenever a foreign word, or a Russian word that had not been assigned its own four-digit code group, was transmitted. The special code group 7810 indicated "begin spell"; a two-digit group, 91, indicated the end of the spelled-out section. Within the spelled-out section, the numerical code groups stood for letters rather than words.

If, as seemed likely, a similar spelling system was used in ZET and ZDJ, that would be a huge wedge to crack this traffic open. Like stereotyped openings, the "begin spell" group itself was likely to appear frequently in messages, offering an entry point. Moreover, any spelled-out section, once the additive was removed, was in effect a simple substitution cipher, in which the same numeral always stood for the same letter. Such simple ciphers are susceptible to the most basic of all cryptanalytic attacks, which exploit the fact that in every language some letters appear far more frequently than others, making it possible to guess the identity of the symbols that stand for each letter by counting how often each appears.* Once the spell table was cracked using this technique, the codebreakers would have a significant handle on reading depths and recovering additive, since it is often easy to guess the additional missing letters of a partially recovered spelled-out word.[6]

The basic point was that if the codebreakers could make a reasonable guess as to what actual code group appeared at a particular spot in one message, then they immediately knew the value of the code group in the corresponding spot of a message in depth with that one, since both messages had been enciphered with the same string of additive key. For example:

message 1	2390
– assumed code group	0234
= recovered additive key	**2166**
message 2	5987
– recovered additive key	**2166**
= new identified code group	3821

Frequently occurring code groups that had been found this way could then be subtracted from every other message group to generate more "hypothetical key" that could be tested against other, unbroken messages to see if stripping this key yielded any other known, frequent code group, pointing to a possible depth.

The Arlington Hall codebreakers meanwhile had made several other important discoveries. It had been known for a while that in the "Red" codebook used in the ZET trade system, the one thousand code groups that stood for numbers all followed a simple pattern. It was a common design feature in codebooks to employ a "garble check" for numerals to

*In English, for example, E has a frequency of about 12 percent, Z about 0.3 percent, versus the 3.8 percent probability that would be expected if all twenty-six letters appeared with equal frequency. The spelling system used in the Russian codes was actually more complicated, based on digraphs—a different number stood for each unique pair of letters, such as CK, EE, LM, and so forth—but the basic principle of cryptanalysis was the same, as digraphs have a characteristic frequency distribution just as do individual letters.

ensure that no mistakes had been made in encoding or decoding; this often involved having the code groups follow a redundant self-checking formula, such as 0102 for the number one, 0204 for two, 0306 for three, and so on. The trade messages clearly had little to offer of intelligence value—most were simply reports of goods being shipped under Lend-Lease—but for the same reason they were full of stereotyped content that made them a cryptanalytic bonanza, particularly when they contained number-laden lists of commodities and shipment quantities in rigid format.

Samuel P. Chew, a professor of English at the University of Oklahoma who knew Ferdinand Coudert from when they were at graduate school together at Harvard, and who had taken the Army's basic correspondence course in cryptology before the war, had made the crucial discovery about the stereotyped format of the trade messages in April 1945. Richard Hallock's earlier key recovery work had been restricted to the first four or so groups of each message, where stereotyped openings were found, but Chew's break made the full length of the one-time-pad pages potentially open to discovery by making it possible to guess the underlying code groups contained within the body of a message. Cracking the spell table would do the same for other messages as well.[7]

Cecil Phillips had made another astonishing discovery the following month. Some of the trade messages, he found, revealed another sort of reused one-time key: the Russian code clerks, when they had a long message to encipher, had adopted the insecure expedient of using a page of key in the normal fashion for the first fifty groups, then using the same key page in reverse order for the next fifty groups. A search for these reuses eventually yielded four thousand such "reverse depths." The work was referred to as the "Red Reverse" problem, and a dozen staffers were assigned just to this task.[8]

It was now becoming clear that multiple pages of duplicate key had been issued by the Russians to users of the various one-time-pad systems; some of the duplicated pages were found in two different pads within the same system, but sometimes a page from a ZET pad also showed up in a ZDJ pad. As careless a blunder as it was to reuse any one-time-pad pages, however, none had been reused more than once: there might be two messages enciphered with the same key, never three or more. The conventional wisdom among cryptanalysts had been that having no more than pairs of overlapping messages to work with—a "depth of two"—was insufficient to break an enciphered code; it took multiple sets of overlapping messages each having depths of three or more to get anywhere. Arlington Hall's Russian section had already proved that wrong, working entirely from depths of two.* By early 1946 some fifteen thousand groups of additive key had been recovered and about two thousand possible code groups had been identified in the ZDJ codebook (which Arlington Hall designated "Jade"), but half of those were considered "doubtful" and only twenty had been assigned even tentative meanings. The path forward was clear in theory, but agonizingly slow in practice: over the following months the section added only about ten code groups a month to the list of known words in the Jade codebook. If the standard work of a codebreaker was looking for the proverbial needle in a haystack, the Russian problem required finding a wisp of straw in a haystack.

In January 1946, Meredith Gardner, a linguist who had worked as a translator on the Japanese military attaché codes, joined the Russian section as its chief "book breaker," whose job it was to work out the code group meanings. Gardner was a shy, slender, self-effacing scholar who had been teaching Spanish and German at the University of Akron; he had a master's in German from the University of Texas, where he had also learned Russian by taking lessons from the Russian-born grandmother of a fellow student. He had similarly picked up Bulgarian, Hebrew, and other languages along the way through his own studies. After being sidetracked for part of 1946 to work on Bulgarian diplomatic traffic, Gardner was back at Arlington Hall's Russian section on November 4.

Over the next two weeks he added eighty-six new recoveries to the Jade codebook, a burst of progress that nearly doubled, to two hundred,

*Trying to solve messages with a depth of two where each message had been encoded with a different codebook, as was the case with depths between the ZET and ZDJ systems, was not completely impossible either, assuming enough progress in code recoveries in each codebook. But it was obviously taking things to a level of unprecedented difficulty. The Red Reverse discovery, though, meant that a sizable amount of key recovered from the reversed-page reuses within the voluminous ZET trade traffic might also be directly duplicated in some of the ZDJ diplomatic traffic, in effect giving a depth of three.

the number of known code group meanings. Then on December 13 he broke the Jade codebook's spell table from a ZDJ message from 1944 that quoted a lengthy report in English. (The message itself was of no intelligence value, consisting of predictions on the likely outcome of the U.S. presidential election.) A week later Gardner broke part of another 1944 message, which also contained large stretches of spelled-out words: they were the names of "scientists who are working on the problem," the message stated, and the list included Hans Bethe, Niels Bohr, Edward Teller, Enrico Fermi, Emilio Segrè, Arthur Compton, Ernest Lawrence, and Harold Urey—all leading scientists who were working on the Manhattan Project, a fact that in 1944 was Top Secret.

It was the first substantial result from the three-year effort to break the Soviet diplomatic codes. ZDJ, it was about to become all too obvious, was not handling the ordinary diplomatic messages of ambassadors and consular staff at all. It was, rather, nothing less than the primary communication channel for spies of the NKGB, the Soviet foreign intelligence service, later to become the KGB. These encrypted telegrams, sent under diplomatic cover, were in fact the principal means by which Soviet agents in the West received instructions from Moscow and filed their intelligence reports back.[9]

Standing over his shoulder as Gardner decoded the atomic scientist message was a coworker, William Weisband. A linguist adviser to the project, Weisband was always a bit mysterious about what he called his "exotic" background. He knew Russian fluently, and could read and speak Arabic. But he spoke English with only a barely discernible accent, had served in the U.S. Army as a translator and cryptologist in North Africa and Italy, and was a friendly, naturally gregarious man who moved easily around Arlington Hall, where he had been stationed since the end of the war.

Gardner did not think anything of it at the time that Weisband was the first person to learn of his break into the Soviet espionage traffic.[10]

The interception of Russian army, air force, navy, and police messages was meanwhile amassing a vast mountain of unprocessed material. By 1946, Arlington Hall was receiving from U.S. and British intercept stations twenty thousand messages a month of Russian military Morse code traffic. Dozens of different code systems were being studied and some were far enough along to be readable, but their intelligence value was slim at best. Nearly all used fairly simple hand encipherment systems such as additive key books that were heavily reused, and thus no great challenge in principle to break; or transposition ciphers in which the order of the figures in the encoded messages was scrambled according to a determined pattern, again a familiar exercise for the codebreakers by this point; or even extremely simple monoalphabetic substitution ciphers in which each letter was replaced by another in an unvarying pattern, which posed almost no real challenge at all. The question was whether the still sometimes tedious and manpower-intensive drudgery required in each case was worth it. Clearly none of these hand systems were carrying high-level communications. If the American codebreakers had been hoping to re-create the triumphs of the war, when they had penetrated the D-Day battle plans of Hitler's generals or read the verbatim orders of Admiral Karl Dönitz to his U-boat wolf packs in the Atlantic, they were so far disappointed; instead of strategic assessments from the Red Army high command or secret orders from the Kremlin, they were reading a report on animal diseases in Siberia from an army veterinarian or an accounting of railcars under repair.

It was likely that the Russians were using machine-generated ciphers to protect their most secret military communications, but what those devices might be had been nearly a complete unknown to the American and British codebreakers during the war. West Coast intercept stations had begun to pick up radio teleprinter signals from the Soviet Far East in late 1944, however, and these seemed potentially important enough that Op-20-G ordered Lieutenant Tordella to drop what he was doing at Station S, attend a twelve-week course at Bell Laboratories in Manhattan on "special processing equipment," and then head back west to set up and run the new experimental test station—designated Station T, it was located at Skaggs Island in San Pablo Bay, just north of San Francisco—that would attempt to apply the "special" equipment to recording the Russian signals.[11]

The interest sparked by the Soviet radio teleprinter traffic came directly from GC&CS's phenomenal intelligence coup of breaking German military teleprinter signals. The German high command employed several teleprinter enciphering machines—the Lorenz SZ40/42 and the

Siemens T52 Geheimschreiber—for their highest-level communications between Berlin and the headquarters of theater commanders and army groups. In one of the most mathematics- and machine-intensive cryptanalytic feats of the war, the Bletchley codebreakers had, without ever capturing or seeing one of the actual German machines, reconstructed the strings of additive cipher generated by these devices (which they called "Tunny" and "Sturgeon" respectively, "Fish" collectively) and developed an electronic special-purpose computer, the Colossus, to match intercepted messages with the correct string of additive key to break them. The messages read at Bletchley included many giving detailed German military plans, dispositions of forces, and orders before and after the Normandy landings. It was a not entirely forlorn hope that the Russians might be using a similar teleprinter-based system to safeguard their most important military signals—and that it might prove equally vulnerable to a concerted attack.

Successfully intercepting the Russian radio teleprinter signals was a challenge to begin with, because the Russians, it was evident, were employing a complex transmission technique known as multiplexing, which combined two or more streams of traffic together in a single radio channel. Teleprinter machines used a system akin to Morse code, known as the Baudot code, which represented each keyboard character as a five-bit sequence made up of "marks" (conventionally denoted as X's) or "spaces" (denoted as dots). The Baudot code thus allowed for 2^5, or 32, different possible characters. In the Russian version, Л, for example, was X X • X X, И was • • X • •, Д was X X X X •; the special character • • • X • triggered a carriage shift from letters to figures to allow some keys to do double duty as numerals, punctuation, and less frequently used letters of the Cyrillic alphabet. The teleprinter could also be operated by

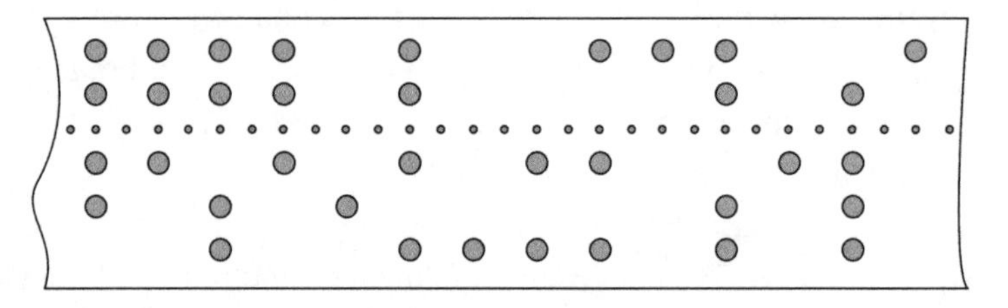

A five-bit paper tape punched with a message in Russian; for details of the Russian Baudot code and teleprinter encryption devices, see appendix B.

a paper tape punched in five rows according to the Baudot system, with a hole representing a mark and no hole for a space.

Once a message was prepared by punching it onto a tape, the entire transmission and reception process was fully automatic. There was none of the tapping out of each letter of a message by hand on a telegraph key while a radio operator at the other end transcribed onto a sheet of paper the dots and dashes he heard coming through his headphones; the paper tape simply was fed into a reader at one end, and the printer at the other end obediently clattered out the complete text. Sent over the airwaves, radio teleprinter signals consisted of a rapid, continuous stream of marks and spaces, with the marks usually represented by a small shift in the frequency of the transmitter's signal. In multiplexing, two or more signals were interleaved in a single stream, like perfectly shuffling two decks of cards together. To separate the signals back out at the other end, a "demultiplexer" with a rotating distributor shunted each successive incoming bit onto a separate wire attached to its own printer.

Some of the Russian signals carried as many as nine multiplexed messages at a time, but those nine-channel signals appeared to be all unencrypted, plaintext traffic dealing mostly with commercial and economic matters. The encrypted military teleprinter traffic was carried by two-channel multiplexed signals, and that was what Tordella was told to focus on. Within a few months he had an experimental demultiplexer operating and was filing the separated-out streams of intercepted teleprinter signals back to Washington. (Only later did he learn that the Army was doing exactly the same thing at its nearby intercept station at Two Rock Ranch in Petaluma, California.)[12]

Keeping the demultiplexer's distributor turning in exact synchronization with the incoming signal proved to be a maddeningly touchy business. The attitude of at least some of the higher-ups at Arlington Hall and Nebraska Avenue was that the whole thing was a phenomenal waste of time in any case, given the cryptanalytic challenges involved. One young Army officer who pressed for more attention to the effort remembered being told that there was no point in throwing away more time and money on a project that would only "add to the growing stack of unprocessed intercepts."[13]

. . .

Help came from an unexpected quarter. A full year before the end of the war, GC&CS had been drawing up plans to send small teams of cryptologic experts along with Allied troops advancing through Germany to locate and seize the equipment and records and, if possible, the personnel of the German signals intelligence services. The aim was to discover any German successes that might be exploited, to learn if the Germans had broken any Allied codes and possibly shared that information with the Japanese, and to secure or destroy any documents that might expose sensitive cryptanalytic techniques. An unstated aim was to get to the German codebreaking experts before the Russians did. In August 1944, General Marshall approved the plan and asked General Eisenhower to accommodate the teams, which would include American cryptologists drawn mainly from the U.S. units already at Bletchley Park. The operation was given the opaque code name TICOM, which stood for the equally opaque Target Intelligence Committee.[14]

Throwing a bunch of very unmilitary civilian linguists and mathematicians in uniform into an active war zone to carry out a James Bond–like mission was undoubtedly a gamble. Art Levenson was a math major from City College who by the skin of his teeth had extricated himself from infantry training at Fort Dix after being called up, and managed to get himself assigned to Arlington Hall, and then to the Signal Corps contingent sent to Bletchley to work on the Enigma project with the British. He recalled the reaction of the regular-Army sergeant in their unit who "couldn't get over this outfit" whose members could barely manage to appear with their uniforms on correctly: when the sergeant heard that Levenson and a few dozen others were going to Germany in April 1945, he said, "Boy, the war must really be over if they're sending you guys." Levenson was assigned to TICOM Team 1, led by a British major, John Tester, who had run one of the two groups at Bletchley that worked on Fish, and one of their key objectives was to try to capture an intact T52 Geheimschreiber. The senior American on Team 1 was Howard Campaigne, a Navy lieutenant commander who had been a mathematics professor at the University of Minnesota.[15]

Theirs was a frequently surreal adventure. A week after the German surrender the members of Team 1 were picking through Hitler's, Goering's, and Ribbentrop's villas at the Nazi leaders' alpine retreat at

Berchtesgaden. The not very convincing cover story to explain their presence was that they were there to check that various German headquarters were obeying the terms of the capitulation forbidding the use of cipher equipment. Getting permission from local American commanders to enter restricted zones or carry equipment out was a constant challenge. Following a report that Field Marshal Albert Kesselring's entire communication "train," four trucks equipped with a T52 machine and radio gear, was located nearby at Zell am See, just south of Berchtesgaden, two members of the team took a jeep over a back mountain road, traversing rocks, meadows, and snowdrifts to get around a blown bridge and evading patrols of the 101st Airborne Division, which had issued a "freeze" order forbidding any movement in the area. When they arrived, the TICOM team found officers of the Luftwaffe high command headquarters still very much in charge: they were all armed, had posted sentries with submachine guns, and "seemed quite unconvinced that the war was irrevocably at an end." Only after a considerable amount of persuading did the Germans reveal the whereabouts of the communications train, which the Americans were then able to secure.

Getting the captured trucks, and the enlisted men who had operated them, out of Germany produced other absurd moments. Levenson and Tester accompanied the convoy, the Germans driving their own vehicles. The men proved far more cooperative than their officers, genuinely eager to be of assistance and happy to have escaped falling into the hands of the Russians, so Levenson—a Jewish kid from Brooklyn—ended up letting one of his German prisoners carry his rifle for him. This led to a tense confrontation with an angry mob of civilians in Belgium, who, seeing obviously German trucks and an armed German soldier, thought the Nazis were back.[16]

The TICOM teams had few leads to go on and for the most part would simply show up at the headquarters of a U.S. Army unit and ask whether they knew of any German signals units or research stations in the area, or any prisoners they were holding who had worked in signals intelligence. A lot of leads went nowhere, and for the most part the occupation forces had just thrown a guard around any German installations they found without bothering to learn what they were. Campaigne remembered following up one report about a possible German research facility high in the Tyrolean mountains; driving up there, he found a lone GI standing guard at the door. Campaigne asked him what the Germans had used the site for. "Aw, them Krauts"—the soldier replied with a dismissive wave of his hand—"they built all kinds of shit."[17]

On May 21, Campaigne received a much more solid tip. The U.S. Seventh Army was holding at the POW camp in Bad Aibling, between Berchtesgaden and Munich, a prisoner who had told the camp authorities that he had worked on intercepting and decoding Russian radio teleprinter traffic in a unit that had most recently been stationed at a barracks in nearby Rosenheim. Campaigne hastened over to interview the man. Unteroffizier Dietrich Suschowk told him that he and the other nineteen men of the unit—all now prisoners in the camp—had buried their equipment and documents on the grounds of the barracks. The next day Campaigne returned with a truck, collected the twenty prisoners, drove them to Rosenheim, and set them to work digging up their cache of cryptanalytic secrets. There was eight tons in all. The Germans were "helpful but afraid" at first, but quickly "showed the greatest willingness to cooperate": the men, all NCOs, viewed themselves not really as soldiers but as "specialists, with a genuine pride in their work," and when Campaigne evinced an interest in the technical aspects, they eagerly offered to reassemble one of the devices and demonstrate how it worked.[18]

On June 5 the equipment was on its way to England by air, and on June 29 six of the prisoners who seemed most knowledgeable about its operation arrived at the village of Steeple Clayton, fifteen miles southwest of Bletchley Park, where their gear awaited them. By the next evening they had one of their prize pieces of apparatus assembled and working. The Hartmehrfachfernschreiber, or HMFS, was a demultiplexing receiver that by inserting different distributors could separate out anywhere from two to nine multiplexed channels. It was an extremely sophisticated piece of equipment. An oscilloscope provided a visual indication of when the receiver was properly synchronized, and an automatic circuit then locked in the synchronization; a short-term memory buffer using relays automatically converted the rapid incoming pulses of the radio signals to the longer, 20-millisecond length that the teleprinter machines operated at. The following day the Germans had fifty teleprinters hooked up, printing out streams of intercepted Russian two- and nine-channel signals.

The work had to be suspended a few days later when the British Post Office authorities called up to report that radio listeners in the area were complaining that they could not pick up the BBC and wanted to know if "any unusual electrical machinery" was being operated: apparently the distributors and related electronics of the HMFS were generating considerable local radio interference. Worried that more nosy inquiries might follow, and satisfied that the secret tests had been a complete success, GC&CS shut down the operation and sent some of the equipment to its intercept station at Knockholt in Kent, which had been collecting German teleprinter traffic during the war; at the end of July, one of the German nine-channel receivers and two of the two-channel receivers were on their way to Washington. The German prisoners stayed in England for four months of further interrogation.

Other TICOM teams meanwhile were producing some other extraordinary finds. U.S. Army divers working to recover the body of a drowned American soldier were dragging the Schliersee, a small lake in Bavaria, when they snagged a large waterproof box. A team of Army engineers was brought in for a methodical search of the deep lake, and over the next several weeks they found twenty-eight more boxes. Inside were four tons of documents, which proved to be the complete archives of the German high command's cipher bureau. Two C-47s flew the entire haul to the RAF's airfield at Biggin Hill, southeast of London, where trucks waited to rush the material to Bletchley Park.[19]

Earlier, another TICOM party, Team 3, stumbled on a German Foreign Office cryptologic unit in a castle in Naumburg, near the city of Leipzig. Warning the staff that they would be shot if any of their files were destroyed, the head of the TICOM team returned a few days later with the rest of the unit to pack up and remove the documents, three hundred thousand pages in all. The trove, filling 170 steel file cases, was spirited out under the noses of Russian forces moving in to occupy the area.

Among the Naumburg material were copies of several codebooks that—only much later—would Arlington Hall realize had been recovered from a burning pile left when the Russians hastily evacuated their consulate in northern Finland during fighting in 1941, and were subsequently passed by the Finns to their German quasi-allies. One was a codebook the NKGB called Kod Pobeda, which had been used for its one-time-pad messages from 1939 until November 1943, when the Jade codebook came into use. It was only in the mid-1950s, when the American codebreakers launched a second round of work on the Russian one-time-pad messages, that they discovered they already had in their possession the actual codebook for this earlier period, whose messages up until then had largely defied solution.[20]

One of the German prisoners rounded up by TICOM Team 1 who stood out as particularly intelligent and helpful was Unteroffizier Erich Karrenberg. He had grown up in Russia, the son of a German manufacturer, and spoke the language fluently; before the war he had been a concert pianist and lecturer in the history of art and music at Berlin University. He obviously had a mathematical mind as well, and from the start the GC&CS experts who spoke with him saw that he was thoroughly conversant in cryptanalytic terminology.

During his time in England, Karrenberg was extensively interviewed and at the request of his interrogators also wrote several detailed reports—"homework," they called it—describing Russian radio call-sign and contact procedures, the organization of the Red Army's radio networks, and the encipherment systems used on the Russian teleprinter traffic. Karrenberg stated that some of the traffic was sent in the clear or with very minimal concealment, such as scrambling some of the multiplexed channels, which he said he was regularly able to break if he had two thousand letters of depth to work from. There were other messages that consisted of three-, four-, or five-letter groups, evidently enciphered codes, some of which the Germans had solved but some they had concluded were enciphered with a one-time pad, and thus unbreakable.[21]

But there was also a considerable amount of text on the two-channel circuits that apparently was being enciphered by an online device akin to the SZ40, which used a bank of cipher wheels to generate a stream of continually changing key that was automatically added to the Baudot code of each character as it was transmitted to produce the outgoing cipher text. The Germans called this device the "Bandwurm." The wheels advanced automatically in step with the outgoing or incoming text, so once the machines on each end were set up with their wheels in matching positions, the entire encryption and decryption process was invisible and seamless.[22]

But Karrenberg reported that the Russian operators were often extremely careless in the way they handled the setup of their messages; they would frequently give away the starting-position indicator of a message during their preliminary unenciphered "chat" as they established a connection, directly revealing instances when two or more messages had been sent with the same starting positions—a huge shortcut to finding depths. Another gaping insecurity was the inclusion of standard preambles and addresses within the enciphered portion of the messages, which meant that there were highly predictable, stereotyped openings that were the same from one message to another, making it fairly easy to locate depths in the traffic by looking for repeated strings of cipher text indicating that the same text had been enciphered with the exact same key. Another rich source of depths came from the fact that the Russians' transmitters and receivers would often get out of sync in the middle of a transmission, requiring messages to be resent, resulting in large amounts of duplicated plaintext being enciphered.[23]

The Germans had not tried to reconstruct the machine's actual wheel patterns and thereby directly predict the key streams it generated, as the cryptanalysts at Bletchley had done to routinely read the German Fish traffic. But even without cracking the key-generation algorithm of the Bandwurm, the German codebreakers, using little more than inspired guesswork, were often able to decode paired Russian messages that they knew from the operator's chat or other clues were in depth. (The method exploited the fact that subtracting two such messages in depth from each other eliminated the key altogether, leaving only the difference of the two underlying plaintexts; see appendix B.) Bandwurm traffic read by the Germans included reports from Army Front headquarters to GHQ Moscow on their position and situation; orders concerning postings, transfers, and promotions of officers; signals intelligence and POW interrogation reports; and other similar information of considerable intelligence importance. Some messages of the Soviet air force and the armed units of the NKVD security forces also passed on the two-channel Bandwurm networks.[24]

The British referred to the Russian Baudot scrambler at first as the "Russian Fish" and then, "pending official designation" of a cover name, "Caviar," which was hardly an improvement as far as security went, but the name stuck, at least informally, for much of the next year.[25] GC&CS and Arlington Hall and Op-20-G began at once a concerted effort to recover long stretches of key, as had been done at the start of the attack on the German Fish. From that it might be possible to deduce the wheel patterns and thereby reconstruct the device's key-generation algorithm. Then, finally, with powerful enough computational machinery, every possible sequence of key could be slid against a received message text and tested for statistical evidence of a likely match. With such a complete system, any message—not just the small percentage that happened to be in depth—could be read consistently.

But, as the Op-20-G War Diary noted the following spring, the work was going slowly: "The greatest need is more long key, which is only available from reading depths." Both the Army and the British decided to temporarily put the project on hold; the Navy was left with only one or two officers, borrowed from other sections, to carry the work alone through the spring and summer; and then in August 1946, Howard Campaigne reported that Op-20-G had temporarily abandoned the project as well due to "lack of personnel."[26]

The main problem, besides a shortage of manpower, was a lack of sufficient intercepted traffic to work from. The HMFS demultiplexers had saved six to nine months' work that would otherwise have been required to design and build equally advanced devices. But the number of devices that had been found at Rosenheim and brought to Britain and the United States was hardly enough to provide complete coverage of the Soviets' teleprinter networks, which numbered ten to fifteen in Europe and three to four in the Far East; at best they would be able to provide "spot checks" on each of the networks. It took the better part of a year for the cryptanalysts' pleas for the means to fully cover the teleprinter traffic to be heard. The Army and Navy finally agreed to budget $200,000 to manufacture additional demultiplexers, and a small assembly line was set up in the basement underneath the cafeteria at Arlington Hall to produce them, but it was not until the very end of 1946 that the sets began arriving at the Army's and Navy's intercept stations.[27]

Although nearly all the details remained classified by NSA even seven decades later, scattered hints in the Op-20-G War Diary and elsewhere strongly suggest that Bandwurm, alias Caviar, was the same system subsequently code-named Longfellow by the British and Americans.[28] When work resumed in early 1947, the British were able to report a "rapid advance" on Longfellow that February. Some twenty "analogs" of the teleprinter encipherment attachment were built (code-named Tan, they employed relays in place of the rotating wheels and were used both for generating key and for decipherment). But it was still not possible to break all of the traffic; doing that hinged on the construction of a planned "superbombe" that would be able to statistically match each message with its encipherment key, allowing continuous reading of the Longfellow messages. (The computer was named Hiawatha, an obvious allusion to Longfellow's famous poem.) The superbombe project faced repeated delays, however. The device was an ambitious gamble: its calculating and logic circuits were to be built around forty thousand vacuum tubes, more than twice the number contained in any of the pioneering digital computers under development in the United States and Britain at that time, and its projected cost was $1 million. Hiawatha was finally approved in late 1947—just a few months before the Soviets abruptly stopped using the Longfellow system, causing the entire project to collapse.

That ignominious end infuriated Campaigne, who observed in disgust that if the Longfellow traffic had been fully collected from the time it first appeared in 1943, the system would probably have been solved in 1945. "And if we had supported this by the analytical machinery recently planned, we could have broken out most of the available traffic. The entire story is one of 'too little too late.' This system was in use for five years, yet we were not ready to read it in quantity until it disappeared."[29]

The huge reduction in staff resulting from postwar demobilization was the most obvious problem causing headaches for the men in charge of the signals intelligence units, though it was actually just one manifestation of a much larger problem they faced in transforming hothouse wartime organizations that had run on improvisation, inspiration, and adrenaline into permanent peacetime establishments that could settle down efficiently to the quiet, steady task ahead, during years or perhaps even decades of peace.

The Army and Navy signals intelligence organizations had both drawn up plans for a permanent workforce of about half their wartime levels: five thousand at what was now called the Army Security Agency (ASA), and twenty-five hundred at Op-20-G, with about half of those totals stationed at their respective Washington headquarters.[30] That looked fine on paper, but said nothing about whom they were actually going to find to fill those jobs once the shooting stopped. Inevitably, the wartime excitement that had drawn in an astonishing number of unusually talented men and women and imbued the work with such a sense of urgency, importance, and necessity was gone. At GC&CS and Arlington Hall in particular, formal lines of authority had never counted for much during the war; getting the job done was what mattered, and in large part because no one planned to make a career of the work, no one was very career-minded about office politics or promotion or pay or protecting their bureaucratic turf. Cecil Phillips remembered wartime Arlington Hall as a true "meritocracy" where a sergeant, who in a considerable number of cases might have a degree from MIT or Harvard or some other top school, and a lieutenant might work side by side as equals on the same problem and no one thought much about it.

At Bletchley Park the culture of informality and individual initiative was even more manifestly a part of the wartime ethos. "No one ever really regarded anyone as their boss," recalled Stuart Milner-Barry, a top-rated member of the British chess team who became head of Hut 6, the army and air force Enigma section. Milner-Barry said he had never once had to actually order anyone to do their work.[31] But by the same token few of the top-level mathematicians and scholars who had put aside their careers to help win the war, and whose intellectual firepower and unmilitary individualism and collegiality had contributed to Bletchley's unique atmosphere, had much interest in being part of a government or military bureaucracy in peacetime. Max Newman and Jack Good, both first-rate mathematicians who had conceived the attack on the German Tunny teleprinter codes and been instrumental in the design of Colossus, left, as did the formidable Alan Turing; all three joined a project at the University of Manchester to develop one of the world's first stored-program, general-purpose electronic computers, the Mark 1. Theirs was just part of a mass exodus that sapped the postwar vitality of GC&CS. William P. Bundy, who had been operations officer of the 6813th Signal Security Detachment, the first American contingent sent to work at Hut 6, returned to England for a vacation in the summer of 1947 and, as he reported in a newsletter that the group's veterans published for a few years, "spent three weeks in England . . . saw almost all of the old Hut 6 people. . . . The trip left me very depressed . . . the wartime spark had gone, and no new urge had taken its place."[32]

At ASA, peace brought a flood of pettifogging orders, policy directives, and procedural instructions, accompanied by a succession of martinet junior officers who rotated in and out and often knew nothing about cryptanalysis but were sticklers for organization, military protocol, and the chain of command. Lengthy interoffice memoranda circulated dissecting the merits of developing a personnel handbook, or analyzing whether a proposed change in policy that would allow civilian employees of Arlington Hall to be admitted to the post movie theater was consistent with Paragraph 10, AR 210-389 of the Army Regulations. "Low pay and too many military bosses" would be a recurring complaint from ASA's civilian workforce over the next few years, along with a sense that no matter how much experience they had or how qualified they were, the top positions in each division would always go to a less qualified Army officer.[33]

The Navy's troubles were exacerbated by personnel policies that prevented most of Op-20-G's members of the Naval Reserve—which included virtually all of the civilian experts it had recruited during the war—from being considered for permanent peacetime positions. Wenger noted that Op-20-G had an unusually high proportion of older men (19 percent of them had completed their PhD studies at the time they became reserve officers), which made them ineligible to apply for a transfer to the regular Navy because of age limits intended to ensure that new officers could devote a full career to the Navy. Wenger proposed creating five hundred new civil service positions to retain Op-20-G's most experienced men, and Louis Tordella and Howard Campaigne were two of the unit's wartime officers who had the experience of going out the door of Nebraska Avenue one day, returning the next day in civilian clothes, sitting down at exactly the same desk, and resuming exactly the same duties. But the bureaucratic machinery was slow to turn, and only three hundred civil service slots had been approved by the summer of 1946.[34]

At Arlington Hall the civilian workforce was down to twenty-two hundred by the summer of 1946, and as at GC&CS, many of the brilliant linguistic, mathematical, and legal minds who had contributed to the triumphs of the war were gone, Richard Hallock, Ferdinand Coudert, Genevieve Feinstein, and Alfred McCormack among them. William Friedman and his original three assistants, Frank Rowlett, Abe Sinkov, and Solly Kullback, had all decided to stay on, however, and they would prove a stabilizing force in maintaining the technical professionalism of the agency through its difficult times to come. They had been able to persuade a number of their best protégés to stay on as well—among them Cecil Phillips; Leo Rosen, an MIT electrical engineering major who had contributed a crucial idea to the reconstruction of the Japanese Purple machine; and Samuel Snyder, who had devised the first method for using IBM card sorters and tabulators to search for repeated groups in enciphered codes in 1937.[35]

But Rowlett, in a long paper he wrote toward the end of the war, pointed out that holding on to the currently trained group of cryptanalysts was only the most immediate challenge; if ASA was going to be set on a permanent footing it would have to face up to the reality that recruiting was a never-ending requirement. Finding people who wanted to make a career of cryptanalysis was an entirely different proposition from what had gone on during the war. One thing the war *had* shown was that screening tests were of little use in identifying those who had an aptitude for the work, and that only lengthy on-the-job training and experience would tell. There was after all no such thing as a pool of professional cryptologic experts in civilian life that the services could draw on: "No cryptanalytic proving ground exists outside of the field itself," Rowlett wrote.

If the Army wanted professional cryptologists it would have to create the profession itself and offer the kinds of rewards that customarily accompanied a professional career; it also would have to provide nearly all of the required professional training. In the war it had been possible to secure the temporary services of senior men, but a permanent organization would have to be built, and continually rebuilt, from the ground up with young men and women at the very outset of their careers. As Rowlett pointed out, the training demanded of a cryptanalyst "requires time and the resiliency of youth. The long hours of false leads and blind alleys demand the faith in ultimate success which characterizes young people."[36]

The new and deliberately vague name for what had been the Army's Signal Intelligence Service hinted at a more fundamental difficulty the wartime signals intelligence agencies were encountering in trying to remake themselves as a permanent part of the government national security establishment. (The Navy matched the Army Security Agency with its own minor masterpiece of bureaucratic obfuscation in the new name for Op-20-G's Nebraska Avenue headquarters: Communications Supplementary Activities, Washington, or CSAW—pronounced "see-saw." The British Government Code and Cypher School, which in April 1946 had moved from its never-convenient emergency wartime location at Bletchley Park to a set of drab government office buildings in Eastcote, a northwestern suburb of London, was rechristened with an equally misleading name, Government Communications Headquarters, or GCHQ.) The very existence of government bureaus that were intercepting and decoding communications in peacetime was a matter of moral murkiness, with the constant potential for diplomatic, legal, and political backlash. Instructions that ASA issued in 1946 decreed the use of the new code word CREAM for "special intelligence," defined as all information "which results from the decryption of the texts or substance of encrypted communications." The aim, the order explained, was "to restrict dissemination of this type of 'TOP SECRET' material within the narrowest possible limits, based on the 'need to know.' "[37] It pointedly added:

> Disclosure of "CREAM" material or its existence causes a two-fold embarrassment to the U.S. Government:
>
> (1) Operational and loss of intelligence
>
> (2) Political

The lack of any clear statutory authority for what they were doing reinforced the imperative to disguise their work. Wartime censorship had provided adequate legal authority for the military agencies to copy cable traffic, and a lengthy brief prepared by the Army's judge advocate general at the end of the war argued that the president, by virtue of his broad powers to conduct foreign affairs, had the authority to continue monitoring foreign communications in peacetime; the brief also cited a 1928 Supreme Court precedent, *Olmstead v. United States,* which held that government wiretapping did not violate the Fourth Amendment's privacy protections, since it did not involve the "searching" or "seizure" of a person or his private effects or the physical invasion of his home.[38] Getting around the language of the Federal Communications Act of 1934, however, required more dexterous legal footwork, since the act pretty clearly made it a federal crime for anyone to divulge to a third party the contents of any intercepted communication.

The cable companies nonetheless agreed to continue their wartime arrangement, allowing an Army officer in civilian clothes to pick up microfilm or paper tape copies of all international telegrams. "The Army came to me and asked for the company's cooperation," the president of RCA Global related years later to congressional investigators looking into the origins of the program, "and, by damn, that was enough for me." But in fact the entire arrangement had plenty of people worried, and ASA went to great lengths to keep it secret. The presidents of RCA Global and the two other major cable firms, ITT World International and Western Union, sought repeated assurances from the government that the program was essential to national security and that their companies would not face prosecution for their actions. In 1947 the executives called on the new secretary of defense, James V. Forrestal, to ask if he could reassure them that the president himself had authorized the program. Forrestal did, though he added that he could not bind his successors.[39]

Yet putting things on a more solid legal footing risked exposure itself, and when a plan was discussed by the signals intelligence agencies the following year to submit formal legislation that would legalize the companies' assistance, the counsel for the U.S. government's interagency Communications Intelligence Coordinating Committee warned that the crucial thing was to head off any "embarrassing line of questions" when the bill was taken up by Congress. "The greatest danger in debate will be ill-advised screams from some members of the press that the bill invades freedom of the press and civil rights." The bill died a quiet death, and the question of the legality of signals interception in peacetime retreated back into its murky shadows. It would fester there for another three decades, until the Watergate scandals at last made it impossible to ignore the hidden toll that excessive secrecy exacted, which by then had become every bit as damaging as the much more obvious price to be paid by excessive disclosure.[40]

3

Learning to Lie

Throughout 1947, Meredith Gardner continued to toil away on Arlington Hall's most secret project, adding to the reconstructed codebook that lay at the heart of the Soviet spy messages. It was becoming clear that the four different one-time-pad code systems used for messages sent to and from Soviet embassies and consulates each belonged to a different organization of the Soviet foreign and intelligence bureaucracy: GRU, or military intelligence; naval GRU; NKGB; and "true" diplomatic, which dealt mainly with mundane consular matters such as visa applications. Up until 1941 the GRU was the main foreign intelligence organization of the Soviet Union, but the NKGB, on Stalin's directive, had since eclipsed the military spies in that role, even though the GRU and naval GRU continued to operate abroad.

Nearly all of the NKGB messages that had been rendered vulnerable to cryptanalysis owing to their encipherment with duplicated key pages were sent during just three years, 1943 to 1945. The Arlington Hall cryptanalysts would eventually conclude that the duplication was the result of wartime disruptions that forced Soviet printing plants to cut corners for a few months in early 1942. Once the duplicate pages issued during this short period were used up, the window on the Soviets' most secret communications left ever so slightly ajar by this fatal slip slammed shut once again: even with the codebooks broken, only messages in depth could be read.[1] But in the meantime thousands of potentially breakable messages enciphered with duplicate key had been transmitted—and successfully reading them might still yield valuable intelligence about ongoing Soviet espionage operations or the identity of agents still in place. When the project, best known by its final code name, Venona, finally ended in 1980, 49 percent of the 1944 NKGB New York–Moscow messages had been read at least in part, 15 percent of the 1943 messages, and 1.8 percent of the 1942 messages, along with 1.5 percent of NKGB Washington–Moscow traffic and 50 percent of 1943 naval GRU Washington–Moscow traffic.[2]

The Kod Pobeda book found by TICOM Team 3 was applicable only to NKGB traffic sent before November 1943, and as of 1947 Gardner remained unaware the book had even been recovered. Thus the effort at Arlington Hall at this point still relied upon pure cryptanalysis and "book breaking" to reconstruct both the one-time-pad pages and the underlying codebooks, all sight unseen of the originals. Gardner's breaking of the spell table used in the post–November 1943 NKGB Jade codebook meant that messages containing proper names were the ripest for picking, and by the summer Gardner had recovered portions of several 1944 and 1945 messages that contained what were clearly Soviet cover names for dozens of agents in America. The results were so fragmentary that issuing them as individual serialized translations and circulating them to Arlington Hall's usual list of recipients in the War and State departments and the White House, the practice throughout the war, seemed slightly ludicrous to Gardner, as well as all but impossible given the extraordinary security restrictions surrounding the work on the Russian problem. Not knowing quite what else to do, he decided to write up a series of special reports that attempted to pull together the available information; Gardner sent them to ASA's Cryptanalytic Branch chief, Frank Rowlett, in the hope that he might get the attention of top officials at ASA and the War Department's military intelligence staff, G-2, who in turn might know what to do with material that was at once too enigmatic and too hot to handle through normal channels.[3]

A draft of Gardner's "Special Analysis Report #1" landed on the desk of Colonel Harold G. Hayes, chief of ASA, on July 22, 1947. Gardner listed about two dozen cover names he had culled from hundreds of partially readable NKGB messages. From context he was able to determine that certain other words spelled out in those messages were cover names for places: TYRE was New York City; SIDON was London; CARTHAGE was Washington. In several messages the code word ENORMOZ appeared, once in juxtaposition with the names of Los Alamos scientists; it seemed likely that this was the code word for atomic bomb espionage. And an agent code-named LIBERAL and ANTENNA was mentioned as a go-between who passed on information gathered by certain other persons working on ENORMOZ. In one of his special reports Gardner summarized the information he had been able to assemble about this agent:

> LIB?? or possibly LIBERAL: was ANTENKO until Sept 1944. Occurs 6 times, 22 October–20 December 1944. Message of 27 November speaks of his wife ETHEL, 29 years old married 5 years.[4]

Another agent who caught Gardner's eye was G or GOMER or HOMER; he had cropped up in a number of Washington–Moscow and New York–Moscow messages. HOMER appeared to be well connected to highly placed British officials, and from the fragmentary translations Gardner was able to make in 1947 and 1948 he seemed to be supplying political intelligence of a distinctly above-average quality: in one message, HOMER provided details on advance planning for Roosevelt and Churchill's secret conference that took place in Quebec in September 1944, while others proved to be verbatim copies of telegrams sent by Churchill to the British embassy in Washington.[5]

Still other messages read by Arlington Hall in 1947 and 1948 unmistakably pointed to well-placed Soviet agents in the U.S. War Department and in the Australian government. Two messages from New York to Moscow sent in late 1944 on the NKGB system contained long quotations of documents supplied by an agent ROBERT: the documents, prepared by the War Department general staff and classified Top Secret, dealt with plans for postwar U.S. troop deployments. In October 1947, Gardner read two long cables in the same system that had been sent from Canberra to Moscow in early 1946; the Australian traffic had been intercepted by the British, who turned it over to Arlington Hall. They appeared to be extracts of a classified report published by the British War Office. GCHQ's liaison man on the project, Philip Howse, contacted London to ask that a search be made for the original publication for comparison with the ciphered cable. GCHQ replied two weeks later: "Papers have been traced and being forwarded forthwith. Wonderful job." This was a huge windfall, providing a spectacular run of matching plaintext—a "crib" in cryptanalysts' lingo—which added a flood of codebook recoveries.[6]

That the Soviets had spies operating in the Manhattan Project and in the highest levels of the British, American, and Australian governments was startling enough, but the traffic revealed something else about America's erstwhile ally and rapidly emerging foe. The Russians were, to put it simply, extremely old hands at this game. To the Soviet spymasters conspiratorial thinking, counterespionage, and countersurveillance were second nature, and the precision with which they spelled out arrangements for dead drops, communications procedures, and shaking tails showed how thoroughly they grasped that the spy game was above all a deeper counterspy, or counter-counterspy, game. A message to a Soviet NKGB officer in London shortly after the end of the war specified the elaborate precautions to be taken for a rendezvous with an American agent, a member of the Communist Party USA who upon returning to Washington from his Army service had "agreed to continue work on the collection of intelligence":

> DAN will go to the meeting and await our man for 10–15 minutes on the pavement immediately at the exit of the Regent Park tube station in Regent Street. DAN will have the magazine "John Bull" in his hand. Our man is to approach first and, after greeting him, will say, "Didn't I meet you at Vick's Restaurant at Connecticut Avenue." To which this DAN is to reply—"Yes, Vick himself introduced you." After this our man is to show DAN a small (19 groups unrecoverable) is to show an exact copy of this label. Then the two men will [talk] business. We recommend ALAN to contact DAN.[7]

To a later generation brought up on spy novels all of this might seem old hat, but in fact these were the hallmarks of a professionalism of a kind that their Western adversaries were neophytes at by comparison circa 1948. The Soviets had none of the Boy Scout bungling and amateurish credulity of their sources that had so tarred the OSS. When Igor Gouzenko defected in Canada, Moscow immediately sent instructions to protect its most important agents in the United States from exposure, even if it meant breaking off most or all contact with them for months, even years, in order to preserve their future usefulness:

> Surveillance has been increased. Safeguard from failure: HOMER, RUBLE, RAID, MOLE, ZHORA, and IZRA. Reduce meetings with them to once or twice a month. Minor agents should be deactivated. Carefully check out surveillance when going to meetings, and if anything seems suspicious, do not go through with them.[8]

The NKGB was playing a long game, for keeps.

As careful students of Russian history like the State Department's George Kennan observed, the Soviet genius for conspiracy, paranoia, and secrecy had deep roots. Russian hostility toward and suspicion of the outside world was not merely a reflection of the Communist system but the stubbornly entrenched legacy of a village peasant society whose backwardness had endured for centuries after its equivalents in Western Europe had been swept away by the modernizing influence of Christian morals, cash-based markets, public education, the rule of law, a politically engaged landowning class, and growing national and linguistic identity. By the end of the nineteenth century even the most remote rural farm worker in Norway or England or France was tied to something in the world beyond family and clan loyalties. The Russian muzhik, up to and even after the time of the revolution, lived in a world that was not even medieval but prehistoric in its horizons. There was virtually no cash economy; most landlords were absentees utterly uninterested in their communities; most of the Orthodox Church's priests were uneducated and corrupt; public schooling was all but nonexistent; peasant families farmed land in a primitive system of shifting allotted strips to which no one ever held permanent title, and thus had no incentive ever to improve. Law was limited to what the government forced one to do, mainly pay taxes and be conscripted into the army. There was scarcely even a sense of Russian national identity or patriotism: the czars feared anything that might imply an allegiance to institutions other than the person of the czar himself, and so actively discouraged the most basic attributes of the modern state.[9]

The result, as Kennan explained in what would become his famous analysis of Soviet intentions and thinking—the "long telegram" that he cabled back to Washington from the U.S. embassy in Moscow following Stalin's belligerent Bolshoi Theater speech in February 1946—was that Russia's rulers had *always* needed an external threat to bolster a government that had *always* been "archaic in form, fragile and artificial in its psychological foundation, unable to stand comparison or contact with the political systems of western countries." During an earlier tour in Moscow, in 1936, Kennan had made the point by composing a report consisting entirely of sentences taken from the dispatches of the American minister in Saint Petersburg in 1853, during the time of Czar Nicholas I: "Secrecy and mystery characterize everything"; "nothing is made public that is worth knowing"; the Russian government possessed "in an exquisite degree the art of worrying a foreign representative without giving him even the consolation of an insult." Or, as Kennan put it in his own words on another occasion, "Russians are a nation of stage managers: and the deepest of their convictions is that things are not what they are, but only what they seem."[10]

Marxist-Leninist ideology had only applied a crueler, pseudo-intellectual gloss to the venerable instincts of Russian rulers. To the "enemies of Russia" the Bolsheviks now added the "enemies of the Socialist revolution"; that ever-present threat, Kennan explained, provided them an ever-present justification "for their instinctive fear of the outside world, for the dictatorship without which they did not know how to rule, for cruelties they did not dare not to inflict, for sacrifices they felt bound to demand." Concessions by the West would not alter Stalin's behavior, because the Soviet Union had *never* been a normal state, interested in fostering international stability.[11]

As his biographer John Lewis Gaddis noted, Kennan was pointing out to Washington the painful truth that the end of the war had "accomplished the only objective—military victory—that the USSR shared with anyone else." Earlier, Kennan had suggested that the only realistic option was to counter the Soviets' implacable self-interest with old-fashioned balance-of-power politics: simply divide Europe in half and accept Soviet domination of the East and forget all of the idealistic hopes for a new international order based on collective security, economic cooperation, and principle above might that the UN and the World Bank were meant to create. But in the long telegram Kennan offered a different prospect, a strategy of patience, of the West's matching the Soviet long view with one of its own; the only thing that would bring about a change in Soviet behavior was if some future Soviet leader finally recognized that the strategy of fomenting perpetual conflict abroad was no longer reaping the benefits to the regime it always had in the past. That required "long-term, patient, but firm and vigilant *containment*" to deny the Soviets any success, particularly when it came to territorial demands such as those Stalin was now peppering the West with for parts of Iran, control of the Turkish Straits, even naval bases on the Mediterranean in North Africa.[12]

Kennan later ruefully remarked that his dispatch reminded him of one of "those primers put out by the Daughters of the American Revolution designed to arouse the citizenry to the dangers of the Communist conspiracy," but the fact was that his five-thousand-word message contained more than a few home truths that Washington had been painfully slow to wake up to. (Stalin, whose intelligence service quickly supplied him a copy of Kennan's classified cable, ordered his ambassador in Washington to send an equally long telegram to Moscow analyzing "American monopolistic capitalism" as the force behind the United States' policy of "striving for world supremacy.")[13]

Just how cruelly barbaric the Soviet state's leader was and how accustomed he was to playing a double game—not only in overt diplomacy but in the ruthless deployment of covert forces abroad and repression at home—was not nearly so widely understood in the late 1940s as it would be a half century later. In Stalin, as Gaddis observed, "narcissism, paranoia, and absolute power came together." A "convinced Marxist fanatic from his youth," he maintained an air of "eerie calm," his short, stocky figure and ostentatious pipe projecting the air of a simple Georgian peasant elder; General Eisenhower, meeting him in 1945, was just one of many to be charmed by the act, calling him afterward "benign and fatherly."[14] But that outward pose, noted Stalin's biographer Simon Montefiore, concealed

> a super-intelligent and gifted politician for whom his own historic role was paramount, a nervy intellectual who manically read history and literature, and a fidgety hypochondriac suffer-

> ing from chronic tonsillitis, psoriasis, rheumatic aches. . . . Garrulous, sociable, and a fine singer, this lonely and unhappy man ruined every love relationship and friendship in his life by sacrificing happiness to political necessity and cannibalistic paranoia. Damaged by his childhood and abnormally cold in temperament, he . . . believed the solution to every human problem was death, and was obsessed with executions. . . . No one alive was more suited to the conspiratorial intrigues, theoretical runes, murderous dogmatism and inhuman sternness of Lenin's Party.[15]

In common with his predecessor, the Soviet leader possessed an utter indifference to human suffering and a complete lack of remorse for the millions of innocent people sent to their deaths on his orders. Neither Lenin nor Stalin was a sadist, in the sense of deriving pleasure from the infliction of physical torment; rather, they saw terror as "an indispensable instrument of revolutionary government," the historian Richard Pipes observed, and seemed genuinely puzzled at the idea that they should be troubled over the fate of individuals who might have to be sacrificed on the way to achieving the ideals of the worker's state. "How can you make a revolution without executions?" Lenin asked in exasperation as he ordered death "on the spot" for "counterrevolutionary agitators." Finding the normal system of justice too unreliable and lenient for his purposes, he created, in the utmost secrecy, the Cheka to carry out the job. Modeled directly on the hated czarist secret police, the Cheka was instructed by Lenin to "exterminate" and "liquidate" the state's "class enemies," subject not to any formal law but constrained only by something he termed "revolutionary conscience."

As one of its first acts the Cheka ordered that nothing could be published about the organization without the approval of the Cheka itself. Its first chief, Felix Dzerzhinsky, was a man thoroughly in the bloodless Bolshevik mold, an ascetic ideologue "capable of perpetrating the worst imaginable cruelties without pleasure, as an idealistic duty," as Pipes described him. In the courtyard and basement cells of the Cheka's Lubyanka prison in the heart of Moscow, executions and torture sessions were carried out while truck engines idled to muffle the shots and screams.[16]

Stalin's Great Terror, which began in 1937, aimed to eliminate once and for all the "anti-Soviet elements" who dared to weaken the state in "even their thoughts." Each region of the country was assigned a quota to meet goals the Politburo set: 72,950 in "Category One," to be shot; 259,450 in "Category Two," to be arrested and sent into internal exile. The actual number eliminated in 1937–38 was 700,000, so enthusiastically was the order carried out. A million children of those executed or arrested, often for nothing more than being a troublesome personal rival to a local Communist Party boss, were taken away and placed in state orphanages; most did not see not their mothers again for twenty years. Stalin himself reviewed the execution lists, writing "for" to signify his approval; a tiny crayon dash mark could save a man's life, but Stalin seldom found any reason to be merciful. "An Enemy of the People is not only one who does sabotage but one who doubts the rightness of the Party line," he told his chief of the NKVD, Lavrenty Beria. "And there are a lot of them and we must liquidate them." To hide the appalling extent of the purge, Stalin ordered the 1937 census falsified, then had all of his top census administrators arrested, many of them shot.[17] George Kennan recalled the crushing sense of fear that gripped Russian society at the time. When the purges began, Russians simply cut off even the most ordinary social contact with foreigners, leaving the staff of the American embassy in Moscow completely isolated.[18]

Stalin may not have been a sadist, but Beria unquestionably was, one of the many in Stalin's inner circle who took advantage of the terrifying official power they wielded to act with personal impunity. The NKVD chief kept a collection of blackjacks in his office so that he could personally take part in the torture of victims when he chose; he also stored there an array of women's silk underwear, schoolgirls' sports outfits, sex toys, and pornography. Accountable to no one but Stalin, Beria routinely had women he spotted picked up and brought to his office, where he seduced or more commonly just raped them, in either case having his driver take the woman home afterward in his limousine and present her with a bouquet of flowers as a parting gift. To encourage one of his victims to submit, Beria promised he would release her adored father from prison, knowing that he had already been executed.[19]

By the late 1940s the NKVD and NKGB (reconstituted as full-fledged ministries after the war, they had become the MVD and MGB by then) were sprawling, all-powerful bureaucracies that encompassed the secret police, the vast network of Gulag slave labor camps, and multiple divisions of armed troops organized as regular military units; they were responsible for internal security, foreign espionage and subversion, cryptology, counterintelligence, and ideological supervision. In conducting foreign intelligence operations they acted with the same ruthlessness and impunity with which they suppressed dissent at home. It would not be until 1956 that the successors of the Cheka would finally be stripped by Nikita Khrushchev of their assumed power to carry out secret executions. "The failure of the Bolshevik government to make public, at the time of its founding, the functions and powers of the Cheka had dire consequences," Pipes observed, "because it enabled the Cheka to claim authority which it had not been intended to have."[20]

In April 1947, fourteen months after Kennan's long telegram, George Marshall returned from a meeting in Moscow profoundly upset and discouraged. The former Army chief of staff, who had quietly and adroitly shaped America's military strategy through World War II, had recently been named secretary of state by Truman. He was reserved, unflappable, courteous, dignified; Churchill called him "the noblest Roman."

But his patience had been sorely tried at the meeting of the wartime allies' foreign ministers. Stonewalled day after day by Foreign Minister Vyacheslav Molotov, Marshall had finally sought a session with Stalin, only to get the runaround there too when he attempted to raise the problems of postwar Europe, in particular the future of Germany, which was still divided into separate American, British, French, and Soviet military occupation zones. Stalin doodled on his notepad, drawing wolves' heads in red pencil, a favorite tactic the dictator used to disconcert foreign visitors, as he blandly suggested that there was no urgency in the matter. "All the way back to Washington," recalled an aide, Marshall spoke of "the importance of finding some initiative to prevent the complete breakdown of Western Europe."[21]

In March, Truman had asked Congress to provide $400 million in aid to Greece and Turkey to counter Communist guerrillas now threatening their independent survival. It was one one-thousandth what the United States had spent to win World War II, but it represented a huge psychological step for the country, since it meant "going into European politics," Truman pointed out. But Marshall had been adamant that the choice was "acting with energy or losing by default."

Now the secretary of state asked George Kennan to draw up a far grander proposal to seize the initiative in Europe. His one instruction to Kennan was "avoid trivia."[22]

The European Recovery Program, which instantly became known as the Marshall Plan, was unveiled in a speech Marshall gave in June at Harvard, where he had been invited to receive an honorary degree. His proposal was cast not as a response to Communist pressure but rather as a means to revive the economies of war-ravaged countries so that free institutions would grow and thrive on their own. The price tag was a staggering $17 billion, and though Americans and Europeans would rightly come to see the Marshall Plan as a historic act of American altruism at its best, it was the darkening realities of Stalin's intentions that finally prompted the Republican Congress, initially scornful of "global New Dealism," to approve the appropriation almost a year later, in April 1948. The most alarming event had been a brutal and swift coup in Czechoslovakia in February 1948, in which the local Communist Party, backed by the Red Army, overthrew the only remaining democratically elected government in Eastern Europe. Almost as worrisome was the upcoming election in Italy scheduled for April, which the Communists seemed poised to win with substantial financial backing from Moscow.[23]

In the end, this immediate threat to extend Soviet control into Western Europe was turned back not by the overt aid of the Marshall Plan but covert aid to Italian political parties from the newly established Central Intelligence Agency. The CIA had been created in the sweeping bill passed by Congress the previous year that overhauled the entire defense establishment. The National Security Act of 1947 unified the military services (including the newly created Air Force) in a single department under a new secretary of defense; set up a National Security Council under the president to direct national security policy; and took the first quiet step into the dark business of covert operations by replacing the Central Intelligence Group, which had proved to be a mostly ineffectual coordinating office, with an independent civilian agency under the direct control of NSC.

Although CIA had no explicit statutory authority to carry out overseas operations, the NSC wasted little time filling that breach, issuing on its own initiative a directive in June 1948 that authorized the intelligence agency to engage in

> propaganda, economic warfare; preventive direct action, including sabotage, anti-sabotage, demolition and evacuation measures; subversion against hostile states, including assistance to underground resistance movements, guerrillas and refugee liberation groups, and support of indigenous anti-communist elements in threatened countries of the free world.[24]

The State Department had urged this vastly expanded role. Kennan acknowledged that in adopting the very tactics of the totalitarian state that the United States hoped to counter, it risked compromising its most basic values, and indeed undermining the very premise of the Marshall Plan, namely that the most effective answer to Soviet influence was for America to show the way with its example of openness, freedom, and the rule of law. But he thought that as long as the State Department kept a close eye on the CIA's operations, and that they were infrequent enough and circumspect enough so that, as the NSC's directive had put it, "the U.S. government can plausibly disclaim any responsibility for them" if uncovered, it would be all right. "It did not work out at all the way I had conceived," Kennan would later admit, with considerable understatement. Much later he more frankly said that supporting the CIA's covert operations was "the greatest mistake I ever made."[25]

"Plausible deniability" had seemed like a clever way out of the dilemma at the time, but the catch was that Americans' belief that their government told the truth was at the very heart of the extraordinary trust they placed in their country's leaders. In the 1950s 75 percent of Americans said they trusted their government to do the right thing just about always or most of the time. The Cold War's greatest casualty would be Americans' faith in their government, which fell to 25 percent by the 1980s—in no small part because of the Machiavellian compromises with the truth that had been made in the covert war against the Soviet Union. The discovery that America, too, had become a "nation of stage managers" engendered a cynicism and disillusionment that were never to be undone once they became habitual features of the political landscape.[26]

Kennan had presciently warned in his long telegram that "the greatest danger that can befall us in coping with this problem of Soviet Communism, is that we shall allow ourselves to become like those with whom we are coping." He stressed, "We must have the courage and self-confidence to cling to our own methods and conceptions of human society."[27] Faced with the irreconcilable contradiction between American "conceptions of human society" and the subterfuges it now increasingly employed, U.S. officials in succeeding years would dig themselves ever deeper. An NSC review of strategy in 1950 by Kennan's successor as head of the State Department's policy staff, Paul Nitze, took a fateful step down the road of moral doublethink by arguing that, in effect, when the United States did it, it was acceptable because it was merely the recourse of necessity; deep down, America was morally in the right, so—unlike the "evil men" in the Kremlin—telling a lie didn't make Americans liars, nor did America's moral values preclude actions that violated those very same values:

> The integrity of our system will not be jeopardized by any measures, overt or covert, violent or non-violent, which serve the purposes of frustrating the Kremlin design, nor does the necessity for conducting ourselves so as to affirm our values in action as well as words forbid such measures, provided only that they are appropriately calculated to that end and are not so excessive or misdirected as to make us enemies of the people instead of the evil men who have enslaved them.[28]

The easiest way to duck the issue was with ever-greater secrecy, and that would be the story of American intelligence and covert operations throughout the Cold War. No president tried to justify publicly what a high-level review of CIA covert action a few years later acknowledged was its "fundamentally repugnant philosophy": that in seeking to counter the Soviets on their own familiar playing ground of deceit, sabotage, subversion, and espionage, "there are no rules in such a game." Secrecy accordingly begat secrecy; it was not merely to protect operations and sources but to avoid having to confront basic questions of legality and morality raised by covert operations that the Cold War intelligence establishment retreated ever further from the democratic norms of open government and constitutional accountability.[29]

. . .

Signals intelligence had always been different. It was a cleaner, technical, noninvasive way to gain an inside advantage over a foe, and it also offered the promise of being much more reliable than the messy and dangerous business of spies and human betrayal. But even signals intelligence could not escape the moral black hole that secrecy drew everything into.

The breaking of the NKGB/MGB traffic presented one moral and legal dilemma almost immediately. One of the first of the American spies identified by name from the messages Meredith Gardner began reading in 1947 was a Soviet agent code-named SIMA, who according to the deciphered cables was working in 1945 in the Justice Department's Foreign Agents Registration section and who had previously been in the department's Economic Warfare section. The FBI, which began working directly with Arlington Hall in October 1948 to follow up on leads from the broken messages, quickly determined that only one person fit that description. Judith Coplon had been a member of a Communist student group at Barnard College, and her job in the Foreign Agents Registration office had allowed her to tip off the Soviets to FBI investigations of suspected agents. The FBI tailed her and arrested her in the act of handing over documents to a Soviet MGB agent in New York who was operating under diplomatic cover at the United Nations. But the FBI agents were ordered not to disclose at her trial the source of the lead that had brought her to their attention—on the witness stand they resorted to elaborate circumlocutions, referring to a "confidential informant" who could not be named—and her conviction was overturned on the grounds that the government lacked probable cause to place her under surveillance in the first place. She was retried, convicted again, and acquitted on appeal again for the same reason.[30]

The episode raised disturbing questions about justice and the fundamental right of a person accused of a crime to be confronted with the evidence against him, but the consequences of the airtight secrecy surrounding Arlington Hall's work proved even more fraught in the arena of national politics. Republicans in the 1948 elections were preparing to make as much as they could out of sensational charges that the government was riddled with Communists; Republican senators led by Robert

Taft had waged an ugly smear campaign against David Lilienthal, Truman's nominee for head of the Atomic Energy Commission, claiming that he was "soft on the subject of Communism" and that the Tennessee Valley Authority, which he had headed, was "a hotbed of Communism," and making thinly veiled anti-Semitic allusions to his parents' "foreign" origins. Far more reckless charges would follow in the witch-hunting hearings of the House Un-American Activities Committee and Senator Joseph McCarthy's Permanent Subcommittee on Investigations, culminating in McCarthy's wild claims to have unearthed a secret Communist conspiracy within the government to do Moscow's bidding, going all the way up to "the mysterious, powerful" George Marshall, who had "lost China" to the Reds.[31]

It was to blunt the charges of Communists in government flogged by the Republicans that Truman reluctantly agreed in March 1947 to order the dismissal of any government employee found by the Civil Service Commission or the FBI to be "disloyal." He told his aide Clark Clifford that the whole thing was "a lot of baloney" and a "red herring," and privately expressed fears that the result would be to make the FBI into "an NKVD or Gestapo," sniffing out citizens' denunciations. The FBI dragnet of "name checks" that the loyalty program triggered found no Soviet spies; the thousands who lost their jobs in the process had committed no crime more serious than belonging to left-wing organizations or subscribing to periodicals the FBI deemed "subversive." The government's refusal to reveal to the public the real evidence from the NKGB's own messages about real Soviet spies, several of whom who did hold high government positions, left the stage to McCarthy. The Republican junior senator from Wisconsin was a loudmouth, a boor, a lush, and a demagogue, but he knew what to do with the opportunity, and made speech after speech filled with ever darker ravings of secret conspiracies and cover-ups, all utterly unsubstantiated.

Ironically, McCarthy and the poisoned politics of the 1950s he so helped create gave the real spies the perfect cover: they were able to claim then and ever after that they, along with the many others falsely accused of being Communist agents, were merely innocent victims of the McCarthyite witch hunt. "I think the greatest asset that the Kremlin has is Senator McCarthy," Truman told reporters during the peak of the senator's accusatory spree. It was a truer statement than he probably knew. Like most Democrats and those on the left during the Cold War, Truman would staunchly defend to his death several top aides accused by ex-Communists during congressional hearings of having spied for the Soviets. Among them were FDR's economic adviser Lauchlin Currie, former assistant secretary of the treasury Harry Dexter White, and most famous of all, Alger Hiss, who would be convicted of perjury for denying he had passed secret documents to Soviet agents while a senior State Department official. In fact, their identities as spies had all been confirmed by Arlington Hall's decodes by the time Truman left office in January 1953.[32]

Truman's ignorance of the truth was astonishing but genuine. It turned out that it was not merely the public, the courts, or Congress to whom Arlington Hall's work could not be divulged. When the question was raised within ASA of informing the president of the Soviet spy traffic, Carter Clarke vehemently objected, insisting that the FBI, and the British GCHQ, with whom ASA was working so closely on the project, were "the only people entitled to know anything about this source." Clarke took the matter straight to General Omar Bradley, the chairman of the Joint Chiefs of Staff, and subsequently reported back that Bradley "agreed with the stand taken by General Clarke and stated that he would personally assume the responsibility of advising the President or anyone else in authority if the contents of this material so demanded." Apparently the military chiefs whom bureaucratic circumstance had placed in charge of U.S. peacetime signals intelligence operations against a foreign adversary decided that the circumstances did not so demand, and the president of the United States remained among those *not* "entitled to know" about it.[33]

Of the five cipher machines the Soviet military employed in the immediate postwar years, the Longfellow teleprinter scrambler was among the most sophisticated: it appeared to have been based on the Germans' advanced SZ40 device, and, as the Americans and British were finding, its cryptanalysis was a far from trivial computational challenge even after extensive progress had been made on key recovery.

Likewise resisting solution were the systems code-named Albatross and Pagoda by Arlington Hall. Little was known for certain about Albatross, but it seemed to be a strong variant of a special version of the Enigma used by the Abwehr, the Nazis' military intelligence service; that model had four rotors that turned in a far more complex pattern than the standard Enigmas.[34]

Pagoda (or sometimes Pogoda) was a double-tape teleprinter encipherment system that, as William Friedman concluded in a 1948 internal report, was almost certainly derived from an identical American machine developed by AT&T thirty years earlier and openly described in a published paper in 1926. The AT&T device, attached to a standard teleprinter machine, used two random Baudot key tapes that were changed daily. One tape was 1,000 characters long, the other 999, and each was formed into a loop that was fed through a paper tape reader. The tapes automatically advanced one position as each character of text was transmitted. By adding the characters from each of the tapes together, an extraordinarily long sequence of nonrepeating key, 999,000 characters long, was generated. Friedman had cracked it in 1919: taking up a challenge from the U.S. Army Signal Corps, which insisted the system was "invulnerable," he was provided 150 enciphered messages representing a typical day's traffic. He found that even without any overlapping depths, it was possible to separate out the two key cycles by assuming a crib of commonly repeated characters, notably those used for spaces and carriage returns. After two months' arduous work, he triumphantly sent a message back to the Signal Corps enciphered in the same key that had been used in the challenge messages. The Soviets' version, however, was "a much more difficult" problem, Friedman acknowledged, and "the odds against our present workers" were much greater than those he had faced in 1919.[35]

The other two main Russian cipher machines were knockoffs of far less complex foreign designs. In April 1946, GCHQ passed on to their American colleagues the "surprising discovery" that the Soviet machine cipher they called Coleridge was apparently a "subtractor device a little like a Hagelin."[36] The Hagelin machine was something the Americans and British knew how to deal with. The invention of a Swedish mechanical engineer, Boris Hagelin, and sold on the commercial market by his firm AB Cryptoteknik in Stockholm, the device had been adopted by the military services of a large number of countries, including Italy, the Netherlands, France, and the United States. It offered only modest security, and the U.S. Army employed its version, known as the M-209, only for tactical communications at the division level or lower. But it was small and portable, operated without electricity with an ingenious, purely mechanical mechanism, and offered ease in changing its key settings from one message to the next. A letter to be enciphered (or deciphered) was selected on a rotating alphabet disk; turning a hand crank operated the cipher mechanism and caused a type wheel to print the corresponding letter on a strip of paper.

During the war Arlington Hall had produced a compendium of technical articles on its cryptanalysis and was routinely reading Hagelin traffic of other countries. The major cryptographic strength of the machine lay in an intricate system of pinwheels that produced an irregular, nonrepeating key pattern more than one hundred million letters long. Its considerable weakness was that at any one of those positions there was a choice only of the same twenty-six different substitution alphabets, and each of those alphabets followed a completely predictable pattern. (The Enigma, by contrast, could in principle generate any of nearly eight million million different cipher alphabets.)* Just as with the Russian teleprinter ciphers, two Hagelin messages in depth could often be cracked by assuming a short crib of likely plaintext and building up a solution from the words that started to emerge in each of the matching pair of messages (see appendix C).

By April 1947, Op-20-G's liaison in London was reporting to Captain Wenger that the British believed that "with the possible exception of Longfellow and the B-211, Coleridge is the most important, high-level system from which intelligence may be produced," and while it "has not, of course, ceased to be a cryptanalytic problem," the feeling was that enough traffic was exploitable to set up a section to concentrate on current production. Hugh Alexander, who had headed Hut 8,

*The Enigma's rotors and plugboard (diagram, page 186) had the effect of swapping the identities of thirteen letters of the alphabet with thirteen others at any given setting: if A was enciphered as F, then F was enciphered as A at that same position. Even with this constraint of "reciprocal" encipherment, a fantastic number of different cipher alphabets is possible; the total number of permutations is given by the expression $25 \times 23 \times 21 \times 19 \times 17 \times \ldots \times 5 \times 3 \times 1 = 7{,}905{,}853{,}580{,}625$.

the naval Enigma section at Bletchley, was placed in charge of the "Coleridge Party." Several special-purpose relay machines were built to aid the process of testing cribs on pairs of Coleridge messages in depth. The Navy built a device called Stork that could "drag" a thirty-letter crib through successive locations and measure the statistical roughness of the resulting plaintext in the paired message: when the crib was correctly placed, the matching plaintext would have the characteristically uneven frequency distribution of letters in natural language, as opposed to the perfectly even distribution of random text. Another Navy device, Piccolo, automatically printed out twenty letters of paired plaintext for each position of a crib.[37]

The B-211 that the British mentioned was another Hagelin machine, and one that Friedman's three assistant codebreakers—Rowlett, Kullback, and Sinkov—had also thoroughly analyzed before the war, even having the Government Printing Office publish a pamphlet in 1939 (it was classified Confidential, the lowest secrecy level) describing the machine's solution. Although it gave a convincing illusion of security, in fact the machine was extremely vulnerable. Friedman's team rather dismissively observed that "it offers no more security than does the cylindrical cipher device with known alphabets"—in other words, a manual device consisting of alphabet disks arranged on a spindle that had been around so long that Thomas Jefferson had once described an identical ciphering scheme. The B-211 employed what was known as a fractionating cipher: as a letter was typed on its keyboard it was replaced by a pair of letters according to a fixed 5 × 5 table. These bigrams were then fur-

	L	**N**	**R**	**S**	**T**
A	L	M	Y	F	X-Z
E	O	J	B	R	S
I	P	U	G	C	W
O	K	N	T	D	Q
U	I	H	V	E	A

The 5 × 5 bigram fractionation table used in the Hagelin B-211 machine to replace each letter with a bigram for further encipherment. The letter G, for example, becomes the bigram IR. The Russian version of the B-211, known to the United States as Sauterne, employed a 5 × 6 table to include most letters of the larger Cyrillic alphabet.

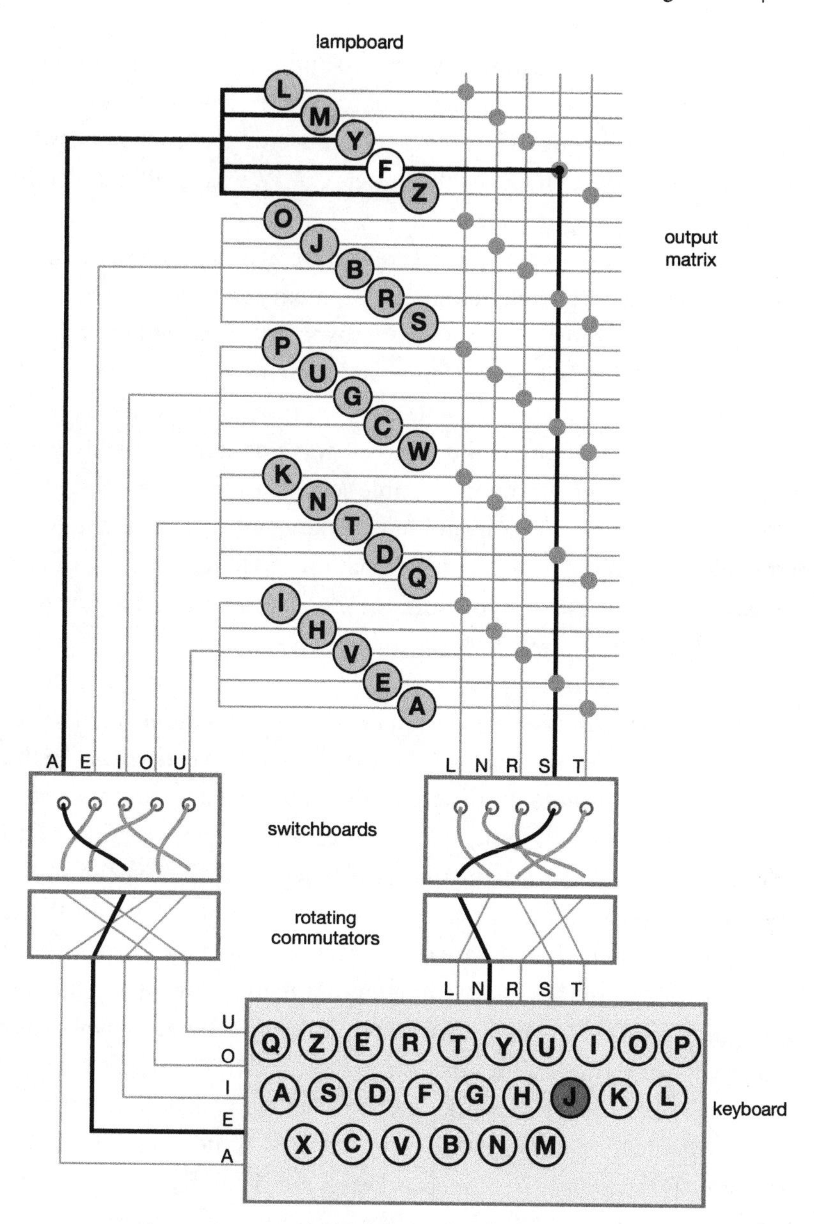

A schematic of the B-211, tracing the encipherment of a plaintext letter (J) to its cipher equivalent (F). Pressing the keyboard connects an electrical voltage to a unique vowel and consonant combination specified by the 5 × 5 bigram table (EN in this case); each letter of the bigram is separately scrambled by a rotor and switchboard to create a new bigram (AS) that is then transformed in the output matrix according to the same 5 × 5 table, causing a bulb to illuminate on the lampboard.

ther scrambled by the machine's plugboard and rotating commutators to generate a new enciphered bigram, which was then transformed back to a single enciphered letter. (The 5 × 5 table covered twenty-five possible letters, so one letter of the alphabet, either X or W, was omitted from the keyboard.)[38]

As complex as the system appeared to be at first glance, the fractionation procedure actually made the problem vastly simpler to solve. Although the rotating commutators moved in an irregular fashion, sometimes stepping with each letter and sometimes not, if the codebreakers had even a short crib—literally no more than a word or two of matching plaintext and cipher text—it was possible to quickly place the crib by ruling out certain impossible sequences in the way the vowel or consonant commutators would have to advance from one five-letter cipher alphabet to the next. The fractionation of a letter into a bigram in fact gave twice as many opportunities for such contradictions to show up, making it a fairly swift procedure to recover the wheel settings and plugboard connections.

Before the war, Boris Hagelin had been "obliged" (in his words, presumably by the Swedish government) to sell two B-211 machines to the Russian Trade Commission in Stockholm. The Soviets then proceeded to produce their own version, replacing the 5 × 5 grid with a 5 × 6 grid to encompass the most commonly used thirty letters of the Cyrillic alphabet. The French and Dutch were also good customers for the B-211. None appeared to be fully aware of its extreme weakness. The Germans had captured one of the Russian versions of the machine during the war and figured out how to break it with a ten-letter crib, but were never able to intercept any traffic. But after the war B-211 traffic transmitted by Morse code on military networks in the Soviet Far East began to appear, and on March 1, 1946, Op-20-G reported the "excitement" of reading the first B-211 message. The codebreakers figured out that the Soviet version, which the U.S. and British code-named Sauterne, had repluggable wheels whose connections could be changed for each message, which was undeniably a significant complication, but with the use of several of Op-20-G's electromechanical analytic machines it was possible to crack that obstacle, and by the next month a regular "watch" working sixteen hours a day had been established to process current traffic.[39]

The U.S. and British codebreakers rightly feared that knowledge of their wartime success against the Enigma would bring about a swift end to the cryptographic naïveté that made these Soviet systems and others like them exploitable into the postwar era. The Russians never succeeded in breaking the Enigma themselves during the war, and although on several occasions the British government cautiously shared with the Soviets intelligence derived from German army and Luftwaffe Enigma traffic—usually disguised as coming from "a well-placed source in Berlin" or some other similar formula, and often warning of strategic troop movements or plans to carry out air attacks on Soviet positions—British intelligence adamantly refused to divulge to the Russians GC&CS's cryptanalytic success in this area. The concern at that time was not so much that it might help the Russians down the road, but rather that since GC&CS knew from the Enigma traffic itself that the Germans were reading much of the Russians' coded messages, any information given the Russians would likely get back to the Germans. Menzies warned Churchill that the result "would be fatal": it would only be "a matter of days before the Germans would know of our success, and operations in the future would almost certainly be hidden in an unbreakable way."[40]

And while the Soviets had at least two spies at Bletchley—that would be discovered much later, in part from the one-time-pad NKGB messages—both worked in Hut 3, which produced the decrypts and translations, not Huts 6 and 8, which did the cryptanalysis.* Thus the Soviets never gained any direct information from GC&CS about the precise cryptanalytic vulnerabilities of the Enigma's design.[41] But the Soviets did capture some Enigmas during the Germans' retreat at the end of the war, and the former chief of the KGB's cipher section told the historian of cryptology David Kahn in 1996 that Soviet cryptanalysts had worked out the mathematics of breaking it, though they lacked the means to produce the equivalent of the bombes required to carry out the task.[42] In the

*John Cairncross, who worked at Bletchley for one year in 1942–43, was one; like a number of other Soviet spies of his generation who would rise to high levels in the British government, he was recruited at Cambridge in the 1930s by his tutor Anthony Blunt, an art scholar and later Soviet mole inside MI5. After the war, as a senior treasury official, Cairncross passed information to the Soviets about the British atomic bomb program. He came under suspicion in 1951, and finally confessed in 1964. The other spy at Bletchley, code-named BARON, worked in Hut 3 in 1941 and has never been identified.

light of subsequent crushing events, it would become all too clear that the Russians had not taken very long after the end of the war to catch on to the inherent insecurity of most existing rotor machine designs in the face of the new cryptanalytic resources available, and how to fix it.

The 1946 agreement between the Army and Navy to "coordinate" their separate signals intelligence operations had merely sidestepped glaring deficiencies in the entire arrangement, which was quickly proving itself unequal to the new technical and intelligence challenges they faced in attacking the Russian problem. The rivalry between the two agencies was a long story of mutual suspicion and secret maneuvering, punctuated by ad hoc compromises that served only to underscore the absurdity of the situation. For over a year before the Pearl Harbor attack, in one particularly bizarre deal, the Army and Navy divvied up responsibility for decrypting Japanese Purple traffic simply by alternating days: the Navy took the odd-numbered days, the Army the even ones. It was a travesty of the bedrock cryptanalytic principle of continuity, but bureaucratic politics trumped even common sense. Wenger later observed that, ridiculous as it was, the arrangement was actually an improvement over the previous situation, in which each service completely duplicated the work of the other, and whenever an important message was broken each would "immediately rush to the White House with a copy of the translation in an effort to impress the Chief Executive."[43]

The Navy was especially suspicious of sharing its most important cryptologic secrets with the Army, taking the view that the large number of civilians at Arlington Hall could not be trusted to handle classified information. At one point the Navy insisted that its rule forbidding civilians to operate or know anything about the SIGABA cipher machine, a joint Army-Navy device used for the highest-level traffic, should apply to the Army as well. The Army pointed out that such a directive would mean that William Friedman, who had overseen the machine's development, would not be allowed to touch his own invention. There were also some basic cultural differences between Op-20-G's traditionally hierarchical military structure and Arlington Hall's more freewheeling, civilian-heavy organization, not helped by what one visiting British naval officer astutely and correctly diagnosed as an ugly strain of anti-

Semitism that was a serious source of friction at times: "The dislike of Jews prevalent in the U.S. Navy is a factor to be considered," he reported, "as nearly all the leading Army cryptographers are Jews."[44]

The 1946 coordination agreement left each service in complete control of its own intercept stations and cryptanalytic staff and preserved the Army's exclusive authority over military and air force traffic and the Navy's over naval traffic; it was only in "joint" areas of responsibility, diplomatic and commercial targets, that the new "coordinator of joint operations" and the interagency board to which he reported (most recently rechristened the U.S. Communications Intelligence Board, or USCIB) had any say. But the coordinator, who alternated annually between the chiefs of ASA and Op-20-G, had no actual authority to order anyone in the other agency to do anything. Knowledge was power when it came to bureaucratic turf fights, and the old habits of hoarding information continued unchanged. Rowlett was not above giving his Navy counterparts deliberately misleading information to avoid sharing ASA's de facto control of the Russian one-time-pad problem; before a meeting with the Navy at Arlington Hall in June 1947 to discuss the project, he instructed Meredith Gardner not to reveal that he had already succeeded in reading some messages from a depth of two, and casually tried to steer the Op-20-G codebreakers into what he knew was a dead-end hunt searching for patterns in the one-time-pad keys, thus leaving the real work to his group at ASA.[45]

The particular absurdity of the arrangement was that the agencies most directly interested in the results of the "joint" activity that came under USCIB's purview—namely, the State Department, FBI, and now CIA—had no actual authority over how much effort would be allocated to these nonmilitary targets, or what priorities would be set among them. Although State and CIA were represented on USCIB, the board's decisions had to be unanimous, giving each of the military services an effective veto over any initiatives they disagreed with. A proposal in 1948 to bring USCIB under the civilian-led National Security Council and give it the power to direct the collection and analysis of signals intelligence in areas of "national" importance met with furious resistance this time from the Army, which protested that this would allow civilians to give orders to its intercept stations, interfering with the military chain of command. In the end, a much watered-down directive from NSC brought the board under its control, but still with the requirement of unanimity in its decisions and with its power limited to "authoritative coordination" rather than "unified direction." The entire attempt to effect "coordination" through such an ungainly mechanism was "totally useless," recalled Oliver Kirby, who worked on the Russian problem at ASA at the time. "I don't know what they 'coordinated.' They didn't bother us at all."[46]

This endless tinkering with feckless bureaucratic mechanisms left completely unaddressed an even more serious weakness in the U.S. signals intelligence system. Almost entirely for reasons that were quirks of history and personalities, both the Army and Navy maintained that the product of decrypting signals was "information," not "intelligence." That semantic invention was the outgrowth of a long fight between the services' communications and intelligence branches for control of signals interception and codebreaking, going back to the 1920s; more by dint of their forceful and politically adept commanders than through persuasive arguments about organizational logic or military efficiency, the Army Signal Corps and Naval Communications always managed to retain the mission, repeatedly fighting off attempts by Army and Navy intelligence to wrest control away from them. Still, some plausible justification had to be invented for such an odd division of responsibilities: thus the euphemistic pretense that signals intelligence was not intelligence.

But bureaucratic fictions often have actual consequences, and as a matter of policy Arlington Hall and Op-20-G were formally limited to supplying translations of their decrypts to the intelligence analysts of other departments, rather than producing a finished "intelligence product" themselves: they could report what an intercepted signal said, but not what it meant. That distinction continued into the postwar period, with the result that the Office of Naval Intelligence, Army G-2, the State Department, and CIA each produced their own analyses of the decrypted messages that the Army and Navy signals intelligence organizations provided. The duplication of effort was one thing; much worse was the utterly schizophrenic separation of the job of translation from analysis.

Bletchley Park by contrast had confronted this problem head-on from the very start. The great elation at breaking the first German army Enigma messages in January 1940 had just as quickly turned to deflation when they turned out to be "a pile of dull, disjointed, and enigmatic scraps, all about the weather, or the petty affairs of a Luftwaffe headquarters no one had heard of . . . the whole sprinkled with terms no dictionary knew," as a veteran of Hut 3 recalled.[47] It was immediately apparent that even to produce a meaningful translation of a message, much less extract useful intelligence from it, was fundamentally a job of intelligence *analysis*. In time Hut 3 would amass a huge card file of cross-indexed names, terms, places—which time and again led to discoveries of the first importance from just such "dull and enigmatic scraps" in decoded Enigma messages. The first identification of the location of the Nazis' rocket experiments, notably, came from an Enigma message reporting the transfer to Peenemünde of a junior Luftwaffe NCO who Hut 3 knew, from earlier signals, had worked on radio guidance systems. The top translators at Bletchley were intelligence officers first, who sifted myriad pieces to assemble an insightful whole.[48]

The U.S. Army and Navy signals intelligence organizations never attempted to perform a similar role. Although Carter Clarke and Alfred McCormack assembled an impressive array of analytic and legal minds in the office McCormack advised setting up in the wake of the Pearl Harbor attack to make sure no signals intelligence warnings were ever missed again (known as Special Branch, it was part of the War Department General Staff's G-2), its analysts were one step removed from the translators who were on intimate terms with the traffic on a daily and hourly basis. Now CIA, with its mandate to serve as the central authority for correlating, evaluating, and disseminating national intelligence, not only began to insist on direct access to the "raw" signals but aggressively objected if the codebreakers added the smallest annotation or interpretation. ("We had people all over us with both feet and clubs," recalled Oliver Kirby, if an ASA linguist dared to add any context to a translation.)[49] It was a fundamental disconnect in the American intelligence establishment that would never be fully resolved, leaving an enduring weakness that would cause officials at crucial moments in the decades to come to treat decoded messages with exaggerated reverence or breezy disdain, with equally fatal consequences in both cases.

4

Digital Dawn

In the summer of 1944, Lieutenant Herman Goldstine, a thirty-year-old mathematician, was waiting for the train to Philadelphia at the railroad station in Aberdeen, Maryland, when he spotted one of the giants of his field walking toward him on the platform.

John von Neumann had joined the faculty at Princeton's Institute for Advanced Study in 1931, the same year as Albert Einstein. Von Neumann was one of a galaxy of astonishing scientific minds who had fled the rising anti-Semitism of Europe for America and Britain in the years before the war. So many were from Hungary—besides von Neumann, the Hungarian contingent included the renowned physicists Edward Teller, Leo Szilard, and Eugene Wigner and the aerodynamicist Theodore von Kármán—that their colleagues joked they must be a race of super-intelligent beings from another planet who had adopted the cover story of being Hungarian to explain away their accented English.

Von Neumann had been an intellectual prodigy as a child, able to divide eight-digit numbers in his head at age six. Throughout his life he could effortlessly recite entire books verbatim after a single reading, and equally effortlessly provide a running translation in any number of languages. Years later, after he got to know him well, Goldstine tried to test von Neumann by asking him how Charles Dickens's *Tale of Two Cities* begins. He was still going fifteen minutes later, without pause, when Goldstine finally stopped him.[1] As a scientist, von Neumann had made seminal contributions to a bewildering array of fields, including game theory, quantum mechanics, economics, topology, and the theory of shock waves. Besides his eminent position at Princeton, he was now also working as a consultant to the Manhattan Project, already looking ahead to the possibility of the "super," or hydrogen, bomb.

At the time of their first meeting Goldstine was the U.S. Army's liaison to a small group of mathematicians and engineers at the University of Pennsylvania's Moore School of Electrical Engineering who were designing what would be one of the world's first digital electronic calculating machines. Called ENIAC—the letters stood for Electronic Numerical Integrator and Computer—the device was being built for the Army's Ballistics Research Laboratory at the Aberdeen Proving Ground, which was hoping to automate the extremely labor-intensive calculations involved in creating aiming tables that allowed gunners and bombardiers to predict the trajectories of artillery shells, bombs, and missiles. That day on the train platform the younger man, with some temerity, approached his world-famous colleague and introduced himself:

> Fortunately for me von Neumann was a warm, friendly person who did his best to make people feel relaxed in his presence. The conversation soon turned to my work. When it became clear to von Neumann that I was concerned with the development of an electronic computer capable of 333 multiplications per second, the whole atmosphere of our conversation changed from one of relaxed good humor to one more like the oral examination for the doctor's degree in mathematics. Soon thereafter the two of us went to Philadelphia so that von Neumann could see the ENIAC.[2]

ENIAC contained seventeen thousand vacuum tubes, weighed thirty tons, and took days to program for each calculation by means of huge banks of patch cords and switches. But the group was already thinking about the next generation of computers that would follow, and von Neumann enthusiastically joined in the discussions. A few months later he sent Goldstine a 101-page draft distilling the ideas for the architecture of the new computer. The key point was that its program instructions would be stored in the computer's memory itself, providing a flexibility and general-purpose adaptability so far lacking in all electronic computing machines. Goldstine later called the paper "the most important document ever written on computers and computing." Von Neumann's enthusiasm for the possibilities of computers led the Los Alamos scientists to submit the very first problem run on the ENIAC, an extremely complex calculation related to the hydrogen bomb's design.[3]

EDVAC, ENIAC's successor, was still on the drawing board in the summer of 1946 when the Moore School decided to invite a select handful of government and academic mathematicians to a special summer school lecture series on the project; as a result, the world's first computer course predated the world's first stored-program computer. The "Course in Theory and Techniques for Design of Electronic Digital Computers" consisted of eight weeks of lectures and seminars presented by EDVAC's developers and outside experts, von Neumann among them.

Howard Campaigne, who was about to take on his new civilian job of chief of mathematical research at Op-20-G, heard about the course from his old boss Howard T. Engstrom. As head of Op-20-G's research section during the war, Engstrom, a former Yale mathematics professor, had always made nurturing talent his chief priority. A courtly and reserved man, he was, several colleagues later recalled, "a conceptualist" who thought about the process of making advances in the field rather than getting caught up in day-to-day crises. Engstrom knew he would never be a mathematical genius himself but felt he could make his most important contribution by encouraging others and creating the environment that would allow them to get on with their work. He often acted as an effective buffer between his research staff and the more hard-charging naval commanders in the organization who demanded instant results. Engstrom had since left the Navy to start his own research company, but kept in close touch with his former colleagues at Nebraska Avenue.[4]

And so when Engstrom called one day and told Campaigne he really ought to find someone, quick, to send to the Moore School course, Campaigne needed no persuading. He chose a young mathematician on his staff, Lieutenant Commander James Pendergrass, who did need some persuading. Pendergrass was on leave, and the whole thing was so sudden there was no time for him to receive written orders, which meant he had to pay his travel expenses out of his own pocket—"always a dangerous thing to do in the Navy," Pendergrass said, because if there was a subsequent hitch in approval of the orders you were stuck.[5]

"I wasn't thinking in terms of computing at all at the time," Pendergrass recalled of his introduction to computers. But the thing that struck him most of all as he sat through von Neumann's and the others' descriptions of their computer design was that it really wasn't a "computer" at all. It was fundamentally a logic machine that manipulated discrete pieces of data; in fact, it had to be "perverted" into doing mathematical calculations at all, Pendergrass thought. "You know," he told his boss on his return to Washington at the end of the summer, "this is just absolutely ideal for our business. It's better for our business than it is for the mathematicians."[6]

Pendergrass quickly wrote up a paper in which he worked out several examples of computer programs that could perform basic cryptanalytic tasks on the new machine. The speed of the digital computer was of course one advantage, but equally important was its flexibility in manipulating data according to a logical sequence of instructions, and the prospect of soon-to-be-available mass storage devices that could hold large amounts of data for ready access and comparison. "The purpose of this report is to set forth the reasons why the author believes that the general purpose mathematical computer, now in the design stage, is a *general purpose cryptanalytic machine,*" Pendergrass began his report. Referring to the large array of bombes and other special-purpose cryptanalysis devices the Navy had acquired at its Nebraska Avenue complex during the war, he continued, "It is not meant that a computer would replace all machines in Building 4, nor is it meant that it could perform all the problems as fast as the existing special purpose machines. It is, however, the author's contention that a computer could do everything that any analytic machine in Building 4 can do, and do a good percentage of these problems more rapidly."[7]

Princeton, Harvard, and the National Bureau of Standards were all working on computers as well, but there was little prospect that they would have time to spare to build one just for the Navy codebreakers. National Cash Register, which had built the U.S. Navy bombes, had already made clear that it was not interested in continuing its wartime collaboration with Op-20-G and was anxious to get back to its commercial business, as were IBM and Eastman Kodak, which had helped with R&D on special-purpose cryptanalytic equipment.

The company Howard Engstrom founded upon leaving Op-20-G was hardly on a par with such industrial technology powerhouses. But it had already begun to show it could deliver on R&D tasks that the Navy codebreakers needed done. That had been the whole idea behind the company, and Wenger in fact had pulled every string imaginable to help set it up as a way to preserve some of Op-20-G's technical expertise amid the postwar downsizing, and to continue the kind of arrangement it had with NCR to build the secret analytic machines it might need in the future. Those machines, the Navy codebreakers now swiftly decided, were going to include a pioneering digital electronic computer.

The whole arrangement with Engstrom's company was more than a little irregular. The founding technical partners were Engstrom; his Op-20-G colleague Bill Norris, an outgoing electrical engineer from Nebraska who had sold X-ray machines for Westinghouse before the war; and Captain Ralph I. Meader, who had commanded the Navy contingent (including three hundred women, members of the Navy's WAVES) that helped build the bombes at NCR's Dayton factory. The Navy had no legal authority to set up a private company, but behind the scenes Rear Admiral Joseph R. Redman, the director of naval communications, "verbally authorized" the project in February 1945 and asked some top Navy officials to see if they could put Engstrom and his colleagues in touch with likely financial backers to get the company off the ground. Secretary of the Navy James Forrestal did not have to look very far: one of his staff aides was Captain Lewis Strauss, partner of the Wall Street investment bank Kuhn, Loeb. But after months of discussions with Engstrom, Strauss's Wall Street associates concluded that the idea would never be a financial success and decided they could not put any money into it.[8]

In late 1945, an officer stationed at the Army Air Forces' Wright Field in Dayton, whom Meader had crossed paths with suggested he talk to a businessman in Minnesota he knew. John E. Parker was a hearty, energetic, wheeling-dealing entrepreneur who had shrewdly turned around a failing light aircraft company, securing a contract from Wright Field to produce wooden gliders in a government-owned factory in St. Paul. Parker was also an Annapolis graduate. A few weeks after being introduced to Meader, he received an invitation to Washington to meet the chief of naval operations, Admiral Chester W. Nimitz. At their meeting the head of the U.S. Navy wasted no words, Parker recalled:

> All Nimitz said to me, as he tapped me on the chest was, "I've looked into your background and there's a job I would like to have you do." And he said, "It may be more important in peacetime than it is in war time." And I said, "Aye, aye, sir." I had no idea what I was going to do.[9]

What he was going to do was be the public face and chief source of capital for Engstrom's new company, Engineering Research Associates. Within a few weeks, in January 1946, the company was incorporated with $20,000 in equity supplied by Parker and his financial partners, plus a $200,000 line of credit they secured. More than three dozen of ERA's new employees came straight out of Op-20-G, recruited by a small office the company set up in Washington and shipped out to Minnesota; others came from the large pool of physicists, mechanical engineers, and electrical engineers freed from defense industries with the end of the war, supplemented by a good number of recent University of Minnesota graduates.

In his role as ERA's president, Parker said he often felt like a man pretending to conduct a symphony orchestra "without knowing a note": "I always used to feel that this group of people sort of felt sorry for me, this poor fellow sitting up in that front office who didn't know what was going on, and didn't understand any of the technical things." But Parker had understood from the start that his job was to provide the money and not ask questions; in any case, all of the real decisions were going to be up to the Navy.[10]

Op-20-G cut more official corners in issuing ERA a series of contracted "tasks" without competitive bidding. The company was left to issue an awkward "no comment" when Drew Pearson, in his tell-all Washington newspaper column, ran a small item a few years later revealing that the Navy had entrusted one of its "most closely guarded secrets" to an inexperienced company that had received a "juicy contract" at the behest of the very Navy officers who later turned up as "highly salaried vice presidents of the company." But in fact ERA performed admirably. The cavernous 140,000-square-foot factory in St. Paul was hardly the ideal place to build sensitive electronic components. The building, originally a radiator foundry, dated from 1920 and climate control was nonexistent; employees wore mittens and overcoats in winter and went shirtless in the summer, and it was frequently necessary to chase out sparrows and swallows that had entered through the screenless windows. One of the early "tasks" assigned the company led to ERA's development, in its very first year of operations, of the world's first magnetic drum memory. It was the diameter of a bicycle wheel, was a foot thick, and spun at only 50 rpm, but it could hold thousands of words of data and access them in any order, regardless of where they were stored on the drum.[11]

To have ERA build a general-purpose computer, however, required negotiating some bureaucratic obstacles that even Op-20-G's friends in high places could not easily waive. All Navy contracts went through the Bureau of Ships, which had already decreed that computer development for the Navy was the responsibility of the Office of Naval Research. Wenger told Pendergrass to try to get ONR's support; Pendergrass drafted a letter outlining their ideas for a computer, but immediately hit another snag as almost no one at ONR was cleared to know about Op-20-G's secret work. The letter finally landed on the desk of Mina, Rees, head of ONR's mathematics branch, who telephoned Pendergrass to ask what she was supposed to do with his letter. "Well, answer it I guess," he replied, and frankly explained that their plan was to have ERA design and build the machine. "She said, 'Fine, write me an answer,'" Pendergrass recalled, "so I wrote an answer saying that this was a great idea. Rees put her initials on it and sent it up to the front office and had it signed by the head of ONR." Von Neumann and Goldstine added their endorsements after Wenger sent Campaigne and Pendergrass to Princeton to line up their support for the project as well.[12]

In August 1947, ERA was assigned "Task 13," the design of a general-purpose, stored-program computer. Six months later the Navy told ERA to go ahead and build it. The cost was to be $950,000. They named it Atlas, after the mental giant in the comic strip *Barnaby*.[13]

The large array of special-purpose machines that filled multiple floors of Building 4 at the Navy's Nebraska Avenue site was an attempt to speed up cryptanalytic tasks that were hopelessly slow using standard IBM punch card methods. The IBM sorters and tabulators could perform basic data compilation tasks useful to cryptanalysis, such as counting how often different words appeared in a large body of plaintext that had been punched onto cards, one word per card; libraries of cards containing the code texts of every message collected in a specific system could be searched for code groups that had already been assigned a dictionary meaning, and a catalog printed out showing what groups preceded and followed that known word to help suggest possible meanings of as-yet-unidentified groups; tables of differences between the values of known, frequently occurring code groups could be prepared to help identify words in messages already placed in depth; and, as in the massive hunts for double hits in the Soviet one-time-pad traffic, a huge corpus of collected cipher text could be combed to try to locate additional depths.

But the IBM machines' lack of any true memory made those latter kinds of hunt-and-compare sweeps, crucial though they were to routine cryptanalytical work, extremely cumbersome. There was no way to store intermediate results, such as the hypothetical additive calculated by subtracting frequently used code groups from every group of a string of cipher text, except by punching out more cards and feeding them back through the sorters and tabulators for each subsequent step of analysis. To do a thorough, exhaustive search for double hits at every possible relative overlap of every possible pair of messages—an "IBM brute-force run"—sometimes required punching literally millions of cards, one for each message at every possible offset, then running sorts to place the cards in numerical order and scanning the resulting printed indexes by eye for repetitions.[14]

The Arlington Hall and Op-20-G machine room groups during the war devised an array of clever add-on contraptions that could be attached to the IBM machines to try to give them the rudiments of a built-in memory, and thus at least a limited ability to search for repetitions directly without all of the initial sorting and repunching. (IBM had a strict rule that its customers were forbidden to touch the inner workings of its machines, which it offered only for lease, not sale, and until the company at last agreed to cooperate the codebreakers always posted a lookout for the IBM service representative before pulling off the cover plate and setting to work with screwdrivers and soldering irons to hook up their latest invention.) Arlington Hall's F Branch developed something they called the Slide Run Machine, which tried to automate the process of placing known additive key against messages. It used banks of relays to subtract a string of five key groups from cards that had been punched with the cipher texts of thousands of messages; other relays were wired to then detect if two or more of the resulting code groups in the sequence matched any of the 250 most frequently used groups in that system.[15]

But aside from still being quite slow—it took about a half second for the Slide Run Machine to test each punch card of cipher text as it was fed in through a hopper—the whole approach, as the cryptanalysts were painfully aware, was extremely crude from a mathematical point of view. It was looking for a few specified needles one haystack at a time when what they really wanted to do was search for every possible needle in every haystack simultaneously. They wanted, first, to be able to test every message against not just five selected groups of additive key but against every possible string of additive; that required some way to rapidly slide tens of thousands of groups of cipher text against tens of thousands of groups of additive key at every offset between them. More important, they wanted to be able to carry out much more sensitive statistical tests on the resulting deciphered sequences of code groups to see if they yielded something that looked like a plausible decoded message text. The technique of using a lookup table of 250 high-frequency words was like trying to determine if a sentence was German or gibberish by testing only if a common German word like *eins* or *und* appeared in a particular five-word sequence. Much better results could be obtained if instead the machine could assess the entire message and test whether it statistically resembled the overall patterns of German word usage.

The measure of what looked like real language versus coincidence involved an application of a statistical method Alan Turing had devised to simplify Enigma bombe runs.* By the end of the war Op-20-G had come up with a behemoth device called Mercury that tried to implement this idea with IBM cards and relays. It represented the apotheosis

*Dubbed "Banburismus"—the pseudo-Latin name was an allusion to the town of Banbury, where the large printed sheets used in the method were printed—Turing's approach involved calculating the odds that a string of repeated letters that showed up in two different messages was random coincidence or the result of the same word being enciphered with the same key in both messages. It was a pioneering application of Bayesian probability to codebreaking that would become a mainstay of modern scientific cryptanalysis; see appendix D for further explanation.

of the punch card–based cryptanalysis machines. Instead of the Slide Run Machine's yes-no hunt for frequently appearing code groups, Mercury used a bank of ten thousand plugboard connectors to assign a frequency weight of 0 to 19 to every single code group in a four-digit code system. As each stripped code group was read from a test decipherment, a bank of relays connected its corresponding weight to an electronic adder, which kept a running tally for the entire message. At the end, if the weighted frequency of all the groups exceeded the average value of actual plaintext, it was flagged as a possible hit.

Setting up the pluggings took days. But the machine proved useful for half a decade after the war because of its ability to carry out at least simple, comprehensive statistical assessments. In addition to placing cipher text against known additive, Mercury offered a powerful tool to directly place cipher text against cipher text in the hunt for depths, exploiting the fact that if two messages of an enciphered code are in depth, the additive key cancels out when one is subtracted from the other, leaving only the difference between the underlying code groups of the message pair. Knowing the plaintext frequency of each code group made it possible to calculate the frequency of every difference between two code groups. So for a cipher text versus cipher text run, the plugs were set up with weighted frequencies of differences.[16]

Still, as William Friedman noted toward the end of the war, all of these adaptations of IBM technology, undeniably ingenious as they were, were "improvisations," attempts to make IBM equipment do "what it was not basically designed to do." The other electromechanical technology that the Army and Navy codebreakers milked for all it was worth during the war was the tape-based optical comparator. Collectively referred to as rapid analytical machinery, or RAM, these offered the most promising approach to vastly speeding up the process of sliding long streams of data against each other. Two superimposed punched paper tapes, or strips of 70mm film exposed in patterns of opaque and transparent spots, could be run at extremely high speeds through an array of photocell sensors that could simultaneously measure the light coming through as many as a hundred groups at a time, making it possible to locate double hits or search for other kinds of repetitions. By forming one tape into a loop that cycled repeatedly as the other advanced one space with each complete revolution of the first, a "round robin" test covering every possible relative offset between the two data streams could be made. Jack Good jokingly said that he had always regarded it as "one of the great secrets of the war" that "ordinary teleprinter tape could be run on pulleys at nearly thirty miles per hour without tearing," and the machines attained some remarkable computational speeds. The Navy's Copperhead, a film comparator used to search for double hits, could scan five hundred messages against five hundred messages in four and a half hours, approximately a hundred times faster than comparable IBM punch card methods.[17]

The optical comparators also provided an efficient means to perform a central calculation used to locate depths in streams of polyalphabetic cipher. Known as the "index of coincidence," this mathematical test was first devised by William Friedman and Solomon Kullback in the 1920s and 1930s (see appendix E for an explanation of the concept). It was a fundamental breakthrough in the cryptanalysis of polyalphabetic ciphers, but it involved a tedious tabulation of the number of identical letters of cipher text that occurred in the same position in each of two messages. The comparators, by shifting two superimposed photographic plates encoded with message texts through all their relative positions, were able to dispense with any electronic counting circuitry altogether, instead weighing up the total number of coincidences in a single analog gulp simply by measuring the total amount of light emerging through the two plates.

The Colossus machines that GC&CS developed to break the German teleprinter traffic grew out of a similar high-speed comparator device, with the refinement that the stretch of key to be tested was stored electronically in an internal vacuum-tube memory rather than on a tape loop. But Colossus was a very special-purpose machine, as well as an electronically temperamental one, and Frank Rowlett surprised the British by spurning their offer at the end of the war to let the Americans have one of their ten Colossi. Rowlett correctly saw that the future lay in a different direction. If anything, there was more promise, as a model for a general-purpose cryptanalytic calculator, in the "statistical bombe" the U.S. Navy had built as a way to recover the rotor settings for Enigma messages that had resisted the normal approach. Instead of depending on a plaintext crib, the statistical bombe (rather like Mercury did in weighting word frequencies of enciphered codes) deciphered a string of message text at every possible rotor setting and then assessed the roughness in the letter frequencies that emerged from its trial decipherments, to see if any matched that of German plaintext. Although the method was applicable to Enigma messages only in the specific case where the basic setting of the machine had already been recovered, with sufficient improvements in speed and power the approach might offer the basis for a general "cipher text only" attack against a variety of machine cipher systems.[18]

As Pendergrass realized during the Moore School lectures, the digital computer could fully implement all of the cryptanalytic functions that IBM and RAM equipment could often only roughly approximate. The flexibility of the computer to pluck a specified piece of data with equal ease from anywhere in a large stored list and compare it with any other piece of data was its most powerful feature for cryptanalytic application. Combined with the ability to perform complex statistical calculations on the fly, the new computers would provide the cryptanalysts a tool that they hoped would continue to give them an advantage into the postwar decades in the never-ending seesaw race between codemakers and codebreakers.

The major technological challenge of the first computers was fast memory. ERA was banking on a promised RCA vacuum tube, the Selectron, but delays in its availability led the designers to settle on an improved magnetic drum made by ERA itself that was only eight and a half inches wide, spun at 3500 rpm, and held 16,384 24-bit words; with some clever programming techniques, it could be made to perform with an access time of as little as 32 microseconds to find and retrieve a word of memory (though it could take a thousand times longer).[19]

Samuel Snyder, who through much of the war led the machine cryptanalysis section at Arlington Hall, belatedly received a copy of Pendergrass's report and, with a combination of inspiration and agony at the thought that the Army was about to be surpassed by the Navy in the field, set out on a whirlwind tour of computer centers; he visited the research groups at Princeton, Penn, and the Bureau of Standards and some of the new companies getting into the business, including Raytheon and a firm recently established by the ENIAC and EDVAC developers, soon to be called UNIVAC. But months of negotiations with possible suppliers led nowhere. Snyder finally decided to take a bold gamble and have ASA just design and build its machine itself. For the memory they purchased a Rube Goldberg–esque system, developed by the Moore School group for EDVAC, called a mercury delay line. The problem in storing a number electronically was to find a way for a circuit to retain a series of electrical pulses until they were needed again. The Moore School memory exploited the fact that sound waves move a lot slower than electricity, especially sound waves moving through a very dense substance such as mercury. An electrical pulse was first changed into an acoustical pulse and the sound then traveled down a two-foot-long glass tube filled with mercury. At the far end, the sound was converted back to an electrical signal, then fed back into the start of the line. At any moment, eight 48-bit words in the form of mechanically vibrating sound waves would be moving in a row down the line. The time it took to access a word from memory depended on how far that particular word was from the end of the line at a given moment, and varied from 48 to 348 microseconds. ASA's computer called for two large cabinets holding sixty-four delay lines each to give 1,024 words of memory: 1K.[20]

The ASA team gamely dubbed their computer with the name of a comic strip character too: theirs was Abner, named after the hillbilly bumpkin Li'l Abner, the joke being that it was "strong but dumb." But Snyder's group, with the benefit of having spent some additional time thinking about the requirements of cryptanalysis and what specific computer architectures were most suitable for those tasks, came up with a number of innovations that promised to offset Abner's brainpower deficiencies with considerable flexibility. A great deal of emphasis was placed on input and output capabilities that could handle large streams of data from a variety of sources and media that the codebreakers were already using to accumulate libraries of intercepted traffic; Abner was to be equipped with devices to read and write on magnetic tape, paper tape, and IBM cards.

It also was designed to carry out a number of special functions unique to cryptanalysis, including adding alphanumeric characters together in modular arithmetic (as was involved in adding or subtracting key-in systems such as the Russian teleprinter ciphers and the Hagelin) and, even more important for cryptanalysis, initiating a running comparison of two data streams and counting coincidences between the two. The group had correctly noted that this fundamental cryptanalytic task was an inefficient one when carried out by the standard general-purpose computer program. Comparison runs involved performing exactly the same steps over and over; but the stored-program architecture's very flexibility was a curse here, since it required the computer to retrieve and execute, one at a time, the same sequence of instructions for every single comparison of two individual numbers. Abner's "Swish" function did away with all that by throwing the computer into a streaming mode that just pointed to two stacks of data and sent them running sequentially into an analytic unit that tallied coincidences without any further instructions; it then automatically placed one of the streams back into the memory a specified number of words away from its original location, ready to be retrieved again, thereby sliding the two sets of data through every possible offset.[21]

In fact, the war-era optical comparators would remain faster than the conventional general-purpose computer at performing this central cryptanalytic task until the third generation of digital electronics arrived in the mid-1950s.[22] The electronic realization of the streaming function that Snyder's group incorporated into Abner would be a key feature in the future mammoth, technology-pushing computers built for NSA by IBM into the 1960s.[23] Only with the advent of supercomputers would the advantages of this kind of specialized cryptanalytic computer architecture be overtaken by brute processing speed.

Atlas did not have a streaming function, but ERA, at Op-20-G's request, built a number of specialized electronic comparators that did much the same thing. The first of them, Goldberg (ERA's "Task 9"), cost $250,000, could perform twenty thousand comparisons a second, and did all the basic cryptanalytic tests, including round robins, index of coincidence counts, crib dragging, and language-weighted statistics. A more specialized machine, Demon, was rushed into production in the meanwhile to tackle one of the Soviet high-echelon teleprinter encryption systems; equipped with a magnetic drum memory taken from Goldberg, it was designed to test a large number of cribs at every location of one of two messages already found to be in depth, and determine whether the resulting plaintext in the paired message yielded recognizable high-frequency Russian words stored in a dictionary.[24] The first Demon was delivered in October 1948, just in time for the whole Soviet problem to come crashing down.

On June 24, 1948, the Soviet military administration in Germany, without warning, issued orders halting all rail traffic into the western half of Berlin. The blockade left the two and a half million German civilians living in the French, American, and British sectors of the former capital completely cut off from the supplies of food and coal that had kept them alive since the end of the war. Lucius Clay, the military governor of the American zone, later called it "one of the most ruthless efforts in modern times to use mass starvation for political coercion." It was Stalin's first move from diplomatic obstructionism to outright military confrontation with the United States.[25]

Exactly what Stalin hoped to gain remains uncertain to this day. The immediate precipitating event was a plan announced by Clay to introduce the new West German deutschmark in West Berlin to replace the inflated and worthless notes the Russians had flooded Germany with, and Stalin may have been trying to keep the West marks out. He may also have been more directly hoping to force the Western allies to abandon their military presence in Berlin altogether, or pressure them into backing away from their moves to end the military occupation in the western zones and establish a new German state. Berlin was in any case the place where the Soviets were confident they could apply the maximal pressure with the least cost to themselves, and so it would remain throughout the Cold War. (As Khrushchev, with his usual earthy directness, would later remark, "Berlin is the West's balls. Whenever I want to make the West scream, I squeeze Berlin.")[26]

Four days later Truman approved Clay's ad hoc decision to respond to the blockade by trying to supply Berlin by air. Recognizing the enormous political risks, but believing that basic principles were at stake—as well as any hopes for the formation of a West German government that could anchor the stability of postwar Western Europe—Truman did not even consult with advisers before issuing his instructions: "We stay in Berlin, period." A month later he overrode objections from his Air Force generals and ordered 160 large four-engine C-54 cargo planes, more than half the force's entire global airlift capacity, to join the operation, then increased that to 200 in early September. No one had thought it possible to supply by air forty-five hundred tons of food and coal a day, the amount Clay calculated was the bare minimum needed to keep the city from starving and its electricity plants running, but C-54s were landing every couple of minutes at Tempelhof Airport and the city's other two tiny airports, idling on the tarmac with two engines running as crews hustled off ten tons of bulging burlap sacks, then were back in the air in less than twenty minutes.[27]

Truman took one other highly visible action, ordering two squadrons of B-29s, sixty of the mighty long-range bombers that, everyone in the world knew, had dropped the atomic bombs on Hiroshima and Nagasaki, to RAF airfields in East Anglia. Not made public was that the aircraft sent to Europe were not in fact equipped to deliver atomic weapons. Truman had impatiently cut short Secretary of the Air Force Stuart Symington at a meeting at the White House when Symington glibly tried to argue that custody of atomic bombs should be transferred from civilian control to the Joint Chiefs. "Our fellas need to get used to handling it," Symington foolishly insisted. Truman shot back, "This is no time to be juggling an atom bomb around." But the deployment of the B-29s was intended as a not very veiled warning to the Soviets, the first gambit in a grim game of nuclear bluff that both sides would play with increasing realism and risk as the Cold War progressed.[28]

The cryptanalysts at Arlington Hall and Nebraska Avenue were absorbed in their own crisis while the Berlin confrontation played out in public. For over a year they had been watching with growing unease a series of changes in Soviet communications security practices. In August 1946 the Russians began enciphering the indicator groups that told the intended recipient of an enciphered code which key pad or additive book page had been used. The change occurred in all of the five-digit codes used by the armed forces, the MVD, and the MGB. In September 1947 there was a noticeable drop in the number of official messages transmitted in the clear on the internal radiotelegraph networks of the Soviet Union, and in December 1947 a notice was issued by the telegraph authorities in Moscow that coded telegraph messages were no longer to be sent over unscrambled radio teleprinter channels, but were restricted to landlines only; at the same time there was an upsurge in scrambled radio teleprinter traffic on certain links. In early 1948 military traffic in the Soviet Far East sent using the Sauterne machine, the Soviets' version of the Hagelin B-211, went off the air; in April 1948, use of the Longfellow teleprinter encryption machine ceased. Throughout 1948, codebooks for a number of Russian enciphered codes, including some naval and MVD systems, were changed from one-part codes to the much more difficult to break two-part codes.[29]

None of these changes in themselves stood out from the sort of routine upgrades that happened all the time, but the tempo of the activity suggested an overall tightening of security. On Monday, November 1, 1948, something the U.S. and British codebreakers had never seen before took place. An urgent report the next day from the Op-20-G traffic analysis section cataloged the sweeping changes that had occurred in virtually every communications system of the Soviet military and MVD:

> Unprecedented Coordinated Russian Communication Changes.
>
> 1. Beginning 1 Nov extensive communication changes, which overshadow all previously recorded changes in type and areas affected, were effected in the Russian Naval, Military, and Police Communications Networks. Major fixed radio stations of all services (Navy, including Naval Air; Military, including Military Air; and Police) are now employing what appears to be the same type of call sign system, similar in appearance to International Berne call signs. Other sweeping changes occurred down the line of the three services affecting radio procedures, call signs, message formats, and in some instances the frequency plans.
> 2. Points of significance which arise from these changes include:
>   a) The efficiency with which the changes were executed in all areas by the three services, which is in direct contrast to previously recorded changes of individual services.
>   b) The coordination exemplified by the three services in enacting the changes.
>   c) The extensiveness of area application: All Naval Fleet areas; European and Far Eastern Police; and European and Far Eastern Military areas.

3. It is not evident that there was any reason for the execution of these changes on 1 Nov other than readiness by all services to do so. The coordinated timing does indicate clearly a central unified control or direction of all service communications.[30]

The Soviets had pulled the plug even more emphatically than was first apparent. All traffic passing over military, naval, and police radio links was replaced with nothing but practice and dummy messages. The Soviets' heavy use of radio teleprinter and other radio links for their internal communications had been a recourse of necessity given the extensive damage to landlines during the war (and the Soviet Far East and Central Asia had never been well served by landlines and had always depended heavily on internal high-frequency radio networks). Reconstruction remained far from complete in late 1948. But apparently the Russians had concluded that their security concerns could not wait, and were shifting whatever they could of their most secret communications to available landlines immediately, even if it meant drastically cutting back the amount of traffic that could be handled.[31]

The November 1 change would become a fabled part of NSA's lore, known within the agency as "Black Friday" (although the changes had actually been made over the weekend of October 30–31 and introduced the following Monday).[32] The full extent of the disaster only became apparent the following spring when real traffic started reappearing on the radio nets, now employing greatly improved—and completely unbreakable—technical and security procedures. The keying errors or other mistakes that had allowed most of the Soviets' machine-enciphered military traffic to be routinely read by U.S. and British codebreakers for the past several years had been corrected, and the much more disciplined systems that now replaced them slammed the cryptanalytic door shut.

The one important Soviet machine system that had remained on the air immediately following Black Friday was Albatross. When Coleridge traffic vanished on Black Friday, following the earlier disappearance of the Sauterne and Longfellow machines, some of that traffic was temporarily taken over by Albatross. But Albatross had stymied the best efforts of Arlington Hall's codebreakers, and would continue to do so for years to come even as the cryptanalysts marshaled ever more powerful special-purpose analyzers, and then the first digital electronic computers, to attack the problem.[33]

The Soviet one-time-pad systems used for communications that passed through diplomatic cables also underwent a comprehensive alteration in indicators and addresses. Although Arlington Hall had not found any duplicate key use on messages sent after 1945 on most of these nets, it had continued to collect, sort, and study current traffic in the hope that reuses might still appear. (Some NKGB/MGB messages sent between Canberra and Moscow had employed the old duplicate key pad pages as late as 1948.) But with the indicator changes, it became impossible even to sort the traffic into the five different systems that had been identified: GRU, naval GRU, trade, consular, and MGB. Cecil Phillips was finally able to break the new indicators to allow that sorting and traffic analysis to resume, even if the messages themselves remained unreadable.[34]

A sudden, across-the-board change in Soviet code systems and communications procedures was so alarming that London and Washington briefly considered the possibility that it indicated preparations for imminent war. At a meeting in November 1948, the U.S. Communications Intelligence Board reported that their British counterparts had definitely ruled out that explanation but could offer only speculation in its place. The changes might be part of a "methodical drive to improve communication security"; they might be a temporary stopgap while security defects that had been discovered were corrected; or they might have been an urgent response to leaks about U.S. and British successes in reading Soviet cryptographic systems. The chiefs of the Army Security Agency and the Navy's CSAW were asked for their views: the Army's Colonel Hayes was "strongly inclined toward the belief that leakage of information had been the primary cause," while Captain Wenger of the Navy leaned toward the more mundane explanation of "methodical" improvement, but acknowledged that a leak could not be ruled out. By the end of the year the board agreed to proceed on the assumption that the Soviets had been tipped off that their codes had been penetrated. USCIB ordered the "need to know" rule tightened and directed that even those specially cleared to receive communications intelligence at "user" agencies such as CIA were not entitled to know anything of the technical details concerning how it was produced. In April 1949, GCHQ and USCIB agreed that henceforth there would be a "complete separation of work on non-Russian from that on Russian," both on the part of the traffic analysts, cryptanalysts, and translators who produced the material and the users who evaluated and disseminated the results for intelligence.[35]

Those were all prudent steps in any case, but they did nothing to reverse the damage that had already been done; nor were the British and Americans any closer to finding out if they did in fact have a Soviet spy or spies inside their signals intelligence agencies, and if so just who he or they might be.

At the start of 1949, 1,073 of the 3,124 workers at ASA and CSAW were assigned to the Russian problem; at GCHQ there were another 389. Hundreds more manned the 524 intercept radios tasked to collect Soviet radio traffic at the 38 field sites the United States operated around the globe. It had been a huge surge of personnel from the handful that had begun work on the Russian problem a few years before. Now it was not clear if any of them had a job to do anymore.[36] On March 24, 1949, Frank Rowlett, now chief of ASA's Operation Division, responsible for all communications intelligence (COMINT) production, and Solly Kullback, chief of its Research and Development Division, reported the one glimmer of hope left in the effort to exploit Soviet communications:

> Russian communications security measures introduced over the period of the last eighteen months, including retirement of certain major cryptographic systems and the virtual cessation of operational radio activity on Armed Forces links, have increased the dependence of the allied Comint effort on the study and analysis of the large volume of traffic passed on the Russian internal civil radio links. These links, as yet unaffected by Russian communications security measures, are now the major source of current economic and current military intelligence information.[37]

These hundreds of thousands of routine, unenciphered telegrams carried each month over the Soviets' civil internal radio networks had not seemed like a very promising source at first, and more than a few of the traditional-minded cryptanalysts in Britain and the United States, Rowlett among them, were initially dismissive of the idea that the Russians were going to hand over anything of intelligence value on a plate that way. "If the Soviet Union considered it important," Jacob Gurin, a Russian linguist at Arlington Hall, remembered being told, "they would encipher it."

But even before Black Friday a small plain-language unit under Gurin's enthusiastic direction, drawing entirely on Russian telegrams that had been sent by radio in the clear, produced several insightful reports on the Soviet ministries in charge of industrial production. Gurin, known as Jack, spoke Russian fluently; born in Odessa to Russian Jewish parents, he had come with his family to America as a young boy but grew up speaking the language at home. Gurin argued that even the most tedious telegrams dealing with coal supplies, railcar loadings, and labor requirements could, when pieced together, produce a comprehensive picture of the state of the Soviet economy. In fact, given the near-total secrecy Stalin imposed on publication of official information (the locations of government ministries and production sites, the full names of officials, even texts of government decrees were often a secret) and the all-pervasive authority of the Soviet government in the country's centrally planned economy, the plain-language messages represented a source of information available nowhere else.[38]

Gurin's group had begun work in November 1947 with a staff of six but added nearly one hundred Russian linguists during 1948, many shifted to the effort when the encrypted B-211 traffic went off the air. By the summer of 1948, USCIB and GCHQ concluded a formal arrangement—it was embodied in a voluminous appendix K to the 1946 BRUSA agreement—extending their cooperation into Russian plain-language teleprinter and radiotelephone material, with the British contributing to the joint effort a special linguistic unit it had been operating since the end of the war: located on the third and fourth floors of a converted apartment building on Ryder Street, in the fashionable St. James's district of central London, it was staffed mostly by Russian émigrés. Working from clues as tenuous as a list of Gosbank account numbers that the analysts were able to link to Soviet defense industries, Gurin's group issued reports identifying centers of munitions production, assessing the capacity of the Soviet transportation system, estimating the output of vehicle assembly and engine plants, and compiling basic production statistics for steel, chemicals, oil, and electric power. For several years, the plain-language effort would also be one of the primary sources of information about the Soviet atomic program, and one of the few means to monitor warning signs that might indicate mobilization for war.[39]

Based as it inescapably was on piecing together information gleaned from hundreds or thousands of messages, the plain-language effort made even more of a mockery of the fiction that ASA and CSAW were not supposed to be producing their own intelligence analyses but only publishing translations of individual messages; Gurin's group, and its smaller counterpart at CSAW, for the most part ignored the stricture and ultimately published only 0.3 percent of the plain-language messages as separate translations, concentrating instead on producing reports devoted to a series of single topics that pulled together all available information on each. Just as Bletchley's translators and analysts in Hut 3 had done during the war, they amassed a huge card file of names, places, and industries mentioned in every message, along with other collateral information to aid the effort. This would become the nucleus of NSA's legendary Central Reference library on the Soviet defense industry, known as C-Ref.[40]

Even this limited foray into intelligence production raised hackles, especially when analysts at CIA noticed that some of the footnotes in the reports cited messages that had not been officially published and circulated. William Friedman a few years later tried, with obviously limited success, to point out the superiority of the British method:

> GCHQ has consistently strived to furnish its consumers with comprehensive reports (compilations of all pertinent plain-text material plus relevant collateral) on subjects of interest to its consumers. The full-message translation method has been reserved for the very few items which can stand alone and tell a good story. Only in the case of special subjects such as Atomic Energy has there been any attempt to publish all available messages on a subject, and in these rare instances the translations are published in book form rather than as individual cards.

But in Washington, even the "right to write reports" was still being contested, and there was constant pressure on the signals intelligence agencies from consumers "to issue as large a number of individual translations as possible." The Office of Naval Intelligence went so far as to insist that the signals intelligence analysts be barred from issuing *any* reports on Soviet shipbuilding and limit their work in that area entirely to message translation, leaving interpretation to the real experts, namely themselves: ONI informed Gurin's boss, Oliver Kirby, that they wanted to see "everything" so that they could take over the job. Gurin protested that it had taken his group years to build up the expertise and experience needed just to identify and interpret the content and meaning of the messages they read. Many of the telegrams in fact were nothing but lists of materials and shipments and other production statistics in standardized formats that would be meaningless to someone who had not already read thousands of other similar messages and could notice connections between them. A naval officer on Gurin's staff, Neal Carson, was so furious about ONI's demands that he threatened to resign his commission if they persisted, which apparently so stunned some of the Navy intelligence officers who knew and respected Carson that they backed down.[41]

The plain-language operation was the first plunge into the massive data collection and sifting that would characterize NSA's standard approach in the decades that followed, and it upended business as usual. At the pinnacle of Arlington Hall's wartime success decoding Japanese army traffic, it had processed a little over 100,000 messages a month. When the Russian plain-language effort began in 1947, ASA and CSAW were each handling a few thousand of those messages a month. The number quickly grew to unprecedented heights: 200,000 a month in 1948, 700,000 in 1949, 1 million in 1950. The peak would be 1.3 million messages a month in early 1951.[42]

Simply recording, delivering, and sorting that much traffic was a challenge on a scale that dwarfed anything previously attempted. Intercepting high-frequency signals was a complex art and science that depended on atmospheric conditions, sunspot cycles, and other factors that affected how far a radio wave would propagate by skipping off the earth's ionosphere. Not all U.S. intercept stations were yet fully equipped with teleprinter demodulating and receiving equipment needed to automatically record the signals. Intercepting Soviet telephone conversations, which also traveled by radio on many of the same links that handled teleprinter traffic between Moscow and Central Asia and the Far East, was even more difficult, often taxing translators straining to make out indistinct recordings.[43]

The teleprinter traffic arrived at Arlington Hall from field sites mostly on punched paper tape, and the job of sorting and printing out those millions of tapes far more resembled working on the floor of a boiler factory than it did any contemplative scene of cerebral Holmes-like codebreaking. In an institutional culture that accorded preeminent status to the mathematically trained cryptanalysts, and where even linguists were looked down on as mere "support personnel," the clerical tasks associated with plain-language processing were the lowest of the low. At Arlington Hall, the job fell almost entirely to a segregated unit of African American men and women.

Up until 1944 the only African American employees at Arlington Hall were messengers and custodians. One of those was William Coffee, who had studied English at Knoxville College in Tennessee, a historically black college founded after the Civil War, and who was working as a waiter at the Arlington Hall girls' school in 1942 when the Army took over the site. Coffee was kept on as a junior janitor and subsequently promoted to messenger. In 1944, facing pressure reputedly from the White House to improve the representation of African Americans in Army specialties—fewer than 1 percent of black selectees had been assigned to the Signal Corps, notably—the commander of Arlington Hall told Earle Cook, then a lieutenant colonel and in charge of B Branch, the cryptanalysis section, that he needed to hire "some Negroes." The only African American Cook knew was Coffee. "I told him I got this problem," Cook recalled. "I got to have about a hundred and some odd people of your race, and I said, 'You're my personnel officer to see that I get the right ones.' Did a marvelous job. Where the hell he got them and how he got them I knew not."[44]

In fact it was not difficult. The District of Columbia had an unusually well-educated African American population, and many of the first black recruits to ASA's cryptanalysis branch were graduates of Washington's historically black Howard University who were working at menial government jobs well below their educational qualifications. The first segregated, all-black unit in B Branch, which Coffee joined in January 1944 and soon became the operating head of, was given the job of working on foreign commercial codes. After the war more African Americans were hired to fill low-level positions at Arlington Hall running tabulating equipment and operating key punches.[45]

After 1947 the rapidly expanding Russian plain-language processing branch became the place where nearly all new African American employees ended up. The work, as a review of the agency five years later found, was "sheer drudgery." Located on the first floor of A Building, the unit had the job of feeding the tapes into printers and stamping documents. The pay was at the very bottom of the government scale; nearly all the positions were at the GS-2 level, $1,440 a year. The one hundred or so men and women of the section, which would be known for most of its existence by its subsequent designation, AFSA-213, sardonically referred to it as "the Plantation." It was noisy, dirty, and mind-numbingly monotonous work, but it was at least a professional position, a step up the ladder from busboys and messengers: a photograph of the unit from around 1951 shows a dozen men of AFSA-213 striking a confident and dignified pose in sharp suits and ties and gleamingly polished shoes.[46]

Jack Gurin tried to improve conditions by providing at least rudimentary training in message analysis. "Their job was to sit there and watch the machine and make sure it didn't jam," Gurin recalled. "If it jammed, you stopped the machine and pulled out the keys or fixed the paper. Then you started it again and waited for the next jam. That was their job. They were college graduates. I looked at them and said this is ridiculous." Gurin taught some of the staff in AFSA-213 how to read the Baudot codes at the beginning of the tapes so that they could act as "scanners," sorting the tapes by their addresses and selecting which were worth printing out. Scanning was not "terribly challenging," Gurin admitted, but it was at least "better than what they were doing before."[47]

Still, even scanning was monotonous and largely mindless production-line work, a point underscored by rigid productivity quotas that required each employee to get through three hundred tapes a day, which worked out to less than two minutes per tape. The scanners were later given a fixed list of thirteen hundred terms they were to look for to decide which tapes were to be kept, which culled the number of incoming messages by 80 percent, but they were not even told what the ultimate purpose of their work was or what the Russian words meant. "So far, no other more humane or less haphazard method of reducing the millions of bits of paper to usable and workable proportions has been developed," the 1952 agency review concluded. Although Arlington Hall had initiated an intensive six-month Russian language course in 1948, none of the African Americans in AFSA-213 were told about that opportunity; William Jones, who had studied Latin and German at school and worked in the section from 1951 to 1955, enrolled on his own in a Russian course at the U.S. Department of Agriculture Graduate School hoping to improve his chances to move up in the agency, only learning much later that the codebreaking agency offered its own course.[48]

A few African Americans were able to find more rewarding positions at Arlington Hall in those years; three black electrical engineers and technicians, Carroll Robinson, Charles Matthews, and W. C. Syphax, were part of the team that built the first Abner computer beginning in 1948. Truman's executive order that year desegregating the armed forces put an end to any legal basis for confining African American employees to separate units, and the overall political atmosphere at ASA in those years had a distinctly liberal cast. But Arlington Hall was still located in the South and segregation was the law outside the gates, including on the public buses that many of its employees rode to work. It would not be until NSA's move to Fort George G. Meade, Maryland, in 1957, and the abolishment of NSA-63—the successor to AFSA-213—in a major reorganization in 1956 that opportunities for African Americans to escape servitude on "the Plantation" of Russian plaintext would noticeably improve.[49]

What did not change was the American codebreakers' conviction that more manpower, more machines, more intercept sites, and more money could make up, through the sheer brute force of mass collection and industrial-scale data processing, for what they could no longer achieve through cryptanalytic finesse alone.

5

Shooting Wars

If signals intelligence was a growth industry, neither the new CIA nor the military services wanted to miss their chance to be a part of it. In peacetime, divided authority usually results at worst in turf battles, bureaucratic inefficiency, poor planning, and duplication of effort. But in the increasingly high-risk stakes of the Cold War, it was about to have literally fatal consequences.

On April 8, 1950, a Saturday, a U.S. Navy PB4Y-2 patrol plane took off from the American air base in Wiesbaden, West Germany, passed over Bremerhaven on the coast of the North Sea, then turned abruptly east and headed to the entrance of the Baltic Sea, flying a course over the water that kept it about twenty-five miles from the coast of East Germany and Poland.[1] The plane, a large four-engine craft, was a Navy version of the B-24 Liberator bomber of World War II fame, modified with a single tail in place of the Liberator's twin vanes and an extra-long fuselage to accommodate a flight engineer's station, radar jamming equipment, and a crew of up to sixteen for extended patrol missions over the open ocean. The Navy called it the Privateer. The crew flying that day had nicknamed their plane the *Turbulent Turtle*.[2] It was part of a squadron usually based at Port Lyautey in French Morocco whose ostensible mission was to act as couriers for U.S. diplomatic posts throughout Europe, Scandinavia, and Western Asia.

The *Turbulent Turtle* had arrived in Wiesbaden a few days earlier for what was in fact a distinctly undiplomatic mission. Like the rest of the Privateers at Port Lyautey, the plane was filled with radio gear designed to pick up radar signals from Soviet air defense units. In addition to receivers that covered the microwave radar spectrum, with frequencies up to 3000 MHz, its equipment racks were crammed with pulse analyzers and oscilloscopes that could identify the particular radar units by the duration and repetition pattern of their transmitted signals, plus direction-finding antennas to take bearings and pin down their location.

Airborne radio surveillance went back to World War II, when the Navy and the Army Air Forces had flown what they called "ferret" flights during the island-hopping advance against Japan in the Pacific. These were extremely dangerous missions, but the information they collected allowed strike aircraft to target Japanese radar installations before each assault, an incomparable advantage in the hard-fought battles for Truk and other Japanese redoubts. After the war, the Navy sold off most of the radio equipment from its ferret squadrons as surplus. In early 1950, scrambling to reconstitute two patrol squadrons in order to begin collecting intelligence about the then almost completely unknown Soviet air defense system, the Navy sent two chief electronic technicians to New York City to see if they could buy back any of the surplus gear. Wearing civilian clothes and carrying huge rolls of cash, the two chiefs scoured the city's Army and Navy stores and bought up all the UHF radar receivers and associated analyzing equipment they could find. The equipment was repaired by Navy technicians and quickly installed in the planes.[3]

The ferret missions against the Soviet Union were carried out in the utmost secrecy, but the urgent reason for them was no secret to anyone by the start of 1950. The previous September, Truman's press secretary wordlessly passed out to reporters summoned to his office a mimeographed statement from the president:

> I believe the American people, to the fullest extent consistent with national security, are entitled to be informed of all developments in the field of atomic energy. That is my reason for making public the following information.
>
> We have evidence that within recent weeks an atomic explosion occurred in the U.S.S.R.[4]

On January 31, 1950, over the unanimous objection of the Atomic Energy Commission (AEC) scientific advisory committee, Truman announced in an equally terse statement that the United States would proceed with development of the "super," or hydrogen, bomb. Up until the moment of the decision Truman held out hope that it would still be possible to place all atomic weapons under international control—such was the faith in the UN as an instrument of international order that was as yet unexistinguished in that less cynical age.

The most nuanced and strategically insightful reservations about going ahead with the superbomb came from the AEC's chairman, David Lilienthal, who accurately foresaw the "false and dangerous" over-reliance on nuclear weapons that would skew and often paralyze U.S. defense strategy and the country's ability to respond with military force in situations short of all-out war. But in the end even Lilienthal acceded to the tough-minded arguments of the otherwise liberal Dean Acheson, Truman's secretary of state (who ironically was about to become the target of McCarthy's most vicious red-baiting smears). Acheson thought the hard facts left no choice and upbraided George Kennan for trying to argue that the United States should lead by example and turn away from an uncontrolled arms race that could only end in ever-greater heights of unimaginable destructiveness. Acheson sarcastically told State Department associates that if that was Kennan's view he ought to resign from the Foreign Service and "put on a monk's robe" and stand on a street corner, but not push his "Quaker gospel" within the department. "How can you persuade a paranoid adversary to disarm 'by example'?" Acheson demanded on another occasion.[5]

The stunningly swift move to near-total reliance on atomic weapons as the cornerstone of America's military strategy for fighting the next war had in fact predated the Soviets' acquisition of the bomb. Facing cutbacks in conventional forces, flush with triumphant claims of victory through strategic airpower over Germany and Japan, and locked in a fierce struggle with the Navy for postwar roles and missions and a looming budget decision between super bombers and super carriers, the newly independent U.S. Air Force seized on strategic bombardment with atomic weapons as the central pillar of its postwar planning. Its Strategic Air Command (SAC), a virtually semiautonomous force commanded since October 1948 by the hard-charging, cigar-chomping Curtis E. LeMay, was training single-mindedly to carry out the mission LeMay referred to as the "Sunday punch." LeMay, the youngest four-star American general since Ulysses S. Grant, had earned the nickname Iron

Ass from the crews he relentlessly drove in the firebombing attacks on Japan. Now LeMay insisted that SAC's only real purpose was to be ready to deliver at least 80 percent of the entire American stockpile of atomic bombs simultaneously in a single massive strike. "The next war will be primarily a strategic air war and the atomic attack should be laid down in a matter of hours," he proposed one month after taking over SAC, and the Joint Chiefs agreed, endorsing his plan and giving SAC the highest budget priority.[6]

SAC's targeting briefing was entitled "To Kill a Nation," and it proposed to carry out the emergency war plan by destroying seventy Soviet cities with 133 atomic bombs. The precise targets were only vaguely specified. But LeMay grandly argued that the effect of the "Sunday punch" would be to make the Soviet Union lose the "will" to wage war, to "stun the enemy into submission," even to cause the entire country "to collapse." Given the huge costs of matching the Soviets' conventional forces in Europe, U.S. Air Force planners, sounding more like actuaries than generals, argued that the ability to deliver a crushing atomic attack in response to any Soviet aggression "was an opportunity to put warfare on an economical, sensible, reasonable basis." All of this required making the threat of atomic airstrikes wholly believable, and that meant that SAC's bomber force had to be "combat ready"—one of LeMay's favorite phrases—capable of flying thousands of miles, penetrating Soviet airspace, accurately locating their targets, and delivering their atomic payloads. Initially appalled at the poor performance of SAC's aircrews, LeMay drove them just as he had harried his B-29 squadrons during the war in the Pacific, making them operate under realistic wartime conditions of high altitude, bad weather, and difficult targeting situations. "We had to be ready to go to war not next week, not tomorrow, but this afternoon. We had to operate every day as if we *were* at war," LeMay wrote.[7]

The Soviet atomic bomb, of course, added a more urgent reason to be sure that American bombers could get to their targets and successfully hit them on a moment's notice: in 1950, the Joint Chiefs of Staff revised the war plan to give top priority to "the destruction of known targets affecting the Soviet capability to deliver atomic bombs." So grave a conundrum did the threat of a Soviet atomic strike pose to American war planners that the Joint Chiefs proposed asking Congress to authorize the president to carry out a preemptive atomic strike on any nation engaged in acts of aggression, defined to include "the readying of atomic weapons against us."[8]

Almost nothing was known about Soviet air defenses, however, and if LeMay was right that the Cold War had to be treated as a real war, taking wartime risks to ensure operational success was justified. In the spring of 1950, Truman approved a plan for surveillance flights to assess Soviet air defenses.[9] Although the ferrets were to avoid actually crossing into Soviet territory, the aim was to fly close enough to the border to provoke a reaction by radar stations and fighter squadrons so that their communications and electronic emissions could be detected and monitored.

Three hours and forty minutes into the *Turbulent Turtle*'s flight early on the afternoon of April 8, 1950, the crew radioed that they had reached the end of their assigned mission path, a point roughly even with the northern tip of Latvia, and were turning back home for the return sweep along the Soviet coast. Half an hour later the plane radioed in again to report its location. That was the last it was ever heard from. At 2330 that night the squadron's headquarters at Port Lyautey received a dispatch from the U.S. naval base at Bremerhaven reporting that the plane was declared overdue. At first light the next morning, three PB4Y-2s were dispatched from Morocco to head east to look for the missing aircraft. Over the next ten days two dozen more aircraft joined the search. A few days after the search was called off a Swedish fishing boat picked up an empty life raft bearing the missing plane's serial number.

An official diplomatic protest by the United States was met by a Soviet statement that "a B-29" had overflown Soviet territory, been challenged by Russian fighters over the Latvian coastal town of Liepāja, and was shot down over the sea when it refused to obey orders to land and opened fire as it tried to flee. As the squadron's history drily observed, "The credibility of the Soviet report was seriously weakened by the fact that the Privateer's only armament was a .45-cal. pistol carried by one of the officer crewmen." The ten-man crew was never found, though years later several prisoners released from Soviet prison camps claimed they had seen or heard about some of the men.[10]

The flight path of the U.S. Navy PB4Y-2 electronic surveillance plane shot down by Soviet fighters off the coast of Latvia on April 8, 1950.

Over the next fourteen years, twelve more U.S. Navy and Air Force ferret aircraft would be shot down over the Baltic, the Sea of Japan, the Kamchatka Peninsula, East Germany, and Soviet Armenia; dozens more would be buzzed or fired on but escape more serious harm. In many cases ferret flights approached as close as a few miles to the twelve-nautical-mile territorial limit, often paralleling the coast for long distances, in what seemed almost a direct taunt to Soviet air defenses. Many of the Air

Force ferret aircraft were modified B-29s, which by external appearances were no different from atomic-capable bombers. The riskiest missions sent a SAC bomber racing toward the Soviet border to deliberately provoke the Russians to scramble fighters and activate their radar networks while a ferret flew nearby to record the reaction. More than ninety men would lose their lives in these electronic intelligence-gathering missions around the periphery of the Soviet empire. When forced to make a public announcement about the losses, the U.S. military would say only that the planes had been conducting "weather reconnaissance."[11]

Signals intelligence was starting to look a lot less like a clean, austere, intellectual exercise and a lot more like the dangerous and dirty business that spying had always been.

The gathering of data from radar signals acquired its own acronym, ELINT, which stood for electronic intelligence, the distinction being that these sorts of radio emissions were not communications per se and contained no messages to be deciphered or read. All three military services considered ELINT a separate business entirely from the work of the cryptologic agencies, adding yet another turf battle to the increasingly messy bureaucratic struggle over the postwar intelligence structure. A chief purpose of ELINT collection was to record technical details of each of the Soviet radar systems—the frequencies they operated on, the kind of pulses they employed—so that jammers and other electronic countermeasures could be designed to block or spoof them effectively. The Air Force also wanted to thoroughly map Soviet air bases and their associated radar stations so that they could be targeted as part of SAC's plan for penetrating Soviet airspace in an all-out attack.

But ELINT was also a potential source of a great deal of operational, strategic, and tactical intelligence. Once the radar systems associated with a particular kind of unit, such as a tank division or a motorized rifle regiment, had been identified, it was possible to begin constructing an "electronic order of battle" that could reveal volumes about the organization of the Soviet military, and even offer immediate warning of the movement of Soviet ground forces if those characteristic signals started showing up in new locations.[12]

With the ongoing blackout of significant cryptanalytic intelligence following Black Friday, ELINT assumed even greater importance, especially when it came to sources of possible strategic warning of Soviet preparations to launch an atomic strike of their own against the United States. Existing radar coverage of the continental United States could offer little better than one hour's warning of approaching Soviet bombers. The planned joint U.S.-Canadian Distant Early Warning, or DEW, line, a chain of radars across northern Canada, Alaska, and Greenland, would extend that to four to six hours, enough time to shift six hundred to one thousand fighter aircraft to the Air Defense Command from other tactical units.

But despite all of LeMay's boasts, a secret investigation in 1957 found that even with a decade of effort to put SAC on a constant war footing, not a single one of its bombers would be ready to get off the ground within six hours on a randomly chosen day.[13] If warning of Soviet preparations for an attack could be extended to three to six days, however, the consequences would be "of tremendous magnitude," a high-level scientific review committee that examined signals intelligence and strategic warning concluded. "In fact, warning of this nature" would permit SAC to fully deploy its forces to be ready for an immediate counterattack if the Soviets then went through with their plans. Moreover, SAC's own preparatory mobilization, "if known to the enemy, might induce him to cease his preparations for an attack."[14]

The scientific committee was chaired by Howard P. Robertson, a physicist at Caltech. The Robertson panel was not very encouraging about the dismal state of efforts to break back into the Soviet high-level code systems. In the absence of being able to read the actual transmitted orders of the Soviet high command, signals intelligence derived from collating plain-language intercepts, low-level tactical code systems, and traffic analysis—a process known as "T/A Fusion"—remained the only hope. In fact, the panel found, "Since late 1948, we have been forced to rely on information obtained by T/A Fusion for almost all COMINT on the Soviet Armed Forces," including almost all of the direct indicators of movements by long-range Soviet bomber forces or other steps suggesting an impending attack.[15]

Still, it was not as forlorn a hope as might have first appeared. Traffic analysis had repeatedly proved its value going back to World War I, when cryptologists at the British Admiralty noticed that sorties of the German fleet from the Heligoland Bight into the North Sea were invariably preceded by a flurry of radioed messages to minesweepers, air patrols, and units in charge of the booms blocking the estuary entrance.[16]

In World War II, Bletchley's traffic analysts were able to map out the radio networks and call procedures used to assemble U-boats into position for "wolf pack" attacks, and even during periods when the naval Enigma messages could not be broken they were frequently able to forecast impending attacks hours or even days in advance just from the pattern of communications and direction-finding fixes on transmitting U-boats. Call signs, transmission schedules, operator "chatter" requesting frequency changes or repetitions of messages or setting up future transmission times, even careful analysis of the distinctive idiosyncrasies of a Morse code operator's "fist" and the unique radio fingerprint of stray noise produced by a specific transmitter, could identify and locate a particular ship or unit even when the messages themselves could not be read. Rapid sending without any request from the receiving party for service was usually a sign that only dummy traffic was being sent; its sudden replacement with real traffic often indicated imminent plans for a military operation.[17]

In view of "our present heavy dependence on T/A" for strategic warning of Soviet military action, the Robertson panel accordingly urged that considerably more attention be given to improving the equipment needed to monitor radar and other signals, perform rapid and accurate direction fixes, and identify transmitters by radio fingerprinting; that trained analysts be stationed at field intercept sites to help notice any suspicious indications in Soviet radio traffic patterns; and that the current system for transmitting "Flash" messages be improved so that warning could reach Washington in minutes if necessary.

If there was anyone predisposed to value the role of top-level scientific expertise in intelligence, it was Howard Robertson. He had served as scientific intelligence adviser at Eisenhower's headquarters during the war and helped to interrogate captured German rocket scientists on the V-2 project. He was also one of the world's leading mathematical physicists, one of the originators of the cosmological concept of an expanding universe. But his panel took pains to emphasize that the jobs the cryptologic agencies still tended to regard as mere "support" to the mathematically adept cryptanalysts had to be given far more prestige, and in particular the disparity in pay and promotion between cryptanalysts and traffic analysts needed to be corrected.[18] In an era when cryptanalysis was no longer king, and might in fact never be again, the heretofore unquestioned place of the cryptanalyst, at the top of Arlington Hall's pecking order urgently needed rethinking.

At the end of World War II, Soviet and American troops had occupied the Korean peninsula almost as an afterthought. Part of Japan's Asian empire since 1910, Korea was suddenly freed of its foreign yoke with the surrender of Japan, and the Allied troops that had been preparing to invade Japanese-held Manchuria and the home islands took swift possession of the Korean peninsula, with the United States and the USSR agreeing on the 38th parallel, which roughly divided the country in half, as the north-south demarcation line of their occupation zones. Elections under UN auspices were to follow to select a new government for the entire country. Stalin, however, refused to allow the Soviet-occupied zone to participate in the vote, instead establishing a Communist government under Kim Il-sung, a Korean who had been trained as an infantry officer in the Soviet Red Army during the war and who had promptly begun erecting a Stalinist cult of personality around himself. The election went ahead in the South in May 1948, choosing as the Republic of Korea's first president Syngman Rhee, a seventy-three-year-old leader of the movement for Korean independence who had spent decades in exile in the United States, where he earned a PhD in political science from Princeton. With the subsequent withdrawal of U.S. and Soviet troops, the Korean peninsula was left split between two rival governments, each claiming to be the sole legitimate representative of the entire nation.

In the spring of 1950, North Korea was literally at the bottom of the list of problems that concerned U.S. intelligence analysts. The U.S. Communications Intelligence Board had tried to offer some guidance to the cryptologic agencies on where to focus their attention, drawing up three lists of priorities. The only item involving Korea to make it to the "A" list (topics "of greatest concern to U.S. policy or security") was "Soviet activities in North Korea," but even that was dropped when the list was revised on May 15. The "B" list, which was for items of "high importance" to be given "expeditious" treatment if possible, included at the very end, "North Korean–Chinese Communist relations" and "North Korean–South Korean relations, including actions of armed units in border areas." But even when a "Watch Committee" set up by CIA to monitor Soviet moves became increasingly worried about North Korean intentions in April, that concern was never communicated to Arlington Hall through the USCIB mechanism.[19]

A U.S. Army intercept station in Japan the previous year had stumbled on some signals of an unidentified network using Soviet communications procedures, and through direction-finding fixes concluded they might be North Korean. After Army G-2 requested a more deliberate search for North Korean signals on April 21, 1950, the intercept stations in Japan were able to copy 220 enciphered messages and sent them back to Arlington Hall for analysis. It was indicative of the all-consuming attention that the Soviet Union held that the staff available to work on this North Korean material consisted of one part-time traffic analyst, one part-time cryptanalyst, and one Japanese linguist who had been trying to teach himself Korean in his spare time over the previous year; the section had no Korean typewriters, Korean-English dictionaries, or reference books.[20]

The situation in the field was even worse, exacerbated by a halfhearted but completely chaotic reorganization instituted at the Pentagon to try to bring the cryptologic agencies of the three military services under centralized control. In May 1949 the Joint Chiefs of Staff had established the Armed Forces Security Agency (AFSA) to consolidate communications intelligence and security activities. But it was the usual bureaucratic compromise that papered over all of the real problems and left the effective power of the Army Security Agency, the Naval Security Group (as Op-20-G was now called), and the newly created U.S. Air Force Security Service (USAFSS) virtually untouched. In the Pacific, ASA remained completely in charge of the three field intercept sites in Japan plus one in the Philippines that the Army operated. ASA also maintained its own regional command headquarters in Tokyo, called ASAPAC—Army Security Agency, Pacific. With a staff of 47 officers and 192 enlisted men, housed in a former Imperial Japanese Army arsenal, ASAPAC not only directed local field operations but even acted as a mini Arlington Hall, processing intercepts and attempting cryptanalytic attacks with its own IBM equipment. Rather than solving the problems of duplication and lack of central direction, the creation of AFSA seemed just to have added a fourth agency to fight for the same spoils, and it was utterly unclear where the dividing line of responsibilities between AFSA and ASAPAC lay when it came to working on the traffic being intercepted in the Far East. Most of ASA's best officers and equipment had been transferred to AFSA in the reorganization, however, which meant that even when it came to activities that fundamentally were the job of field offices, the work suffered from inexperienced leadership, overstretched staffs, and plunging morale.[21]

Reflecting both overall intelligence priorities and their parochial service interests, all three of the services' intercept units in Japan were focused almost exclusively on the Soviets. ASAPAC was collecting Soviet military and plain-language commercial traffic; the Air Force had a small unit, the 1st Radio Squadron Mobile (RSM), at Johnson Air Base outside Tokyo, which monitored Soviet air and air defense signals; and the Naval Communication Unit at Yokosuka was targeting the Soviet Far East Fleet's communications. Although ASAPAC did break a few low-level Soviet military codes, the pickings were slim, which added to the frustrations and low morale. The men who operated the radios at the field sites were mostly enlisted personnel, and the stations in Japan suffered constantly from low reenlistment rates and high turnover. At times their most notable distinction seemed to be the records they set for disciplinary infractions, drinking problems, and the number of VD cases reported.

As an Air Force report would later acknowledge, the radio intercept capability at the outbreak of the Korean War was "pitifully small and concentrated in the wrong places."[22]

Kim Il-sung had repeatedly sought approval from his Russian patrons for a military blitzkrieg on the South that, he confidently predicted, could seize control of the entire country in three days. With large stocks of weapons seized from the Japanese and a hundred-thousand-man army equipped with tanks and artillery, the North Korean army stood ready to carry out Kim's plans. Stalin was wary of being drawn into a conflict, but in January 1950 he gave Kim his assent, believing that it was an inviting chance to chip away at American prestige and that the

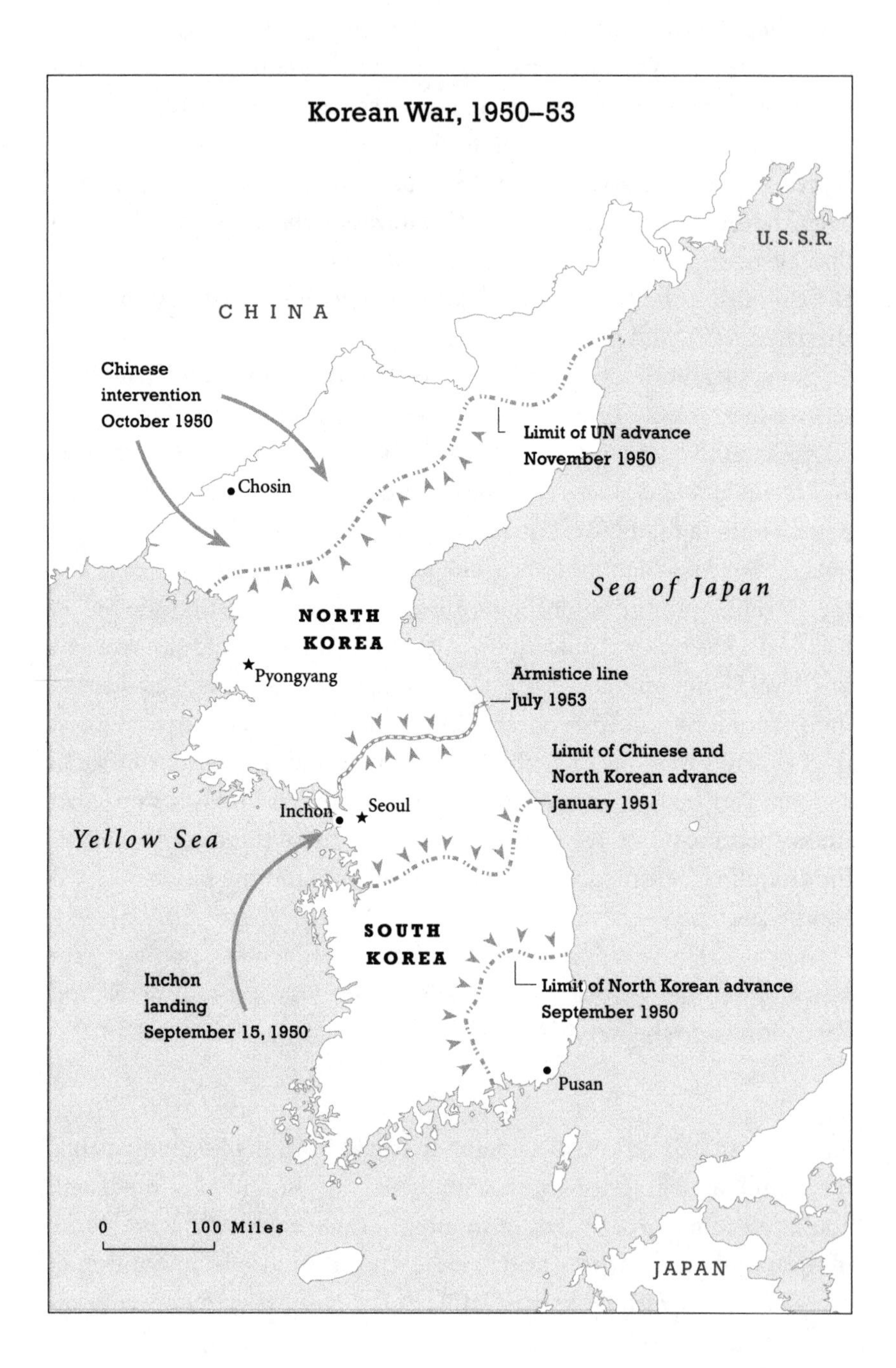
Korean War, 1950–53
U.S.S.R.
CHINA
Chinese
intervention
October 1950
Limit of UN advance
November 1950
Chosin
Sea of Japan
NORTH
KOREA
Pyongyang
Armistice line
July 1953
Limit of Chinese and
North Korean advance
January 1951
Inchon
Seoul
Yellow Sea
SOUTH
KOREA
Inchon
landing
September 15, 1950
Limit of North Korean advance
September 1950
Pusan
0
100 Miles
JAPAN

United States would be presented with a quick fait accompli that it could not or would not respond to.

In the event, the North Korean attack that began on June 25 brought surprise both to Washington and Moscow. Just as with the Berlin blockade two years earlier, Stalin fatally miscalculated American reaction; within hours, Truman, seeing the last hope for the collective security arrangements that the UN was supposed to bring to the postwar world about to crumble away, decided that the United States would come to the defense of South Korea's government, and do so under the authority of the United Nations. "We can't let the UN down," he insisted to his advisers.[23] A Soviet boycott of the Security Council at the time allowed a UN resolution authorizing the use of force to escape a Soviet veto. General Douglas MacArthur, who commanded both the military occupation of Japan and the U.S. Far East Command from his headquarters in Tokyo, ordered the Eighth Army under General Walton Walker rushed to South Korea to try to halt what was indeed becoming a juggernaut of Kim's forces, who in three days had swept unopposed through the South Korean capital of Seoul. By July 31 the Eighth Army had been pushed into a small defensive perimeter surrounding the port of Pusan on the southeast coast, bounded by the Nakdong River to the west and a line of rugged mountains to the north.

For two weeks after the start of the war the U.S. intercept stations in the region kept their focus riveted on Russian radio traffic, so large did fears loom that the North Korean attack was but the initial gambit of a Soviet plan to launch a general war with the United States. (AFSA's sense of where the greatest threat lay never really did change throughout the conflict: of its two thousand new military and civilian personnel added in response to the crisis, only eighty-seven would end up working on the Korean problem, even by the end of the war.) But that also reflected an almost total lack of translators to work on Korean material. None of the intercept units in Japan had a Korean linguist; after weeks of frantic search, ASA found only two Korean speakers in the entire Army available for duty in the theater. Both were teaching at the Army Language School in Monterey, California, where they were obviously needed to train new translators, but so great was the immediate need that both were shipped off to ASAPAC in Tokyo.[24]

But AFSA was quickly able to set up hourly delivery to Arlington

Hall via teleprinter of North Korean intercepts from the monitoring stations in Japan, and by July 3 had made the first translations of North Korean military messages. Astonishingly, they had been sent in the clear. It would later be known that in the chaos of the rapid advance, exacerbated by equipment losses and poor technical training, the communication discipline of North Korean radio operators had all but collapsed, and even the most top-level messages discussing troop movements, battle plans, and daily situation reports were sent with little or no regard to security.

Eleven days later the AFSA cryptanalysts at Arlington Hall solved their first North Korean code, and by the end of July had deciphered one-third of all encrypted North Korean messages received. During the Eighth Army's desperate forty-five-day fight from August to mid-September to hold the Pusan perimeter, intelligence reports from Arlington Hall's codebreakers and translators repeatedly allowed Walker to shift his exhausted and greatly outnumbered troops to meet the North Korean thrusts: messages containing attack orders down to the battalion level, often specifying the exact objectives and weapons to be employed, were now flowing in at a steady pace. Other messages confirming the identity of North Korean units, revealing the location of airfields and artillery ammunition dumps, and reporting the state of supplies had also been read by the end of August. It was, in the words of an officer on Walker's staff, an "utterly amazing" turnaround in the intelligence picture, and probably saved the UN forces from complete and swift annihilation in the first two and a half months of the war.[25]

AFSA also, in hindsight, would be seen to have acquitted itself well in the greatest debacle of the Korean War, MacArthur's disastrous failure to anticipate the intervention of three hundred thousand Chinese troops who came careering across the border in human-wave attacks just when, to most American troops, the war seemed to be over. On September 15, in what is rightly seen as one of the great masterstrokes in military history, MacArthur launched an amphibious landing behind North Korea's lines at Inchon, west of the South Korean capital. Seoul was recaptured two weeks later; the thirteen North Korean divisions besieging the Pusan perimeter retreated in disarray; and by early October MacArthur's troops swept past the 38th parallel almost unopposed into the North. Stalin bluntly advised Kim Il-sung that he would have to evacuate his remaining forces from North Korea. Khrushchev recalled Stalin's resigned acceptance that the war was all but lost, and that U.S. troops would soon be right across the Soviet border. "So what," the Soviet dictator said. "Let it be. Let the Americans be our neighbors."[26]

China's Communist leader, Mao Zedong, had different ideas. For months a hastily expanded group at Arlington Hall under the direction of a twenty-nine-year-old Chinese linguist, Milt Zaslow, had been sending out reports calling attention to the movement of large numbers of Chinese forces. The information came from dozens of civilian telegrams sent in the clear, many of them personal messages to members of military units addressed to them at locations that proved to be stations along rail lines leading to Manchuria, near the North Korean border.

On July 17, based on thirty-one messages, Zaslow's group reported that elements of the Fourth Field Army, considered to be the Chinese army's most capable combat force, had deployed north just before the North Korean invasion. On September 1, based on forty-one more messages, AFSA reported additional Chinese units on the move in late June and again in late July, including a newly formed artillery division. By the beginning of October indications pointed to a massive deployment under way, with twenty troop trains heading to Manchuria. At the same time, AFSA's deciphering of diplomatic cables sent by foreign embassies in Beijing revealed that Chinese foreign minister Chou Enlai had told the Indian ambassador and other foreign diplomats that China was prepared to intervene if American troops crossed the 38th parallel and entered North Korean territory.[27]

The subsequent military disaster would show MacArthur at his worst: his supreme refusal to accept facts that contradicted his prior assumptions; his inability to admit he ever made a mistake; and his reliance on a coterie of devoted sycophants who took pains to shield him from anything he did not want to hear or anything that might "disturb the dreamworld of self worship in which he chose to live," as one historian put it, in the splendid isolation of the Dai Ichi Building, his headquarters in downtown Tokyo. "If he did not want to believe something," said General Matthew Ridgway, MacArthur's successor at Far East Command, "he wouldn't." Once when speaking to General Marshall, MacArthur started to make a point by referring to "my staff." Marshall interrupted him: "You don't have a staff, General. You have a court."

No one in that court outdid the devotion and sycophancy of MacArthur's fifty-eight-year-old intelligence chief, Major General Charles Andrew Willoughby. Emigrating to the United States in 1910 at age eighteen, Willoughby eventually shed the name Adolf Charles Weidenbach that he used as a young officer in the U.S. Army but never lost the heavy German accent of his birth. An exponent of far-right-wing political ideas and Communist conspiracy theories—he was, among other things, an ardent admirer of the Spanish fascist dictator Francisco Franco—Willoughby cut a strange figure in MacArthur's inner circle, where he held the position of G-2 throughout World War II and after. MacArthur facetiously called him "my pet fascist," but those under Willoughby loathed him for his haughty rages and imperious manner, referring to him behind his back as "our Junker general" or "Baron von Willoughby."[28]

MacArthur had made effective use of signals intelligence in the Pacific campaign during World War II but had also insisted on rigidly controlling the information himself. When Carter Clarke tried to set up the usual secure liaison channels so that Arlington Hall could directly send Ultra intelligence to commanders in MacArthur's theater, MacArthur issued orders forbidding the liaison units to send radio messages in any area under his command. In 1944, Clarke finally flew out to see the general personally, bearing a letter from General Marshall explaining the procedures that had to be followed. Clarke got as far as Hawaii before MacArthur's headquarters ordered him back to Washington. In the first months of the Korean War, Willoughby had similarly denied General Walker direct access to ASAPAC's intelligence reports and turned down his request to have an official AFSA liaison attached to his command to receive special intelligence directly from Washington. It was ostensibly out of concerns that the Eighth Army might be captured by the North Koreans, but it ensured that no one could upstage MacArthur's brilliant feats of seemingly omniscient generalship that signals intelligence in fact provided. It also meant that no one would be in a position to challenge his delusional disregard of intelligence that contradicted his decisions until it was too late.[29]

Truman had become so concerned by intelligence pointing to possible Chinese intervention that on October 15 he flew to Wake Island to personally meet with his military commander and discuss the situation. MacArthur, who frequently boasted of his understanding of "the mind of the Oriental," blithely assured the president that there was "very little" chance of the Chinese coming into the fight. "We are no longer fearful of their intervention."[30] Even after the first Chinese troops struck ten days later, Willoughby and MacArthur continued to insist that there was nothing to worry about. The general and his intelligence chief were personally briefed on the evidence coming from the Chinese signals by Lieutenant Colonel Morton Rubin, who had previously commanded ASAPAC and was now MacArthur's signals intelligence liaison, but Rubin made no apparent impression on either man, who habitually received the colonel's reports in stony silence. When the CIA station chief in Korea filed a cable to Washington reporting that he had personally interrogated Chinese prisoners captured in the fighting in the North, Willoughby promptly issued an order to the Eighth Army to "keep him clear of interrogation."[31]

In their initial attack on October 25, the Chinese sent four full armies, 120,000 men, on the offensive; Willoughby asserted that only "battalion-size elements" had been involved. That first assault may have been intended as a notice of China's determination to enter the war if the United States did not pull back to the 38th parallel. On November 6 the Chinese broke off their offensive, withdrew into the mountains facing UN positions, and waited. Throughout November signals intelligence indicators poured in pointing to Beijing's preparations for full-scale war: a state of emergency was in effect throughout China, air defenses were being ordered onto high alert, troops were being vaccinated for diseases prevalent in North Korea, thirty thousand maps of Korea had been shipped to Shenyang, near the border.[32]

On November 24, MacArthur resumed his offensive push toward the Yalu River, which separated North Korea from China. "If they go fast enough, maybe some of them can be home for Christmas," he grandly announced. Two days later thirty Chinese divisions counterattacked, crushing the Eighth Army's right flank and trapping the First Marine Division near the Chosin Reservoir. With the Eighth Army in a headlong retreat that would carry it 120 miles in two weeks through bitter cold, MacArthur was at last forced to acknowledge, at least privately, the truth he had consistently ignored. "We face an entirely new war," he informed the Joint Chiefs of Staff. To the press, however, MacArthur and

Willoughby serenely maintained that they had known all along about the Chinese, explaining that the bring-them-home-by-Christmas final offensive was not in fact one of the worst miscalculations in American military history but a shrewdly successful "reconnaissance in force" that had unmasked the Chinese positions. "We had to attack and find out the enemy's profile," Willoughby told reporters in Tokyo.[33] The Chinese advance would continue until January, when UN forces at last halted it on a line that once again ran south of Seoul.

In July 1951 the North Koreans instituted their own version of Black Friday, making sweeping changes in communications procedures that brought an end to AFSA's triumph in reading the enemy's signals. The cryptologists at Arlington Hall had been decrypting more than 90 percent of North Korean enciphered traffic by the end of 1950, but all of these readily broken codes were now replaced with unbreakable one-time-pad systems. Mimicking Soviet radio procedure, call signs were encrypted and frequency changes made more often. The North Koreans' careless transmission of high-level messages in the clear abruptly stopped. For the rest of the war, low-level intercept of voice and other tactical communications by U.S. Army and Air Force field units deployed close to the fighting would be the only significant source of useful signals intelligence other than traffic analysis.[34]

The effort even to find such a field unit that could deploy to Korea got off to a rocky, at times ludicrous start. During World War II the Army had created a large number of "signal service" or "Y" units that could move with ground and air forces to provide direct coverage of enemy signals sent in the clear or using low-grade code systems. This was traffic whose intelligence valuable was highly perishable, sent by voice or Morse code in the midst of battle, and had to be handled on the spot to be of any use at all. But at the outbreak of the fighting in Korea, ASA could find only one Y unit, the 60th Signal Service Company at Fort Lewis in Washington State, even close to being ready for deployment. It took them three months to arrive in Korea.[35]

The Air Force's 1st Radio Squadron Mobile near Tokyo should have been in a better position to move quickly. But the unit was as green as they come. The jittery commander of the 1st RSM reacted to the news of the North Korean attack by ordering his men to park the unit's vehicles in a circle, bumper to bumper, on the base football field and take cover behind them with whatever weapons they could find, in case the Communists attempted as their next move a surprise parachute landing in Japan. It was "like we were preparing for an attack by hostile Indians," recalled a bemused intelligence officer in the unit.[36] A USAFSS officer sent to take matters in hand, First Lieutenant Edward Murray, assembled a detachment of equipment, men, and several radio vans from the 1st RSM and flew to Korea on July 15, only to find that the Fifth Air Force had already set up its own renegade signals intelligence unit under a warrant officer named Donald Nichols, who had lived in Korea for a few years and was assigned to the Air Force Office of Special Investigations in Seoul when the fighting broke out.

Nichols had no known background in signals intelligence but cultivated a James Bond–like air, not entirely without foundation. During the chaotic first few days of the war, he led a small party on a daring foray behind enemy lines to destroy important documents left at Suwon Air Base, earning him the attention and commendation of the commander of Far East Air Forces, Lieutenant General George E. Stratemeyer. Nichols soon had carte blanche to run secret operations for the Air Force in Korea. Finding two Korean military officers to serve as translators, he quickly began supplying the Fifth Air Force with useful information from North Korean radio messages, even as his ad hoc signals intelligence system violated every established security procedure for handling special intelligence. Secure of his support in high places (Nichols would soon be promoted to captain, Stratemeyer commenting at the time, "This fellow is a one-man army in Korea"), Nichols airily rebuffed Lieutenant Murray's attempt to bring the operation under the control of the signals intelligence professionals, simply claiming for himself the equipment Murray had brought and sending the lieutenant back to Japan. Murray tried again the next month, this time bearing a letter of authority from the intelligence director of Far East Air Forces. The Fifth Air Force responded by handing him an order to leave the country on the next plane out. Only with the arrival of additional trained mobile intercept detachments at the end of the year did the situation begin to resemble the professional, smooth-running Y operations of World War II. By that time Nichols had faded from the scene.[37]

In late March 1951, the 1st RSM, still operating in Japan, picked up Russian ground controllers in voice communication with Soviet MiG fighter aircraft operating over North Korea. It was an astonishing development that the Soviets were actively engaged in the fighting, but it was also a potential intelligence windfall, not only because of the number of trained Russian linguists already available to cover these communications but because Soviet air doctrine called for tight control of fighters by stations on the ground tracking the location of friendly and enemy aircraft on radar throughout the battle. By April, eight Air Force Russian linguists were operating out of a mobile intercept hut in central Korea, passing on information of approaching MiGs to the Air Force tactical air command center, which in turn relayed the warnings to U.S. pilots, disguising the information to make it appear that it had come from U.S. radar stations tracking enemy air movements. In fact, the radio intercepts extended the warning distance well beyond what U.S. radar stations could see. Additional teams at a variety of locations were soon intercepting Korean and Chinese ground-control-to-fighter voice channels and Chinese Morse traffic that continually reported radar tracks of both friendly and hostile aircraft to Chinese air defense units. Security rules as well as technical considerations of the best location for intercepting these different signals kept these operations at separate sites until September 1951, when a decision was made to centralize the processing of all of the air tracking signals at a single USAFSS facility set up on the campus of Chosen Christian College in Seoul.[38]

For decades, standard histories of the air war in Korea attributed the sudden improvement in mid-1951 in the kill ratio achieved by American fighter pilots against Chinese MiG-15 jets to the arrival of the new and more capable American F-86. During the final year of the war U.S. fighters shot down 345 MiGs in air battles with a loss of only 18 F-86s, a kill ratio of 19 to 1. In fact, the real breakthrough had come from pulling together all of the signals intelligence sources in one center so that they could be rapidly correlated and passed on to fighters in the air. "The present top-heavy success of the F-86 against MiG-15s dates almost from the day of the inception of the new integrated [signals intelligence] service," reported an officer involved in the operation. On one day, a visiting ASA colonel observed the system in action as 15 MiGs were shot down without a single loss by U.S. F-86s. With more enthusiasm than originality, the colonel said it was "just like shooting ducks in a rain barrel," but it was an unmistakable demonstration of the incredible force multiplier that the signal interception and reporting system had provided: not a single one of the MiGs was tracked on U.S. radar during the course of the battle; all of the information passed to U.S. pilots had come from listening, in real time, to the communications of the enemy controllers and planes.[39]

An analysis of ground control traffic in June 1952 concluded that more than 90 percent of MiGs engaged in air operations over Korea were being flown by Russians. That the Soviets had engaged in a shooting war with Americans remained classified Top Secret for a quarter of a century. It, of course, could hardly have been a secret to the Russians: it was only the American people who could not be trusted with such information, in an era when nuclear weapons risked turning any spark into a conflagration. "The two superpowers had found it necessary but also dangerous to be in combat with one another," John Lewis Gaddis observed. "They tacitly agreed, therefore, to a cover-up."[40]

Armistice talks had begun in July 1951, but the war dragged on for two more years, settling into a grinding stalemate that resembled the trench warfare of World War I and ultimately taking two million Korean, six hundred thousand Chinese, and thirty-seven thousand American lives. Stalin urged dragging out the negotiations as a way to tie down the United States, keep the Truman administration off balance, and give American military prestige a black eye. Only with Stalin's death in March 1953 did the Soviet regime agree to a cease-fire. The agreement, which went into effect in July, left the two Koreas with almost exactly the same territory had they held when the war began, but with their lands devastated.

Those who had been in a position to see firsthand the miracles that Ultra intelligence regularly performed in World War II were dismayed at what happened to this vital source during the ensuing half decade of peace. "It has become apparent that during the between-wars interim we have lost, through neglect, disinterest and possibly jealousy much of the effectiveness in intelligence work that we acquired so painfully in World War II," observed General James Van Fleet, who had commanded a corps in Patton's Third Army and succeeded Ridgway as commander of the Eighth Army in Korea. "Today, our intelligence operations have not yet approached the standard that we reached in the final year of the last war."[41]

Despite some remarkable successes under difficult conditions, the experience in Korea laid bare the complete unworkability of AFSA and exposed its near-total failure to bring order and central control to the chaos of the sprawling American signals intelligence bureaucracy. By 1952, AFSA plus the three service cryptologic agencies had 32,500 military and civilian personnel and an annual budget of $400 million, but the rivalry and strife that the reorganization was supposed to put an end to had if anything grown worse.[42] At the behest of the new CIA director, General Walter Bedell Smith, Truman in late December 1951 ordered a complete review of the tangled communications intelligence structure. Smith had a well-deserved reputation as a take-charge executive, and the fix was in from the start; by the time the Joint Chiefs of Staff even knew about Truman's order the review committee had already been appointed and its staff, drawn entirely from CIA and the State Department, was in place.

Headed by a prominent New York attorney, George A. Brownell, the committee swiftly concluded that the creation of AFSA had been "a step backwards" in solving the chronic problems that beset the military-run signals intelligence establishment. Even in achieving the most basic goal of consolidating cryptanalytic processing into a single, centralized organization and eliminating the pointless duplication and interservice rivalry that stubbornly resisted earlier efforts at "coordination," AFSA had been less than a roaring success. AFSA's first director, Rear Admiral Earl E. Stone, had been able to bring about what at least had the outward appearance of a merger in the Washington-area Army and Navy headquarters, moving all communications intelligence activities to Arlington Hall and communications security to Nebraska Avenue. But AFSA's seventy-six-hundred-person staff and $35 million budget remained a small share of the total enterprise, and both the Army and Air Force cryptologic agencies continued to grab important projects for themselves. ASAPAC and USAFSS both duplicated AFSA's work on Soviet and Chinese codes throughout the Korean War, and simply ignored attempts by AFSA to take charge of field processing within the theater.

The Air Force had meanwhile established its headquarters of USAFSS at Brooks Air Force Base in Texas, a not too subtle attempt to escape from the Washington orbit altogether.[43]

Control of intercept was more tangled than ever. Joseph Wenger, now a real admiral and deputy director of AFSA, frankly told the Brownell Committee that AFSA made a mockery of the fundamental management principle that "authority must be commensurate with responsibility." Under the pre-AFSA arrangement, the coordinator of joint operations was at least allowed to tell the services' intercept stations what signals to monitor in areas of "joint" responsibility, which included diplomatic targets. The director of AFSA was reduced to begging. Admiral Stone conducted an arduous negotiation with the Air Force, ending with an agreement that the service would place under AFSA control all of its "fixed" intercept stations. The Air Force then declared all of its stations a "radio group mobile." (Most, an Air Force general later admitted, were "as mobile as the Eiffel Tower.")

When AFSA organized a small "Field Activity, Far East" to try to bring some order to the three services' intercept stations supporting AFSA's work on North Korean, Chinese, and Soviet targets during the Korean War, the Navy grudgingly offered some cramped space at its station in Yokosuka, but all three services resisted following its orders, taking the line (as the Brownell report summarized their attitude) that to do so was "inconsistent with normal command relationships and with the responsibility of each Service to provide combat intelligence for its own operations." AFSA was powerless to prevent even the most obvious duplication of effort: for over a year the Army and the Air Force both insisted on intercepting Russian and Chinese air communications, and it was not until March 1952, after months of negotiations, that ASA finally agreed to leave the job to the Air Force. The Navy meanwhile flatly refused to put its worldwide network of direction-finding stations—which provided the single most important source of information on the location and movement of Soviet surface ships and submarines—under central control.[44]

The worst problem was that although it had the outward appearance of following the model of "unification" that had been behind the establishment of the Department of Defense, the creation of AFSA was actually a bureaucratic Frankenstein that left the individual military services once again calling all the shots. Deliberately or not, the result was a near-perfect exemplar of that wonder of organizational theory, the circular chain of command. The service elements answered to the director of AFSA, who answered to a committee called the Armed Forces Security Agency Council, which answered to the services. Just as in USCIB, the decisions of AFSAC had to be unanimous; as the Brownell Committee noted, the council spent most of its time "safeguarding individual Service autonomies."

The internal structure of AFSA similarly ensured that the services had an effective veto over every decision; the director was required to have three vice directors, one from each service, and even at lower management levels every service was represented as a coequal. Hoping to improve relations with the "customers," Stone allowed analysts from the separate military services (along with CIA, State, and FBI) each to maintain "beachhead" offices at Arlington Hall where they could directly access the product. But the effect was only to further undermine AFSA's authority and embolden the services' claim to their "sovereign powers" over COMINT. "Since they felt we couldn't process the stuff fast enough," recalled Oliver Kirby, these "roving intelligence representatives" would each grab the same raw data and write separate, usually mutually contradictory reports. "We had to spend inordinate amounts of time trying to figure out how to get something done within the system," Kirby said.[45]

At the same time, the establishment of AFSA, and its new council made up solely of representatives from the three military services, created a complete parallel command structure alongside the existing USCIB arrangement that was meant to bring signals intelligence under the control of the National Security Council, with representation of all interested parties, civilian and military alike. Who AFSA's director actually reported to, and how it was to serve the needs of State and CIA, was a hopeless muddle.[46]

The Brownell Committee delivered its final report in six months, calling for a complete reorganization that would give the director of the agency real power and break the military services' hold, terminating what it called the failed "experiment" that had put AFSA under the Joint Chiefs' auspices. On October 24, 1952, Truman issued an order accepting nearly all of the committee's recommendations, declaring communications intelligence a "national" function, and establishing the

National Security Agency to replace AFSA. The Department of Defense was named the "executive agent" to carry out these duties on behalf of the government, but the new agency was no longer directly answerable to the military authorities. The director of central intelligence became the permanent chairman of USCIB, and from now on a majority vote would suffice to make decisions—and that majority was now firmly in civilian hands, with State, CIA, FBI, and the secretary of defense each having two votes to the military services' one vote apiece. The NSA director was given "operational and technical control" over all communications intelligence "collection and production resources" of the Department of Defense, and was also made a voting member of USCIB, further strengthening his clout.[47]

With Soviet high-level communications still unreadable, Meredith Gardner and the Russian group at Arlington Hall continued to work on the 1943 to 1945 one-time-pad messages that were vulnerable owing to the Soviets' duplicate key use. In October 1949, Kim Philby arrived in Washington to become the British Secret Intelligence Service's station chief, and within weeks was on the best of terms with Gardner and the FBI agent now working with Arlington Hall to try to identify the Soviet spies named in the messages, Robert Lamphere.[48]

Philby, like his fellow Cambridge recruits of the 1930s to the Soviet cause, was a scion of the British upper classes who managed effortlessly to combine a commitment to Marxism with a sense of privileged entitlement that the normal rules of society simply did not apply to people like him. His father had been a colonial administrator in India who exemplified the breed: Hillary St. John Bridger Philby was a world traveler, amateur ornithologist, and well-connected member of the English old boys' network who in his later years was a renowned Arabic scholar and convert to Islam. He took as his second wife a Baluchistani slave girl presented to him by the first monarch of Saudi Arabia, King Ibn Saud, in gratitude for his service to the kingdom.[49]

In obedience to Moscow's instructions, Kim Philby had feigned sympathy for the fascists in the Spanish Civil War as a newspaper correspondent in the 1930s and had moved from there to ever more responsible positions in the British secret service. Whatever Philby's original ideological convictions, he came to find the dangerous life of a double agent an intoxicating and addictive drug, matched by a relentless appetite for drunken binges, sexual liaisons, and exercising his duplicitous charm on all around him. "You didn't just like him, admire him, agree with him," said Sir Robert Mackenzie, the security officer at the British embassy in Washington, "you worshipped him." In London during the war he had charmed the hopelessly anglophilic James Jesus Angleton, in 1949 already a key figure in the newly established CIA, and on Philby's arrival in Washington he and Angleton became bosom friends, meeting for weekly, and then almost daily, very alcoholic lunches at Harvey's Restaurant in downtown Washington, beginning with bourbon on the rocks, then lobster and wine, then brandy and cigars. At his home at 4100 Nebraska Avenue—just next to the Navy's signals intelligence center, as it happened—Philby and his wife hosted famously drunken parties for his new CIA and FBI colleagues.[50]

One of the first things the FBI brought to the attention of the new British liaison upon his arrival in the fall of 1949 were decrypted NKGB messages from 1944 and 1945 referring to the activities of the Soviet agent HOMER at the British embassy in Washington. There were hundreds of possible candidates who fit the scant information available about HOMER's true identity, but Philby recognized his fellow spy at once. Donald Maclean had in fact been recruited by Philby at Cambridge, had shrewdly allayed doubts about his political past when interviewing for a job in the Foreign Office by acknowledging his involvement with left-wing organizations as a student and disarmingly admitting that he was trying to shake off his Communist views but had "not yet entirely succeeded," and had risen swiftly in the ranks to become first secretary in the Washington embassy, where he was posted from 1944 to 1948.[51]

That fall Lamphere also read a newly decrypted 1944 NKGB cable from New York to Moscow that contained a summary of a theoretical paper produced by the Manhattan Project on the gaseous diffusion process of enriching uranium. The author of the paper was Klaus Fuchs, a refugee from Nazi Germany who had worked on the British and then the American atomic bomb projects during the war. Fuchs was known to have joined the Communist Party in Germany as a young man, fleeing the Nazis' mass arrest of political opponents following the 1933 Reichstag fire; what the FBI did not yet know was that since 1941 Fuchs had been volunteering his services to the Soviets as a spy. A week after reading the message containing Fuchs's theoretical paper, Lamphere was sure he had identified Fuchs himself as the agent CHARLZ who had supplied the document: another cable mentioned that CHARLZ was a British subject whose sister was attending an American university, which fit Fuchs.[52]

Now back in Britain, Fuchs was interrogated by an MI5 officer who patiently waited out his halfhearted evasions and after a few weeks extracted a full confession from the scientist on January 24, 1950. Among other things, Fuchs admitted that in the summer of 1945 he had turned over to his contact Harry Gold a comprehensive technical report containing everything he had learned about the design and construction of the bomb. Like the other atomic spies, Fuchs insisted he was motivated by idealistic aims, to aid a heroic ally that was bearing the brunt of the fight against Hitler: he had been furiously insulted when Gold tried to hand him an envelope stuffed with $1,500 in cash supplied by Moscow in payment for his services.

Later a Los Alamos physicist colleague and close friend, Rudolf Peierls, with whom Fuchs had lodged while working in England, asked him how, as a scientist, he could have swallowed the doctrinaire orthodoxies of Marxism. Peierls was stunned by the "arrogance and naïveté" of Fuchs's answer. "You must remember what I went through under Nazis," Fuchs replied. "Besides, it was my intention, when I had helped the Russians to take over everything, to get up and tell them what is wrong with their system." Peierls's wife, Genia, who had been something of a mother figure to their young lodger, wrote him a more personal rebuke. Hadn't he at least thought about the betrayal of his friends he had committed, and the harm he had done them? she asked.

"I didn't, and that's the greatest horror I had to face when I looked at myself," Fuchs wrote back from prison. "I thought I knew what I was doing, and there was this simple thing, obvious to the simplest decent creature, and I didn't think of it." He told another friend, "Some people grow up at fifteen, some at thirty-eight. It is more painful at thirty-eight."[53]

Matching together Arlington Hall's decrypts and FBI files of suspected Communist agents was yielding a slew of more names. Harry Gold was arrested on May 22, 1950. Julius Rosenberg, identified as LIB-

ERAL and ANTENNA, was arrested on July 17, and his wife, Ethel, the following month. Julius Rosenberg and Harry Gold were unquestionably part of the Soviet spy ring that had made contact with Klaus Fuchs and other Los Alamos scientists and passed their reports on the atomic bomb to Moscow, but Ethel's involvement was limited at most to putting the group in touch with her brother, David Greenglass, an Army technician at Los Alamos.[54]

In the panicky atmosphere following the Soviets' first atomic test and the shock of the Korean War, both Rosenbergs were condemned to death by a federal judge for espionage, a grotesquely barbaric sentence that the Rosenbergs accepted in defiant martyrdom, spurning offers of leniency in return for their cooperation. They went to their deaths in the electric chair at Sing Sing refusing to confess or provide any information on their espionage activities.

By Kim Philby's second year in Washington he knew the net was closing around him. It was only a matter of time before Maclean was identified by a newly decoded message, and Philby tried to keep an eye on the progress of the work by paying a personal visit to Meredith Gardner at Arlington Hall; it may have been with the imagination of hindsight, but Gardner would later recall the silent, rapt intensity with which Philby stood observing the work of the Russian section. If Maclean were caught, the trail would almost certainly point to Philby, as their earlier associations would not be hard to turn up. In June 1950, Arlington Hall read a message referring to a valuable agent named STANLEY operating in Britain in 1945: that was Philby himself.[55]

Philby had promptly alerted the Soviets when Klaus Fuchs was identified in the cables, giving Moscow time to warn some of their agents in the atomic spy network to flee to Mexico. He was also passing on developments in Arlington Hall's reading of the HOMER cables. The immediate result of whatever warnings Moscow gave Maclean, however, was only to reveal how close Maclean was to cracking under the strain of imminent exposure.[56] The son of a former cabinet minister, Maclean was another of the Cambridge spies whose place in the British establishment seemed to have put them beyond suspicion, but his behavior was now becoming wildly erratic. He was drinking more than ever, and in May 1950, stationed at the British embassy in Cairo, he went on a completely unhinged spree, tearing up the apartment of two secretar-

~~TOP SECRET~~ VENONA

Reissue(T83)

From: NEW YORK

To: MOSCOW

No: 915

28 June 1944

To VIKTOR[i].

Your No. 2712[a].

SERGEJ's[ii] meeting with GOMMER[iii] took place on 25 June. GOMMER did not hand anything over. The next meeting will take place on 30 July in TYRE[TIR][iv]. It has been made possible for G. to summon SERGEJ in case of need. SIDON's[v] original instructions having been altered,

[34 groups unrecoverable]

travel to TYRE where his wife is living with her mother while awaiting confinement. [C% From there] [2 groups unrecovered] [C% with STEPAN[vi]]

[11 groups unrecovered]

The NKGB message read by Arlington Hall in April 1951 positively identifying the Soviet spy HOMER ("GOMMER") as British embassy official Donald Maclean. TYRE was New York; SIDON, London.

ies at the American embassy, shredding their underwear, and smashing a huge mirror into the bathtub. The Foreign Office, with remarkable understanding, merely sent him home for psychiatric treatment—then promoted him to run the American desk of the department in London.

In April 1951 the moment Philby had been fearing arrived: Arlington Hall read a message from June 1944 which mentioned that HOMER was traveling to New York to visit his wife, then pregnant, who was staying with her mother.[57]

That narrowed the field to one—positively identifying HOMER as Donald Maclean.

The information went to London and back to Philby in Washington through official U.S.-U.K. channels; Philby at once informed his Soviet handler in New York and said Maclean had to be hustled out of Britain before he could be interrogated and compromise the entire network.

Whether engineered by Philby or just by chance, the British embassy in Washington chose just that moment to send back to London another member of Philby's circle of Cambridge spies, one whose drunken escapades put all others of the group to shame. Guy Burgess, assigned as the public information officer in the Washington embassy, had been living in Philby's basement and misbehaving on a spectacular scale, showing up to work in disheveled clothes, drinking like a fish, insulting and picking fights with everyone he met, flamboyantly parading his promiscuous homosexuality, and at one completely out-of-control martini- and whiskey-sodden party at the Philby home, sketching a viciously obscene cartoon of one of the female guests, whose husband, the CIA's counterintelligence chief William Harvey, then threw a punch at him.

At the time the crucial evidence against Maclean fell into place, Burgess had just managed to provoke an even more serious diplomatic incident, tearing through the outskirts of Washington in a Lincoln convertible and insolently telling off the patrolmen who stopped him three separate times during his wild progress through the Virginia countryside that he had diplomatic immunity, which prompted an official protest from the State Department.[58]

Maclean now was under constant surveillance in London as MI5 hoped to catch him meeting his Soviet handlers. But as Philby correctly calculated, a visit to Maclean by his friend Burgess, who had not yet been implicated, would not arouse suspicion. Philby quickly hatched a plan for Burgess, as soon as he arrived back in London, to make arrangements with the Soviet embassy to spirit Maclean out of England.

Burgess secretly decided it would be a good idea for him to make an exit too. By the time the Foreign Office realized the men had slipped away and sent an urgent cable to its embassies throughout Europe to have Maclean stopped and brought back "at all costs and by all means," they were already in Moscow, having traveled with Soviet passports under false names. On June 7 the news of their defection to the Soviet Union became public. Philby—whose close association with Burgess was impossible to hide—was horrified at Burgess's precipitous flight, which threatened to give away the whole game.[59]

Philby was recalled to London shortly afterward and under pressure from Washington dismissed from the service. Bill Harvey wrote up a devastating indictment pulling together all of the evidence pointing to the conclusion that Philby was a Soviet spy, and CIA director Smith made clear that unless Philby was fired, the American intelligence relationship with London was over.

For a decade Philby avoided arrest by coolly keeping up the boldest of fronts, maintaining he was guilty of nothing worse than an "imprudent association" with Burgess, whom he said he had never suspected of being a Communist. He was more than a little aided in his deception by the utter inability of most of his former colleagues, his faithful CIA drinking companion Angleton included, to believe such a betrayal possible by one of their own. Menzies, the head of SIS, insisted that Philby could "not possibly be a traitor." Only in 1963, after he was confronted in Beirut with new evidence by an SIS official who told him "the game's up" did Philby escape to the Soviet Union.[60]

The spring of the year before Philby's departure from Washington, another message read by Gardner and Lamphere identified several Soviet agents who had been gathering information on aircraft and aircraft engines at defense plants on the West Coast in the 1930s and 1940s. James Orin York, an aeronautical engineer who worked at the Northrop and Douglas companies, was interviewed by the FBI and related how in mid-1941 he had been turned over to a new controller, a man named "Bill" who had given him $250 to buy a camera. York used it to copy specifications of the P-61 Black Widow night interceptor, among other classified projects. He was eventually paid $1,500 by Bill, whom he met regularly to hand over information and obtain lists of specific material the Russians were interested in. Bill once let slip that he knew Arabic, and eventually also mentioned his last name, which York recalled was something like "Villesbend."

In May 1950, AFSA suspended William Weisband, its linguist adviser on the Russian project who had worked so hard to be helpful and ingratiating to Meredith Gardner and others at Arlington Hall. A few months later, York positively identified Weisband as his handler Bill, pointing him out to two FBI agents on the street outside the federal courthouse in Los Angeles where York had just testified before a grand jury.[61]

Notes made by a Russian journalist and former KGB officer, Alexander Vassiliev, who from 1994 to 1996 was permitted to examine files in the KGB archives, later confirmed that Weisband—one of whose code names was ZHORA—was, as many in NSA had long suspected, the source of the leaks that triggered the Black Friday code changes. Following Gouzenko's defection, Weisband had been deactivated as a precaution along with most of the other Soviet agents in America, but in February 1948 the Soviets reestablished contact, and that was when Weisband passed word to Moscow of Arlington Hall's cryptanalytic successes against Soviet code systems.

In August 1948, Weisband asked to be given asylum in the USSR, but the MGB kept him in place, providing a steady flow of payments and vague promises of future assistance: he was too valuable an asset to give up quite yet. At Arlington Hall he continued to supply documents, which he smuggled out under his shirt and passed to Soviet officials at rendezvous points around Washington. In July 1949, Weisband told his handlers he was worried that suspicions might lead to him, and asked that the Russians not be "overly hasty" in introducing cipher security changes that might expose him. In fact, by then the deed had already been done, as a 1949 MGB report noted:

> On the basis of materials received from ZHORA, our state security agencies implemented a set of defensive measures, which resulted in a significant decrease in the effectiveness of the efforts of the Amer. decryption service. As a result, the pres. volume of the American decryption and analysis service's work has decreased significantly.[62]

When first questioned by the FBI in 1950, Weisband denied everything. In 1953 he finally admitted knowing York but refused to either admit or deny that he had engaged in espionage for the Soviets. Convicted of contempt of court for refusing to testify to a grand jury, he served a year in prison, but was never indicted for espionage: NSA officials adamantly opposed a prosecution that risked exposing any of the agency's cryptanalytic successes.[63] It would not be the first or the last time that NSA sought to keep secret from the American public what even the Soviets already knew, or to hide behind the shield of national security its own blunders, scandals, and bureaucratic miscalculations.

6

"An Old Mule Skinner"

The man chosen to put NSA on its feet following the shake-up of the signals intelligence organization in the midst of the Korean War neither asked for nor wanted the job. Lieutenant General Ralph J. Canine, who had succeeded Admiral Stone as director of AFSA in July 1951, had no prior experience in signals intelligence or cryptanalysis and by his outward air had no interest in such esoteric disciplines. Chief of staff of XII Corps under Patton's Third Army in World War II, Canine was as regular Army as they come. In his five years in charge of the agency the only times he ever missed his weekly Thursday morning nine o'clock haircut were the few occasions when he was out of the country or was summoned to the White House.[1] He was an avid golfer and a four-handicap polo player, which placed him just below the professional level. When faced with long technical explanations he would complain about "the long-hairs coming in and giving me a lot of baloney," or observe that what qualified him to be director of NSA was his long experience in the Army dealing with mules, having commanded pack artillery units in his early days during and after World War I. (He had originally intended to become a doctor, graduating as a premed student from Northwestern University in 1916, but, as he put it, "the First World War rescued me and my future patients.") He later claimed that William Friedman tried to teach him the elements of cryptanalysis after he became director. "I gave up after the first lesson."[2]

Canine—it was pronounced kuh-NINE, and woe to anyone who got it wrong—had, in his words, been "violently against" the assignment at first. At the time of his appointment he was serving as deputy G-2 at the Pentagon and was fully ready to retire after thirty-five years in the Army.

Canine went straight to the Army chief of staff to demand "why in the world he had sent me. . . . I knew a little about managing people after managing a good many soldiers," Canine said, but "I didn't have the least idea what this problem was."[3]

But for all of his wisecracks about just being "an old mule skinner," Canine was a skilled organizer, troubleshooter, and string-puller with few equals in the Army. His year and a half as director under the AFSA system had quickly convinced him of the impossibility of that arrangement and the need to shake up the existing organizational dysfunction that was crippling the agency's work. But he also saw at once the importance of recruiting and retaining top-level civilian cryptanalysts and having NSA take the lead in developing the newest and most powerful computers, and he made those goals his top priority and most enduring legacy. A CIA officer who had worked in cryptology during World War II and had no particular reason to pay the rival agency any compliments was one of many who later gave Canine full credit for the transformation he effected during his tenure: "He raised NSA from a second-rate to a first-rate organization."[4]

One of the new director's first acts was to tear down the old Army-Navy divisions within the organization and make it clear to everyone that for the first time they were going to have to do what the director, and not their military services, told them to do—and if they had a problem with that, they needed to take it up with the president of the United States, because that was where the order had come from. If they did not want to make the new system work, he said in his first address to the workforce of the new NSA, delivered from the stage of the post theater at Arlington Hall a few weeks after Truman's October 1952 directive, "Come in and let me know and I will get you a job with CIA."[5]

Canine had already issued what would become a legendary order requiring that all of the furniture in each office—there was a promiscuous assortment of green and brown filing cabinets that had been inherited from ASA and Op-20-G—match. This, of course, sounded exactly like the kind of Army chickenshit that the many ex-GIs among Arlington Hall's civilian workforce thought they had escaped with the end of the war, and was greeted with unconcealed derision—someone leaked the story to the *Washington Post,* which ran a mildly sarcastic item about the redecorating "color scheme decided upon by the top brass"—but it proved to be a shrewd move.

"I knew that the way you got people to do things was to know the fellow that was giving the order," Canine later explained. "And I knew if I made them move all their files, that they'd be mad at me and they'd know who issued the order. . . . I had done this before, at G-2."[6]

His other way of getting people to know who he was was by constantly dropping in to their offices and asking about their work. To prevent these informal visits from taking on the stilted air of an official inspection he played a regular cat-and-mouse game with the director of the production section, Brigadier General Woodbury M. Burgess, who would always insist on escorting Canine when he came to talk to the staff. Canine tried to slip out of his office at Nebraska Avenue, show up unannounced at Arlington Hall, and quickly enter A or B Building by a side door. When he discovered that Burgess had left orders with all of the guards to notify him the instant the director appeared, Canine threatened the PFC at the gate on his next visit. "Don't you dare call Burgess," he barked, "or you'll lose that stripe on your arm."[7]

Canine joked that he aimed to run the agency like "a dictator" ("not like DeGaulle, but a reasonable dictator") and told his staff that everyone had a vote—but he did not have to count the votes. At the party for his retirement in 1956 he was presented with an album of cartoons illustrating some of his favorite hard-ass maxims ("You guys give me a hard time and you'll wind up on an island so far out that it'll take you six months to get a message back"; "What did you do today to earn your pay?"; "What does he do for the man who shoots the cannonballs?"), and no one could forget the way he could turn an academic title like "Doctor" or "Professor" into something that sounded like a dirty word.

But the fact was he was no anti-intellectual, and while he professed to have been apprehensive about managing civilians, something he had no experience with, he went to bat for the civilian workforce repeatedly, having quickly grasped that finding people who were willing to make a normal career out of the peculiar talents required for NSA's mission was paramount to its effectiveness in the long term. He secured the first "supergrade" civil service positions for the agency, creating a cryptologic career path that could allow nonsupervisory professionals to rise to the

GS-16, -17, and -18 levels, earning $12,000 to $14,800 (compared with the average NSA salary at the time of about $4,000); the three top veteran civilian cryptanalysts in the agency, William Friedman, Solomon Kullback, and Abraham Sinkov, received the first three supergrade promotions. Canine similarly made a point of selecting civilians to attend the National War College after fighting to get NSA an annual place in the class. Louis Tordella was the first, Sinkov the second.[8]

The new director also did not hesitate to use his Army connections to jettison the useless "military bosses" that the civilian NSA employees chafed under. "I had some military guys around here who were just breathing. That was all—and taking up space," he recalled. But he had an old World War I buddy at the Pentagon who could make personnel assignments. "All I had to do was call him up and the guy didn't come to work in the morning. And I was not bashful about calling him up."[9]

His knowledge of how to do an end run around military regulations was equally adroit. When NSA was preparing its move to Fort Meade in Maryland, the base commander, Floyd Parks, called up Canine one day and told him that if he could get the approval of the secretary of defense, he would build a new golf course that NSA employees could use. The catch was that there was an Army rule that no post could have more than one golf course, and Meade already did have one. Canine made an appointment to see the defense secretary, Charles Wilson.

"Mr. Secretary, I've got something here I want you to sign," he said, and handed Wilson a letter of approval, already typed on Wilson's stationery, which Canine had managed to secure a small supply of just for occasions such as this.

"Why don't you send it in here the regular way?" Wilson asked suspiciously.

"I want to be here when you sign it," Canine said.

Wilson read the letter, asked a few questions, picked up his pen and said, "Now quit bothering me," signed it, and started to put it in his out tray. Canine smoothly intercepted the signed approval: "Don't do that. Let me take it. I'll deliver it to General Parks." By the time anyone else in the Army learned about it the golf course was built.[10]

His most effective tool was that he had nothing to lose. On his way out the door to a meeting at the Pentagon or Capitol Hill, Canine would regularly tell his secretary, "I may not be director when I come back."

But he always was, invariably returning with the budget increase, personnel policy exemption, or new project he had insisted he needed to do his job. Once at a congressional hearing Canine was grilled by a congressman who demanded to know why the agency was requesting 111 management engineers in its budget. General Motors, the congressman pointed out, had only 10 such positions in its entire huge organization. Canine launched into an impassioned defense of NSA's needs. The congressman quickly withdrew his objections. Only after the hearing did the agency's embarrassed comptroller inform his boss that the number was a typographical error: they had actually meant to ask for 11.[11]

Trying to make a quotidian job out of spying was rife with incongruities that no amount of support from the top could remove. To put it simply, NSA was a very strange place to work, and it became all the stranger as it grew into a veritable espionage city in the expansion boom that the Soviet atomic bomb and the Korean War set off. Throughout the 1950s the agency's workforce tripled, from a little over four thousand in 1950 to twelve thousand by the end of the decade. (The service cryptologic agencies, with their ongoing responsibility for operating intercept sites and providing tactical signals intelligence units and secure field communications, expanded even more rapidly, from six thousand to more than sixty thousand in the same period.)[12]

To be sure, NSAers always viewed themselves more as technocrats than spooks; Thomas R. Johnson, a longtime NSA employee who spent part of his time in the agency's history office before later doing a tour at CIA, recalled that there was "barely a recognition that what you are doing is part of a foreign intelligence operation" on the part of NSA staffers—who in fact, when Johnson directly put the question to them, flatly refused to accept the idea that what they were doing was even "spying" at all. Their day-to-day activities took place in an atmosphere that resembled "a laboratory or a high-tech clean room" and admitted none of the sordid moral ambiguities of the back-alley world of deception and betrayal into their "sterile" work.[13] Allen Dulles offered a wry introduction to the new moral landscape that CIA employees, by contrast, entered the moment they walked through the doors. One new CIA recruit in the 1950s, part of a group of recent college graduates starting together at the agency, arrived two days late for his initial training session owing to an unusual winter storm in the South that delayed his flight to Washington. "Today we are joined by Mr. Calhoun," the director of central intelligence told the assembled group the next morning, "who says he was unable to be here earlier because of . . . a snowstorm in Alabama." He paused significantly. "When we get through with him, he will be able to lie a lot better than that."[14] At NSA, Johnson noted, one always spoke of "techniques"; at CIA, it was "tradecraft." A CIA officer quoted to Johnson the tried-and-true adage for dealing with embarrassing questions: "Admit nothing, deny everything, and make countercharges."[15]

There was another cultural difference between the two spy agencies. CIA had an aura of Ivy League and East Coast sophistication, but NSA was unmistakably Middle America. NSAers were surely the squarest spies on earth, and the NSA personnel newsletter from the 1950s is almost unbelievably hokey, with its reports of employee clubs and events: square dances, talent contests, hobby shows (prizes awarded for best fishing lures, stamp collections, and model trains), the NSA Men's Glee Club, the NSA Duckpin Team (congratulated for setting a new season record), and an upcoming package trip to New York City (two nights at the Warwick Hotel, dinner and dancing and floor show at the Copacabana nightclub, and a tour of Radio City Music Hall and the United Nations building, all for $42.50; and best of all, "You don't have to use any annual leave," as it was taking place over the Veterans' Day weekend). One issue opened with a large page-one photograph of NSA employees attending the annual meeting of the NSA Federal Credit Union, accompanied by the exciting news that the membership had voted to approve a 4 percent dividend on share deposits.[16] A photograph in NSA's historical files from this period showed the finalists in the annual Miss NSA beauty pageant, the contestants in evening gowns and each wearing a sash bearing the number of the section they worked in.

If it were not for articles bearing titles like "Be On Your Guard Against Espionage" and "List of Subversive Publications" interspersed with the explanations of sick leave and vacation policies, pension options, and adjusted overtime pay rates, there would be nothing to suggest that NSA did anything more exciting for the government than log disability claims or distribute pamphlets containing poultry management tips. But, of course, that was the rub. However outwardly normal NSA's managers tried to make life for their thousands of employees, the all-pervading secrecy surrounding the work guaranteed that it could never truly be anything of the kind. Canine said he hoped that some of the improvements he made in career opportunities and employee morale "might make them forget for a while that they work like convicts," having to pass a double row of fences patrolled by armed guards every day, and one of the first things he did was to change the rule forbidding agency employees even to say where they worked.[17]

But security policies became ever more convoluted, intrusive, pettifogging, self-contradictory, and frequently self-defeating as bureaucratic ossification set it. Along with imposing ever-tighter physical security restrictions involving roving guards, searches of briefcases and bags, partitioned-off areas, and an elaborate system of collecting and disposing of confidential trash in centralized incinerators, NSA's security office produced multi-hundred-page security manuals detailing procedures in which common sense was routinely trumped by proliferating rules and regulations. Friedman was one who argued in vain for a bit of common sense, noting that it was ludicrous to try to pretend that the very existence of cryptanalysis could still be a secret after the congressional Pearl Harbor investigation in 1945–46 had disclosed American success in reading Japanese codes during the war. As he would later write:

> I think it is a bit late to assume that the degree of secrecy about cryptology of any World War II days can be maintained indefinitely. When we dropped the bomb on Hiroshima, the nature of warfare was changed forever, and when the Pearl Harbor Investigation bomb was dropped, the nature of crypto-warfare was changed forever. I think we should therefore face up to the facts. . . . Nobody would even dream of attempting to hide the fact that there are such things as nuclear bombs, guided missiles, etc. Why should anybody nowadays think it sensible to try to deny or hide the fact that there are such things as codes and ciphers and that there are ways of making and breaking them—without telling them exactly *how*?[18]

Aside from the Pearl Harbor revelations, the burgeoning scope of the postwar signals intelligence empire was simply impossible to keep completely hidden. NSA's budget was tucked away inside the Army's appropriations, but it was hardly possible to conceal the construction of a 1.4-million-square-foot building right next to the new Baltimore–Washington Parkway, where ground was broken in July 1954 for the new NSA headquarters at Fort Meade. The expanding network of intercept stations around the world, including the construction (beginning in the early 1960s) of massive direction-finding antenna fields at NSA sites in the Philippines, Turkey, Japan, Thailand, England, Italy, and Germany, were hardly inconspicuous either. These "elephant cage" arrays, designed to provide accurate directional fixes on high-frequency signals from thousands of miles away, consisted of ninety-six 120-foot-high towers arranged in a circle a third of a mile in diameter.

As NSA personnel assigned to these stations immediately learned, their facility was invariably referred to by the locals as "the American spy base." James V. Boone, who did a tour at an NSA site in Germany, recounted what happened when his wife went to register their children for the local German public school. The German secretary went through the form, asking the American woman the usual questions, including her husband's place of employment. She replied with the standard answer for NSA employees overseas: "Department of Defense." The secretary wrote, "NSA."

"Why did you do that?" the American asked, startled.

"Because," came the German's unfazed reply, "if it had been CIA, you would have said 'U.S. Government.' "[19]

It was hard not to see many of the rote but rigidly enforced security practices as thoroughly beside the point under such changing conditions, but NSA's security office was if anything becoming more rigid and paranoid in its zeal to plug real or imagined leaks. A 1950 law—whose constitutionality was never definitively tested even six decades after its enactment, although legal scholars have raised considerable doubts as to its validity—made anyone who "knowingly and willfully communicates, furnishes, transmits, or otherwise makes available to an unauthorized person, or publishes" any classified information concerning codes, code machines, or the communication intelligence activities of the United States or any foreign government liable to a fine or imprisonment of up

NSA's massive "elephant cage" antenna arrays, designed to intercept and precisely locate HF radio signals from Soviet military forces, were erected at sites ringing the Soviet Union in the 1960s.

to ten years.[20] The fact that it seemed to apply not just to government officials who had been granted security clearances but to anyone who "publishes" this information was a sweeping assertion of the secrecy powers of government, and NSA's security office began seeing dangerous and possibly Communist-inspired security breaches even in the most innocuous places. The office launched an absurd investigation in 1953 after a TV series, *Dangerous Assignment,* aired an episode titled "The Venetian Incident" in which a trenchcoat-clad U.S. secret agent flies to Italy to recover the missing parts of a code machine. The histrionically melodramatic plot offered nothing more genuinely secret than the fact that the vital parts were "rotors," which in the TV version appeared as clear plastic disks not remotely like the wired wheels of any actual cipher machine—not that those were a secret either. But NSA, warning that "the practice of showing films of this type is particularly dangerous to communication security," proposed that "a full investigation should be conducted into the background of the writer of the screen play and his associates to determine if, in the past, they have been associated in any way with cryptographic operations."[21]

In the mid-1950s Friedman discovered to his endless irritation that NSA's by-the-book security czars were classifying or reclassifying centuries-old material dealing with codes used during the Civil War, even the American Revolution. In 1958, after Friedman's retirement but while he was still working as a consultant to the agency, two officials from NSA arrived one day at his home on Capitol Hill with a truck and two helpers and proceeded to confiscate dozens of items from his library that they claimed were now classified Confidential—including Friedman's own lecture on the Zimmermann Telegram of World War I, which NSA had declassified five years earlier; his 1922 published article on the index of coincidence; and the correspondence courses on elementary cryptography he had prepared for the Army, which had *never* been previously classified.

"I am hampered by restrictions which are at times so intolerable and nonsensical that it is a wonder I am able to retain my sanity," Friedman wrote a few years later of the classification rules. NSA's insistence on keeping information about obsolete paper-and-pencil cipher systems classified seemed particularly absurd to him. "Automation in cryptography began more than a dozen years ago," he wrote one correspondent, and noted that even the "smallest nations" did not "care a fig" about hand ciphers anymore. Overclassification was just as much a threat as underclassification, Friedman observed, when it becomes "a handicap rather than a help in National Defense."[22]

. . .

At the same time that NSA's security office was offering this parody of flat-footed official obtuseness, the number of people throughout the government cleared to know about NSA's work was expanding on an unprecedented scale. By 1955 there were forty-two thousand people cleared to receive Category III COMINT, the highest security level.[23] The fiction that it was possible to keep up the World War II–era practice of cloaking the origins of signals intelligence with such euphemisms as "a usually reliable source" was wearing thin.

To keep up with the thousands of clearances that needed to be processed during the rapid expansion of NSA during the Korean War, Canine took a fateful misstep that even more deeply confused the illusion of security with genuine security. The polygraph, or so-called lie detector, was one of those quack effusions of American turn-of-the-twentieth-century inventors that might understandably have suckered a gullible public in an earlier era of electrical wonders, but that by 1952 was obviously pure bunkum to anyone with even a modicum of scientific knowledge. J. Edgar Hoover refused to allow the machine to be used in FBI investigations, noting its complete unreliability in detecting truth or falsehood. (Repeated studies since, including a review by the National Academy of Sciences, have affirmed the elementary fact that there is no physiologic response unique to lying and that for all of their pseudoscientific poring over squiggly traces recording pulse, respiration, blood pressure, and skin conductivity, polygraphers did little better than flipping a coin in concluding when a subject had been "deceptive.")[24]

But the CIA had already become enamored of the polygraph—the agency would also invest embarrassing sums in mediums and clairvoyants who claimed to be able to locate Soviet missile sites by telepathy—and since 1948 had been administering the tests to investigate "major loyalty or security risk matters" in COMINT-cleared personnel. In late 1950, CIA began asking all new job applicants to "volunteer" for a polygraph interview. It was as voluntary as anything else in such a coercive situation: by 1955 only six of the agency's twenty thousand applicants had declined to submit to the test. The CIA's arguments for the polygraph were based not on any scientific proof of its validity—there weren't any—but rather that it was an "extremely valuable aid to any investigation." That was a roundabout way of saying that people who were wired up to a machine and told by an examiner with a persuasive manner that it had shown they were lying sometimes could be pressured into revealing something they had been concealing.[25]

Canine became equally captivated by the notion that a gadget could solve NSA's security problems. The immediate problem was that conducting a thorough background investigation took time: it required sending agents out to interview neighbors and associates and investigate financial records and employment histories. By December 1950, 39 percent of AFSA's employees were waiting for their clearances, new hires left to cool their heels for months at the agency's training school, located in a former warehouse at 1436 U Street, NW, in Washington, or perform make-work tasks to fill their time. In January 1951, Canine decided that new employees who passed a polygraph interview would be given an "interim" security clearance that would allow them to start work immediately on classified projects, pending completion of their full background check.

Canine's faith in the magical device had all the blindness of a true believer. In December 1953 he ordered that all new civilian employees submit to polygraph testing as a mandatory condition for employment at NSA. "The Director has repeatedly emphasized his firm conviction that the polygraph is more reliable and more protective of security than the background investigation," his deputy for administration wrote in a 1956 memorandum that argued for periodically polygraphing existing civilian employees as well, to probe for "membership in subversive organizations," "association with known or suspected subversives," and unauthorized disclosure of classified information. (The military wisely used its authority at this time to bar the administration of the test to any service personnel.)[26]

The trouble, aside from the abuse of privacy and due process inherent in the whole business, was that conscientious but perfectly innocent people tended to show a "deceptive" response in the standard polygraph examination, while pathological liars sailed through. In their zeal to clear the initial backlog of pending clearances, NSA scoured police departments and private detective agencies around the country to hire supposed polygraph experts to administer the tests, which took place in hastily erected soundproof rooms at the U Street building. NSA examiners frequently asked intrusive or embarrassing personal and political questions—"Did you sleep with your husband before you married?" "Are you now or have you ever been in sympathy with leftist ideas?"—and while the process was certainly speedy for those who "passed," it became an Orwellian nightmare for the 25 percent whose clearances were held up because of their "unresolved" polygraph results.

Some of the more scientifically knowledgeable NSA officials tried in vain to halt the program. Howard Campaigne warned that it was sure to "get out of hand," provide a false assurance of security, and "disclose information the Agency does not want to have," tarnishing the records of capable employees with "minor derogatory data" that had nothing to do with their performance or loyalty. William Friedman's disdain for the polygraph was apparent in a critical article he clipped and placed in his private files: an investigative reporter who had interviewed several employees about their experiences with "the NSA Chamber of Horrors" quoted a victim of the process saying, "Halfway through, I felt like someone being tried in a Moscow purge." The article drily observed that the polygraphing unit was located in a "heavily guarded building between a gas station and an undertaker's parlor"; perhaps the more apropos geographical fact was that the U Street building had begun its life at the turn of the century as the factory for a famous quack patent medicine of the day.[27]

As subsequent events would make all too clear, the touching faith that a piece of Edwardian pseudoscientific electrical gadgetry could safeguard the nation's most important secrets would prove farcically mistaken, for almost every one of the real spies to betray NSA in the ensuing years passed a polygraph interview with flying colors, while obvious signs that in retrospect should have set off alarm bells about their behavior were blithely ignored, largely due to such misplaced confidence in hocus-pocus.

It was thus oddly fitting not only that the first major public disclosures of NSA's function and activities came from a spy scandal that all of the agency's obsessive security efforts failed to avert, but that the real damage done by the affair was a result of NSA's own handling of the matter:

the injury was almost entirely self-inflicted. On October 9, 1954, the FBI arrested Joseph S. Petersen Jr. at his modest apartment not far from Arlington Hall. Petersen, a thin, lanky, bespectacled forty-year-old former college physics teacher, had taken the Army cryptology correspondence course in 1941 and joined the Signal Intelligence Service during World War II. At Arlington Hall he gained a reputation as a cryptanalytic troubleshooter and often rotated through various sections that needed assistance.[28] In the spring of 1942, while working in the Japanese section, he met a Dutch army colonel and cryptologist, Jacobus A. Verkuijl, who had just arrived in Washington after fleeing from the Netherlands East Indies when Japanese troops marched into the Dutch colony.

Verkuijl had shared with William Friedman information he had gathered on Japanese diplomatic codes and had been welcomed at Arlington Hall, at first as an official liaison of the Dutch government-in-exile. But by the following year Verkuijl was already in bad odor with the American authorities; many of his coworkers at Arlington Hall found him arrogant, not particularly hardworking or competent, contemptuous of Americans, and much more interested in collecting information than sharing it. In August 1943, the War Department suggested to the Dutch government that Verkuijl's skills might be better employed in the Southwest Pacific theater; when that tactful attempt at diplomacy failed to produce results, the U.S. Army's G-2 bluntly informed the Dutch ambassador two months later that "Verkuijl had seen in Arlington Hall everything he could see" and it was time for the arrangement to end.[29]

But one of his American coworkers who had taken to the Dutchman was Petersen, who for a while was assigned as his assistant. Before leaving Arlington Hall, Verkuijl put his new American friend and colleague in touch with the head of the coding section at the Netherlands' Washington embassy, Giacomo Stuyt: explaining that the Netherlands was trying to set up its own signals intelligence organization and needed American help, Verkuijl suggested that Petersen could continue to act as a liaison between the two countries. Over the next decade, Petersen supplied his Dutch contacts with a steady stream of documents, meeting several times a month in a restaurant or car to hand over material he had removed from Arlington Hall.

Stuyt was particularly interested to learn of the Americans' considerable efforts directed at reading the diplomatic traffic of neutral nations, the Netherlands included. As early as 1945, Stuyt was warning his government that the Hagelin cipher machines that Dutch diplomats and the Foreign Ministry relied upon for high-level communications had been broken by Arlington Hall: one of the many documents Petersen turned over was Friedman's 1939 paper on the cryptanalytic solution of the Hagelin B-211. Petersen's roving commission gave him access to a considerable array of materials, but what the Dutch were most interested in by the late 1940s, and frequently pressed Petersen to supply, were documents relating to the cryptanalysis of Belgian, French, British, Italian, and Norwegian diplomatic code systems.[30]

It was not until 1953 that suspicion fell on Petersen, and then only because a naval officer dismissed for homosexuality named him to investigators as a fellow homosexual. When the FBI began questioning Petersen, he acknowledged his contacts with the Dutch and consented to a search of his apartment, which uncovered a large cache of classified documents and letters from Verkuijl and other Dutch officials requesting information and thanking him for his help. Against the objections of USCIB, Canine insisted that Petersen be prosecuted, and specifically wanted him charged under the recent communications intelligence amendments to the Espionage Act.[31]

Petersen's indictment on October 20, 1954, was front-page news in the *New York Times,* which identified the National Security Agency as "the closely guarded code-breaking organization of the Pentagon." An Associated Press report carried across the country further explained that NSA

> is essentially a radio monitoring service. It has a network of radio receiving stations and other equipment, some of which is based overseas. It listens to the world's radio traffic, both conventional messages and coded material. . . . Secrecy even tighter than that shrouding the Central Intelligence Agency surrounds the National Security Agency. It is not listed by name either in the Washington directory or in the Pentagon phone directory.[32]

Though a subsequent Top Secret damage assessment by USCIB concluded that Petersen had practically given away the store ("the catholic scope of the information supplied to the Dutch was enough to keep them, at least, well informed as to the level of competence of the U.S. Comint effort up through 1951"), both governments quickly conspired to keep the true extent of Petersen's spying a secret. The U.S. State Department said it had "no reason to question the good faith" of the Netherlands government—whose ambassador in Washington, Jan Herman van Roijen, had hastened to issue a statement that his government thought Petersen was acting all along "in accordance with an authorized arrangement between the two countries." Meeting privately with CIA director Dulles, Van Roijen made a show of fully cooperating with the U.S. investigation, providing a precise accounting of the 538.50 Dutch guilders the embassy had spent on dinners, lunches, and drinks entertaining Petersen. He omitted to mention the $5,000 a year they had directly paid him for his services beginning in 1947, later increasing that to $7,500 a year, which almost doubled his $7,700 NSA salary. A classified internal history of the Dutch signals intelligence service a decade later candidly acknowledged that were it not for the trove of information supplied by Petersen, their cryptanalytic bureau could never have attained its current status and capabilities, which far exceeded those that might be expected of such a small country. The author wryly concluded, "Petersen never even received a thank-you note."[33]

Dulles for his part strongly pressed the Justice Department to avoid introducing any evidence at Petersen's trial that would reveal just how much he had handed over. Prosecutors agreed to limit their case to a few of the more innocuous documents he took, including a copy of the Chinese telegraphic code—a standard system used to render Chinese characters as numbers in commercial cablegrams that NSA, ridiculously, classified as Secret. Expecting a lenient sentence after agreeing to plead guilty to one count of misusing classified communications intelligence information, Petersen was stunned when the judge threw the book at him, ordering him to prison for seven years on the grounds that the offense was "not *what* the defendant withdrew, but *that* he withdrew, records from the National Security Agency."

If Canine was hoping to make an example of Petersen as a deterrent to other NSA employees, that backfired as well; most of his colleagues, aware only of the charges that had been publicly revealed at his trial, thought he had been railroaded by overzealous prosecutors. Petersen was paroled after four years, and was subsequently hired by a former Arlington Hall colleague from the Japanese section, David H. Shepard, whose company, Intelligent Machines Research Corporation, was working to commercialize optical character recognition technology; at Petersen's trial, Shepard had testified on his behalf and vowed to give him a job as testimony of his belief in his innocence.[34]

Though U.S. officials briefly worried that some of the information Petersen passed on about American work on Soviet codes might have subsequently leaked from the Dutch to the Russians, hastening the Black Friday code changes, USCIB never could find any evidence to substantiate that. Far more damaging than any direct loss of information to the Dutch—who, however duplicitous they were in spying on their ally, were after all members of NATO—was NSA's own handling of the case, the board concluded:

> Last, but by no means least, the publicity attending the arrest and prosecution of Petersen has very materially increased the already damaging aggregate of information in the public file on the nature and extent of the U.S. COMINT effort in general. . . . It must be assumed further that all nations of the world have once more been alerted to the fact that if their communications can be read they very likely will be read. This can hardly fail to reduce the extent of cryptologic naiveté which has been so helpful to the U.S. COMINT effort in the past.[35]

USCIB's damage assessment also insisted that "the polygraph screening procedure now used by NSA (and CIA) would have detected one major weakness of Petersen's and caused his rejection at the time." That was a not very veiled allusion to his homosexuality, but it was also a logical absurdity. Although it was not far-fetched to worry in the social milieu of the 1950s that homosexuals might be subject to blackmail by foreign agents, Petersen's homosexuality had nothing whatever to do with his espionage. Petersen told his NSA and FBI interrogators that "he did not know why he did it." Those who knew him thought it had simply been the sense of importance Verkuijl's friendship gave him. At Arlington Hall, Frank Lewis and his wife had tried to befriend Petersen, too, recalling him as "such a strange guy," socially awkward and lonely.[36] But that was hardly a particular sign of anything out of the ordinary at a place like NSA or GCHQ. As a British security commission would later conclude, "because of the nature of GCHQ's work and their need for staff with esoteric specialisms, they attracted many odd and eccentric characters," and trying to spot a security risk by keeping an eye out for unconventional behavior at a signals intelligence agency was like trying to find a KGB agent by looking for someone with a Russian accent.[37] The very nature of the work, the burgeoning growth in the size of the staff, and NSA's unquestioning embrace of the mission to "get everything" had created a situation ripe for exploitation by even the friendliest foreign power.

Simply trying not to drown under the torrential flow of data from its field collection sites was becoming a major, and self-justifying, driving force behind NSA's relentless expansion since the start of the Korean War.

By 1955, two thousand intercept positions were sending in thirty-seven tons of printouts and paper tape a month, most of it shipped by air; more urgent traffic was forwarded by radio teleprinter at the rate of thirty million words a month. One single Soviet system was producing two million messages a month, mostly plaintext. It was not unusual for the machine section at Arlington Hall to punch a million IBM cards a month on just one problem. In June 1955, Canine noted that even with the additional 1,065 civilian positions he had recently been authorized to fill, "we . . . are still unable to process fully all the material we receive," and current expansion plans for field intercept were going to increase the amount of incoming traffic by 60 percent when completed in 1958.[38]

Canine was appalled upon his arrival as director to discover what an internal study had just described as the "deplorable and deteriorating state" of the agency's communications systems and its inability to handle the flow of incoming traffic. It could take five to six days to deliver a routine message. Even the most urgent messages took at least five to six hours to get from the field to the cryptologists at Arlington Hall: basic security demanded that intercepted traffic had to be reenciphered using NSA's own codes before being sent over the airwaves to avoid letting a target know which of his messages were being collected and studied, and if it was coming from a distant field site it might have to be sent through as many as six relay centers on the way to Washington. At each of those centers a teleprinter would punch a tape from the incoming signal; it would then have to be decoded by a cleared operator, reencoded, and a new tape punched and placed in a box stacked with other messages awaiting transmission on an outgoing circuit to the next radio station down the chain. By the mid-1950s, NSA was sending or receiving 70 percent of all encrypted traffic flowing in and out of Washington. Canine pushed for a system of dedicated cryptologic radio channels manned only by cleared personnel and equipped with automated switches at relay stations that would seamlessly handle the forwarding of the traffic, and work had begun on the system, known as the COMINT Comnet. But funding was slow to come and technical disputes with the services kept delaying basic decisions.[39]

Innovations in radio technology were meanwhile filling the airwaves with new types of signals to be intercepted. During the Korean War, in early 1952, the voice channels connecting Russian and Chinese ground controllers with MiG pilots in the air suddenly went silent; after several months of searching, U.S. intercept stations discovered that the communications had all shifted from HF to VHF channels. Very high frequency radios, operating at frequencies of 30 to 300 MHz, had been experimented with during World War II, but this was their first major operational appearance. VHF could carry more information in a smaller bandwidth, but it also was limited to line of sight, which meant that it was no longer possible to place an intercept station hundreds or even thousands of miles away and pick up a signal that had "skipped" off the earth's atmosphere, as was the case with HF. Radars, and radio signals in airborne IFF systems—used to identify friendly aircraft to one another and to antiaircraft units to prevent accidental shootdowns of one's own planes—also used these much higher-frequency, line-of-sight signals in the VHF or even higher UHF bands.

So, too, did the new microwave links that communications companies were beginning to use to handle large volumes of short-range, point-to-point telephone and data traffic as an alternative to laying expensive landline cables. In the mid-1950s, Western listening stations in Berlin discovered that the Soviet military had installed a spiderweb of crisscrossing microwave links in East Berlin, carrying multichannel signals. Covering all of these new kinds of transmissions required not only entirely new types of receiving equipment but a proliferation of intercept stations located considerably closer to their targets than in the past.[40]

This was one area where the British retained an important advantage over their American cousins: that was a chief reason NSA remained willing to continue the BRUSA agreement into the postwar period. West Berlin offered an incomparable outpost right in the heart of the enemy camp, but elsewhere in the world, thanks to its foothold in the former empire, Britannia still ruled the airwaves. Even in many former colonies that had since achieved independence, Britain had negotiated an ongoing military presence. British intercept posts located at military bases in Ceylon, Singapore, Malta, and Cyprus as well as naval bases at Kiel in Germany and in Turkey and RAF bases in Europe and the Middle East (from which the RAF also operated its own ferret flights) were well positioned to collect Soviet signals. Britain also turned over to the Americans the World War II–era radio intercept station at Chicksands, an RAF base in southeast England, which gave NSA a major outpost of its own and would later house one of the elephant-cage antenna arrays.[41]

Although the British dominion countries, which included Canada and Australia, were not parties to the 1946 BRUSA intelligence-sharing agreement, their signals intelligence bureaus were heavily controlled by London, and GCHQ's governing body, the London SIGINT Board, had the authority under BRUSA to disseminate any shared communications intelligence to the dominions without specific U.S. approval. Gouzenko's 1946 revelations of Soviet spies in Canada and the discovery the following year of the NKGB Moscow–Canberra messages pointing to Communist infiltration of the Australian government prompted the United States to cut off intelligence cooperation with those countries. But after the Conservatives won Australia's elections in 1949 and replaced the previous Labour government, concerns diminished and Australia, along with New Zealand and Canada, were ultimately accepted as full members of BRUSA, whose official name was changed in 1954 to UKUSA to reflect the inclusion of the United Kingdom's dominions. Australia's signals intelligence service subsequently helped staff a listening post in Hong Kong that added to the coverage of the periphery of the Communist countries.[42]

That still did not necessarily solve the problem of intercepting short-range, line-of-sight signals, but here the British with their tradition of naval daring and reputation for eccentric pursuits were also able to make a significant contribution. Apparently no one thought much of it when yet another team of British archaeologists showed up and set up camp in some remote and inhospitable location, and more than once a British undercover team using that eminently believable story carried out missions to intercept Soviet radar and missile-test signals from spots such as northern Iran along the borders of the Soviet Caucasus. The Royal Navy also began conducting signals-gathering missions with submarines that could slip in undetected near major Soviet naval bases. Although the Soviets had already demonstrated their willingness to confront "unfriendly air intrusion," noted the director of British naval intelligence, Rear Admiral John Inglis, "no difficulties were placed in the way of submarine visitors." Submarines operating off Murmansk had collected "considerable VHF voice, IFF, and radar" traffic from Soviet air and coastal defense units in the area. The Royal Navy was preparing to expand these operations when a bungled spy mission executed with almost unbelievable ineptness by the British SIS brought it all to a screeching halt.[43]

The facts still almost seem to defy belief, the storied British spies acting more like Inspector Clouseau than James Bond. On April 17, 1956, the Soviet leaders Nikita Khrushchev and Nikolai Bulganin arrived in Britain for a state visit. It was the first-ever visit to the West by any Soviet leader, and Prime Minister Anthony Eden viewed it as affirmation that Britain still counted as one of the Big Three world powers. The Russians traveled on the latest cruiser of the Soviet navy, the *Ordzhonikidze,* accompanied by two destroyers. The day before the warships were to arrive at Portsmouth harbor, two men checked into the city's Sally Port Hotel. The first gave his name as "Smith" in the hotel register, an alias that might have been criticized for a certain lack of imagination had the point not been rendered moot from a tradecraft standpoint when he added in the address column, "Attached to the Foreign Office."

The second man registered under his own name. Lionel "Buster" Crabb was a minor celebrity in Britain. A World War II "frogman" who had carried out a number of courageous diving exploits, Crabb had earned the George Medal for heroism. He was now forty-seven, retired from the navy, and not the man he had been after years of heavy smoking and drinking. The following night he rang up an old navy diving buddy in the area, and when they met at a pub, Crabb asked his friend if he could help him with his gear in the morning, as he was going to carry out a "secret" dive in the harbor to "take a dekko at the Russian bottoms." Crabb then proceeded to down five double whiskeys, each followed by a beer chaser. At seven the next morning he met his friend, donned his gear, slipped into the harbor, and was never seen alive again.[44]

With the press hot on the trail of the mysterious disappearance of a national hero, and a major diplomatic disaster in the offing, SIS mounted a frantic cover-up, sending a police officer racing to the hotel to rip the page out of the registration book and warn the hotel owner to keep quiet, which had the predictable opposite effect. Khrushchev carried on with his itinerary, touring the Houses of Parliament, the Royal Naval College, and the British Industries Fair, and attending a special performance at Covent Garden. A poorly educated peasant's son who had been a coal miner and factory worker before rising in the ranks of the Communist Party, Khrushchev alternated throughout his visit between anxious efforts to avoid making a fool of himself and touchy and insecure braggadocio: at one dinner party he boasted that Soviet missiles "could easily reach your island," and he erupted in fury at a private meeting with British Labour Party officials when several pressed him on the arrests and liquidation of social democrats and trade unionists in Eastern Europe. "If you want to help the enemies of the working class, you must find another agent to do it," Khrushchev shouted, leading one of the Labour leaders later to describe their guest as "a simple-minded man" who might possibly be qualified to hold "a secondary position in a British trade union." But the eight-day "B & K Tour," as the British press dubbed it, concluded without any major embarrassments on either side. Khrushchev's jocular inquiry of Eden at one formal dinner whether he had any "missing or lost property" was shrugged off as just some obscure Russian joke.[45]

It took the head of British intelligence four days to finally inform the British prime minister the facts of the matter: that Crabb had been sent on a bungled attempt to examine the *Ordzhonikidze*'s propeller and rudder. Two weeks later the entire story spilled out, with the Soviets lodging a formal protest of "this operation aimed against those who had arrived in Britain on a friendly visit" and a furious Eden, facing questions in the

House of Commons, vowing that "disciplinary steps" would be taken and stating that his explicit instructions forbidding any espionage activities during the Russians' visit had been disregarded.

Exactly what happened to Crabb remains unknown to this day, but a decomposed body in a diving suit was pulled from the harbor a year later, and in 2007 a retired Soviet sailor who claimed to have been a combat diver assigned to the *Ordzhonikidze* said he had spotted the British frogman while carrying out an underwater patrol to protect the ship, and had slit his throat as the intruder attempted to attach a device to the ship's hull.[46]

One of the steps Eden took was to order a complete review of intrusive intelligence operations, and in particular to reassess "the balance between military intelligence on the one hand, and civil intelligence and political risks on the other." An early casualty of the review was the Royal Navy's signals-collecting missions near Soviet naval bases. British naval officials in Washington complained to London of their "embarrassment" at not being able to "make good our part of the bargain" to supply the promised intercepts from these missions to their American colleagues and the loss of British prestige that would result "unless we resume these activities," but the decision had been made. When it became clear that the U.S. Navy was intending to fill the void by launching its own submarine operations off Murmansk, the British Admiralty quietly dispatched Commander John Coote, who had made the Murmansk run several times, to brief the first American crew, but only on the strict understanding that he was not to tell any other British naval officers in Washington what he was up to.[47]

In spite of these unaccustomed political restrictions on its activities, GCHQ was able to remain a considerable force in the U.S.-U.K. intelligence alliance. Although it would always operate under the shadow of its far larger and lavishly funded American counterpart, GCHQ underwent a similarly explosive expansion throughout the Cold War, and for much the same reasons. The organization had been down to a virtual skeleton crew in the first year after World War II, but was soon on its way to becoming the largest and most important part of the British intelligence establishment. In 1946, GCHQ freed itself from the control of SIS and its chief, Stewart Menzies, and became a quasi-independent part of the Foreign Office. The agency would grow to more than 11,500 employees by the mid-1960s, which was not only more than SIS and MI5 put together but more than the entire British diplomatic service, including the rest of the Foreign Office staff in London and every British embassy and mission around the world. In 1952, GCHQ moved its headquarters to Cheltenham, a bucolic spa town in Gloucestershire, a hundred miles northwest of London.[48]

Just as at NSA, British cryptology in the Cold War was perhaps inevitably becoming part of a muscle-bound and hidebound bureaucracy, hemmed in by rules and social and political strictures that left little room for the eccentric geniuses who had achieved their idiosyncratic triumphs against the German Enigma and other enemy code systems in World War II. Hugh Alexander had tried to lure the mathematical genius Alan Turing back to GCHQ with an offer of £5,000 to consult for a year. But in January 1952, Turing naïvely reported to the local police a burglary in his home, and under questioning about inconsistencies in his account even more naïvely revealed that the principal suspect was basically a young male prostitute he had paid to have sex with. Consensual homosexuality—"gross indecency"—was still a crime in Britain at that time, and Turing was arrested and pleaded guilty. In October of that year he confided to a friend that he knew he would never be permitted to work at GCHQ again. In June 1954, the man who had conceived the most brilliant machine-based attack on an impenetrable cipher in the history of cryptanalysis, the inventor of the logical basis of the modern digital computer, died from cyanide poisoning, either by accident or suicidal intent, at age forty-one.[49]

Among the administrative nightmares of the explosively growing, disjointed, and highly technical top-secret organization that Canine inherited was a notable lack of skilled managers. That was a failing common to creative and technical enterprises, which always tended to attract people more at home dealing with abstract ideas than with their fellow human beings, but it was especially acute in the very abstract world of cryptanalysis. "I had a terrible time finding people that could manage," Canine related. "We were long on technical brains at NSA and we were very short on management brains."[50] The splintering of the work into hundreds of separate problems, each isolated technically and for security reasons from one another, exacerbated the difficulties of trying to assert managerial control on an organization made up of thousands of individualistic thinkers who marched to no identifiable drum known to management science.

With some of the same energetic naïveté he showed in instituting the polygraph as a cure-all for NSA's security problems, Canine enthusiastically brought in high-profile business consultants including Arthur Andersen and McKinsey and Company, commissioned organizational studies, and instituted a "management development" program that produced flow-chart-crammed publications describing in organization-science jargon the wonders of applying such concepts as "unavailability-forced decision making," "inventories of personnel potential," and "planned programs of experiences" to develop "the large reservoir of executive material" within the agency.[51]

It was all well-intentioned but in the long run had dismal consequences, creating within NSA a large class of professional managers whose only job was to *be* managers, and who often had no real knowledge of the technical problems the men and women they were supposedly in charge of were working on. Jack Gurin would years later write a rueful article for the agency's in-house publication *Cryptolog* entitled "Let's Not Forget Our Cryptologic Mission," lamenting the accumulation of a large number of positions within NSA that were "only distantly related" to its core work. A pseudonymously bylined article by another NSAer ("Anne Exinterne") more pointedly noted the considerable overabundance of managers at the GS-13 level and above who were continually shunted from one position to another: that had been another of Canine's hobbyhorses, as he believed that a regular "rotation" of managers was key to developing leadership skills. Those "left standing when the music stops" in this game of management musical chairs were invariably "assigned to a staff position with vaguely defined duties, or given responsibility for a low-priority problem, or made deputy to someone who operates very effectively without a deputy."[52]

The worst consequence of this, in the view of old cryptanalytic hands like Frank Lewis, was not just that a supervisor was all too often put "in the awkward position of making decisions affecting problems completely beyond his understanding," but that those in such a position tended to cover up their ignorance by the familiar bureaucratic expedient of sweeping problems under the rug or putting on a Potemkin village show of efficiency. "This has reached the nadir," Lewis complained, "when section analysts are called on the carpet for seeking the advice of staff specialists, thereby making the section 'look bad.'"[53]

The other nightmare Canine inherited was Arlington Hall. The hastily constructed wartime A and B Buildings had always been uncomfortable, but now they were becoming literally uninhabitable, structurally unsound firetraps. By 1954, 30 percent of the workforce had been assigned to an evening shift to relieve the impossibly overcrowded conditions. Bringing the buildings up to reasonable standard, including air-conditioning, was going to cost $2.5 million, and replacing them altogether with a new building $9.5 million, but there was a mountain of red tape involving construction of military buildings in the Washington area, as well as a new policy requiring critical facilities to be located away from possible targets of Soviet atomic bombs. In the meanwhile, the best that NSA's employees could hope for in the sultry Washington summers was the appearance of the staff member whose job it was to whirl a wet- and dry-bulb thermometer to measure the dew point; if the combination of temperature and humidity exceeded certain maximum values (95 degrees Fahrenheit at 55 percent humidity, 98 degrees at 45 percent, 100 degrees at 38 percent), everyone would be sent home early.[54]

After months of desultory discussions by the military authorities of alternative locations—Kansas City; St. Louis; Tulsa, Oklahoma; Brooks Air Force Base, Texas; Cheyenne, Wyoming; and Birmingham, Alabama were among the dozens of sites proposed—the Joint Chiefs of Staff finally issued a decision in 1951 that the agency would relocate to Fort Knox, Kentucky. Six hundred thirty miles by road from Washington, the site would presumably offer safety from an atomic attack on the nation's capital if not much else by way of attractions. Already hinting at the discontent brewing from the agency's civilian workforce about the proposed move, a site survey team sent to Fort Knox in August 1951 more than a little defensively insisted that "the region is neither a wilderness, nor undesirable," and that "any normal Washingtonian can be as comfortable and happy in this area as anywhere." The survey party likewise brushed aside the little matter of racial segregation in Kentucky, noting that "this is accomplished without noticeable friction as an accepted principle of long-established social order" and "appears to be no problem for either the whites or the negroes native to the area"—though acknowledging that it might "necessitate adjustments" for Arlington Hall's current "colored employees," who were not used to this quaint practice on quite the scale that it existed in Kentucky. Their report also warned against releasing "partial or incomplete" information to the staff about the move, as "a single erroneous fact or misleading statement may destroy faith and confidence in the reliability of the whole effort."[55]

That was a little late, as the faith and confidence of Arlington Hall's employees in the whole effort was already hovering near zero. Determined to undo the decision, and with a survey of his civilian staff in hand showing that most intended to quit rather than move to Fort Knox, Canine was able to get the Army chief of staff General Omar Bradley on his side. A new site-selection board was named; as usual when Canine went to work on a problem the outcome was foreordained, and the site he had wanted all along, Fort Meade, was chosen in February 1952.[56]

It was not until 1957 that NSA's new building was fully ready for occupancy. The final cost of construction was $35 million, a 75 percent cost overrun. It was everything Arlington Hall was not: fully air-conditioned, modern, spacious, with a fourteen-hundred-seat cafeteria, five thousand parking spaces, three floors of offices along a double corridor and two wings, a basement space covering 4.9 acres to house the agency's vast array of computers and related machinery, all on NSA's own 950-acre site with room for expansion—which was needed almost immediately, as NSA's growth had already exceeded the capacity of the new building by the time it was finished, requiring the communications security group to stay behind at Nebraska Avenue (where it remained until 1968).

It was also in the middle of nowhere. Located in the still-rural Maryland countryside halfway between Washington and Baltimore, the main thing Fort Meade had to offer was space; the post, which had served as a major training base in both world wars, had hundreds of wooden barracks but not much else in the way of permanent facilities. There was only limited housing available on post, requiring most of NSA's staff to commute. The new Baltimore–Washington Parkway eased the twenty-mile drive from downtown Washington, but the other roads leading to Fort Meade were tiny, winding two-lane rural routes, and there was only

A cartoon from the NSA newsletter; the "Meademobile," parked at Arlington Hall, provided employees information about the Fort Meade area in advance of the agency's move there in 1957.

limited public transit available, the only real option being a one-hour-and-twenty-three-minute slog by train from Washington's Union Station and connecting bus from Laurel, Maryland.

To win over still-reluctant NSAers, Canine was able to have the relocation declared a "permanent change of station," permitting the government to cover the full cost for civilian employees to move their household effects. A trailer dubbed the "Meademobile" was parked in the picnic area between A and B Buildings at Arlington Hall and staffed by cheery uniformed young women attendants who offered information about houses and apartments, schools, churches, and shopping in the Fort Meade area. In the end only 2 percent of NSA's staff quit because of the move. Canine retired just before NSA took up quarters in its new home. The myriad details of the relocation had almost fully occupied his final two years as director.

Canine's successor was Air Force lieutenant general John A. Samford. He was virtually Canine's opposite, a cerebral intelligence professional who fully understood the world of signals intelligence, "more of a pedant than a pilot, more of a philosopher than a fighter," in the words of an internal CIA history. Samford probably wished he had something as trivial to worry about as the logistics of relocating ten thousand people and hundreds of tons of sensitive equipment and classified files, compared to the challenges that greeted him as he walked into the new director's office on the second floor of the NSA Operations Building for the first time: above all, the codebreakers' continuing inability to do what they were supposed to do to justify their half billion dollars a year, namely to actually break the codes of America's chief adversary, the Soviet Union.[57]

7

Brains Versus Bugs

NSA's failure to read much if any Soviet encrypted traffic since 1947–48 was obviously becoming more than just a temporary setback: something fundamental had changed in the nature of the Russian cryptographic systems, and in the eyes of some scientific experts called in to assess the situation, NSA had failed to keep up with the times. The Brownell Committee, whose report back in 1952 had led to the creation of NSA, circumspectly observed that "efforts in certain important parts of the cryptanalytic field have not been crowned with success, to say the least," and suggested that a completely new approach was called for. "The attack is timid, parsimonious, and too bound by the remembrance of past accomplishment to make the fresh and untrammeled start that is demanded," the committee reported, citing cryptanalytic experts outside the agency they had interviewed. The Robertson panel reiterated the point the following year: "Top priority should be accorded to the solution of high-level Soviet cryptographic systems." And in 1955 yet another outside panel, this one headed by retired Army general Mark Clark, part of a comprehensive review of government that Eisenhower asked former president Herbert Hoover to carry out, urged that given the "vital" importance of communications intelligence as insurance against a surprise atomic attack, an immediate effort "at least equal to the Manhattan Project" was justified to regain the ability to read high-level Soviet systems.[1]

Despite the millions of dollars invested in computers and special-purpose analyzers, no breakthroughs had been achieved. The very first computer program written for Atlas was designed to search the Russian one-time-pad traffic for repeated key use, and a GCHQ special-purpose digital comparator, Oedipus, joined that effort in 1954. Oedipus was basically an electronic improvement on the U.S. Navy's war-era relay and punch card Mercury machine. It had a stored read-only memory that used thousands of semiconductor diodes to assign a weight to each of four thousand known code groups, while a magnetic drum memory held ten thousand strings of hypothetical key derived from the first three words of accumulated message traffic. A paper tape containing the first three words of a series of new messages to be tested was read in, each was stripped in turn of each of the ten thousand possible sequences of key, and the resulting code group values were looked up in the dictionary and the weight score recorded to see if any represented possible key-page reuses. This machine-aided analysis added to the discoveries of additional depths in the mid-1940s messages, but despite extensive searches no later instances of exploitable one-time-pad traffic were uncovered at all.

"The present outlook . . . is very bleak," an NSA review of the effort concluded. "There can be no doubt that [the Soviets] are aware of the basic principles of correct pad usage and that they are now capable of producing key properly. . . . There is every reason to believe that they are now operating their pad systems correctly."[2]

The Soviet military's Albatross machine was proving equally unyielding. CIA analysts who were consulted by NSA's scientific advisory group stressed that solving Albatross was so important from an intelligence standpoint that it would be worth doubling NSA's budget if that was what it took to achieve it. Fifteen Robin comparators, built by ERA on a rush order in 1950, had been at work for five straight years searching through one million Albatross messages trying to find a handle on the problem. Long streams of messages were punched on paper tape and each tape was then formed into a loop, one of them ten characters shorter than the other; they were then fed at a rate of five thousand characters per second past photoelectric readers that compared the two data streams, ten characters at a time, counting coincidences between the two until every message on one tape had been run against every possible position on the other. To complete all possible comparisons within a set of 480 messages required almost two weeks of continual runs on a Robin machine.

These massive "round robin" searches were hoping to turn up depths, or more particularly what were known as "busts," messages that had been enciphered with the machine set improperly or the mechanism malfunctioning. One kind of bust occurred when the rotors of the machine failed to turn as the keys were pressed, which resulted in the machine generating an easily broken monoalphabetic substitution cipher that would not only render the immediate message readable but might give away some general details of the machine's internal wiring. Another type of bust that could lead to a message being read was when the operator retransmitted a message that the recipient had failed to receive properly and reused the same rotor settings but slightly changed the wording or spacing the second time, producing a long and exploitable depth.

But those five years and one million messages located only 138 busts, and Albatross continued to resist routine solution. Arlington Hall was not even sure how many rotors the machine had, and the agency's scientific advisory group, which included Engstrom, von Neumann, Robertson, and other mathematical and computing experts, glumly concluded that by far the best hope for a quick solution of Albatross "lies in the direct approach"—that is, stealing one.[3]

Weisband's leaks hastened the Soviets' adoption of much more secure and resilient systems, but perhaps not by much: the weaknesses of the Enigma and other first-generation rotor machines were readily apparent to any reasonably sophisticated, mathematically minded cryptologist. The larger technical reality was that in the seesawing competition between codemakers and codebreakers, the former were surging ahead in the postwar era. Even before the war, the United States had recognized the vulnerabilities of the Enigma's architecture, and William Friedman's SIGABA machine—which remained unbroken by the Germans throughout the war and, as far as is known, was never broken by the Soviets after the war either—incorporated a number of design features intended specifically to thwart the kinds of cryptanalytic attacks that ultimately defeated the Enigma's apparent security.

A basic measure of the statistical resilience of a cipher is the keyspace, which is basically the number of different possible permutations that have to be tried to find the one key sequence that will successfully decipher a given message. The keyspace of the four-wheel naval Enigma was in theory about 10^{25}, or 10 million billion billion, which understandably inspired confidence on the part of the Germans. But much of that apparent security was illusory, owing to design choices that had no doubt seemed inconsequential but had subtly undermining effects. The art of cryptanalysis of a cipher is fundamentally an attempt to discover shortcuts that can reduce the keyspace to a small enough size that it becomes feasible just to try all of the remaining possibilities. Alan Turing's brilliant solution of the Enigma effectively reduced the keyspace that had to be tested to half a million or less—well within the capabilities of even a primitive electromechanical calculator like the bombe. A major flaw of the Enigma that made Turing's method possible was its reciprocal substitution, which also meant that no letter could ever be enciphered by itself: Q could never stand for Q. That design permitted the same setting of the machine to be used for both enciphering and deciphering. But among other things it also greatly simplified the cryptanalysts' challenge of placing a plaintext crib in the proper position alongside its matching cipher text, the first step in Turing's process: the crib could be slid along the cipher text until it was in a position where no "crashes"—the same letter occurring in both texts in the same spot—occurred.

The predictable, odometer-like sequence in which the Enigma rotors stepped, with the moving middle and left rotors advancing only when the rotor to the right of them had made a full revolution, also proved a fatal weakness. This meant that for long stretches of text the middle and left rotors did not move at all; only the right rotor, where the electrical pulse entered and exited the entire scrambler unit, moved with every letter. Had the Germans designed the Enigma so that its "fast" wheel was instead in the left or middle position, a cryptanalytic attack would have been vastly more complicated.

The SIGABA had a keyspace considerably smaller than the Enigma's (effectively a little less than 10^{15} as it was actually used during the war), but it resisted cryptanalysis by avoiding this fundamental flaw of regular, cyclic rotor movement. Its most notable feature was a bank of control wheels that themselves changed position with each enciphered letter; the electrical impulses that emerged from this second scrambler bank controlled whether each of the five cipher rotors advanced, or not, at each successive step, thereby generating a seemingly random and highly unpredictable pattern to the cipher wheel movements.[4]

The follow-on to the Albatross machine adopted by the Soviet army

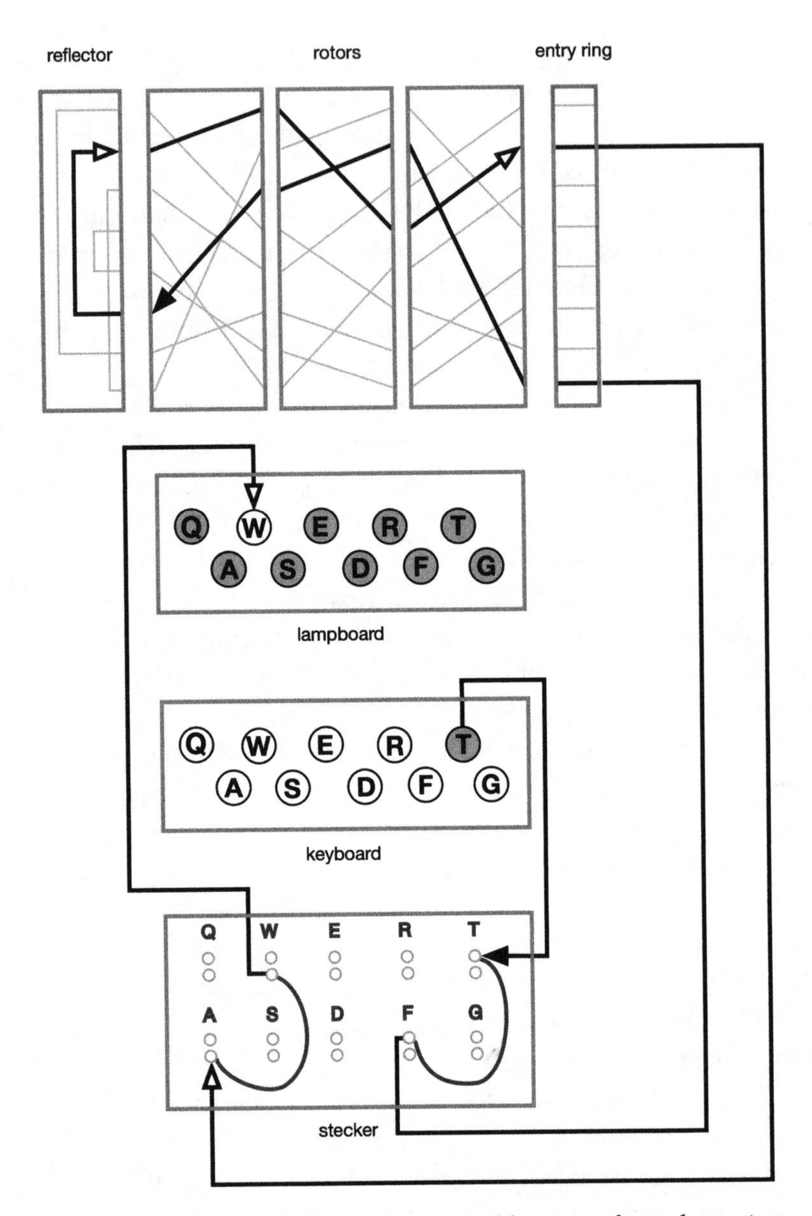

A simplified schematic of the German Enigma (showing only ten letters instead of the complete twenty-six), tracing the path from keyboard to lampboard for one letter. Because of the reflector, a letter could never be enciphered by itself. Alan Turing's method for recovering the daily setting of the Enigma shrewdly exploited this and other design flaws to eliminate the effect of the plugboard, or stecker, altogether, cutting the number of possibilities to be tested by a factor of about 10^{14}.

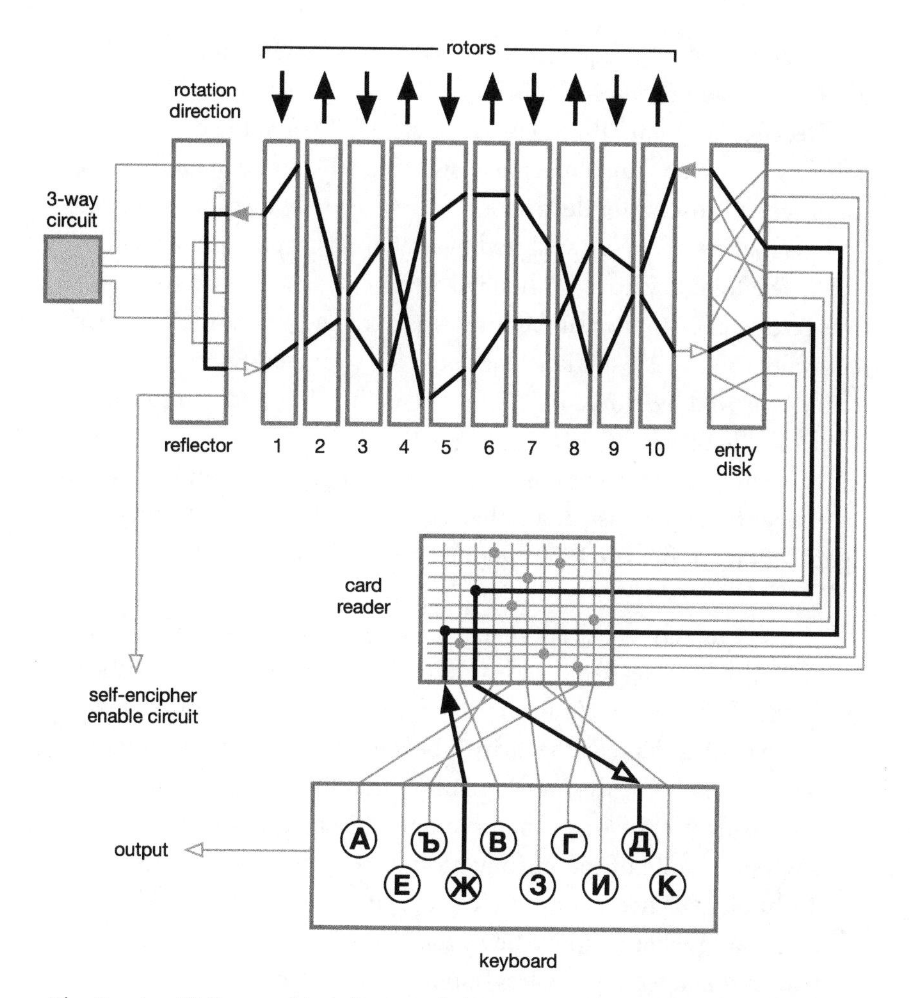

The Russian Fialka machine eliminated the major insecurities of the Enigma by allowing a letter to be enciphered by itself and by employing ten rotors that turned in a complex fashion, as opposed to the Enigma's far more predictable stepping pattern. A punch card took the place of the Enigma's plugboard.

and Warsaw Pact closed even more of the Enigma's cryptographic loopholes. Named the Fialka—Russian for "violet"—the machine had ten cipher rotors, the odd-numbered wheels turning in one direction, the even-numbered wheels in the opposite direction. Multiple turnover pins allowed each wheel to cause the next odd or even rotor down the line to step, or not. Most shrewdly of all, a semiconductor circuit caused one letter at each position to be enciphered by itself, while three other letters were scrambled in a sequential, nonreciprocal pattern (for example, И becomes Ы, Ы becomes Д, Д becomes И).

Decades later, after the fall of the Soviet Union and the demise of the Warsaw Pact, a few Russian, Czech, and Polish Fialkas that had not been returned to Moscow for destruction as ordered in 1989 came on the collector's market and were analyzed by academic cryptologists outside of NSA. The total keyspace of the Fialka was calculated to be 10^{75}, which was getting close to the number of atoms in the universe, and was nearly equivalent to the 256-bit key length of the Advanced Encryption Standard, approved by NSA in 2002 for protecting Top Secret information in electronic data.[5]

Faith in American know-how was one thing, but scientific objectivity was another, and as yet another external review of NSA's cryptanalytical efforts in the late 1950s pointed out:

> Our cryptanalysts believe that some of our own cipher machines are entirely unreadable with foreseeable technology, even if the enemy has a complete machine, and we have no reason to feel that a similar degree of security is beyond the capabilities of other countries. . . . It is the Panel's opinion that the advantages will be increasingly in favor of the cryptographer as against the cryptanalyst, in spite of the introduction of computer techniques. . . . Technology is irresistibly making the situation worse rather than better, and what is now true of the [Soviet Union] may become true of other, technically less-sophisticated countries.

This panel was headed by an even more distinguished outside scientist, William O. Baker, vice president for research at Bell Labs, and their final conclusion was delivered with blunt scientific objectivity indeed: "No national strategy," the Baker Panel warned, "should be based on the hope or expectation that we will be able to read" high-level Soviet encrypted traffic in the near future.[6]

If the problem would not yield to ingenuity, it was always possible to steal the answer, as von Neumann and company had proposed.

The Russians were past masters at this game. As far back as the 1930s, foreign diplomats in Moscow assumed as a matter of routine that their rooms were bugged and their telephones tapped. In May 1937, an assistant at the American embassy discovered in the attic of the ambassador's residence, Spaso House, heaps of fresh cigarette butts, "several piles of human excrement," and a fishing pole strung with thin wires that had apparently been used to lower a microphone into the wall behind the ambassador's study, where he was accustomed to dictate much of his correspondence. George Kennan spent days trying to catch the eavesdropper, including one sleepless night lying in wait armed with a nonfunctioning flashlight and an unloaded revolver, but the intruder never made a reappearance.[7] In 1944 a Navy electrician brought in to make the first thorough electronic sweep of the embassy found 120 hidden microphones. After that, recalled a member of the embassy staff, "They kept turning up, in the legs of any new tables or chairs that were delivered, in the plaster of the walls, any and everywhere."[8]

By the 1950s the efforts of the MGB's Second (Counterintelligence) Chief Directorate against embassies in Moscow were often directed specifically at acquiring cryptographic information that might help solve foreign diplomatic code systems, either by scooping up the texts of messages before they were enciphered (as with the American ambassador's dictated correspondence) or by directly tapping into code rooms. The electronic surveillance was supplemented by break-ins, spies planted among the Russian housekeeping staff, and a veritable assault division of prostitutes, comely ballerinas from the Moscow Ballet, and other seductresses who worked tirelessly to compromise the marine guards, code clerks, even CIA officers stationed at the U.S. embassy.

During Charles Bohlen's four years as ambassador from 1953 to 1957, a dozen U.S. officials, including the CIA's first Moscow station chief, Edward Ellis Smith, confessed to having been confronted by the KGB with graphic photographs of their sexual escapades with Russian women. "All of these people were out of the country in twenty-four hours," Bohlen said. But those were only the ones he knew about. The maids, cooks, housekeepers, and other local employees who worked for foreign missions were all supplied by a Soviet agency, Burobin, which was little more than a wholly owned subsidiary of the Second Directorate. As George Kennan recalled from his brief tour as ambassador in 1952, "I and all the rest of us were substantially helpless" to control their activities on embassy premises. The longtime caretaker at Spaso House, Sergei, occupied his own apartment, which he always kept locked; when Kennan finally demanded he turn over the key to U.S. officials, Sergei stalled for a few weeks, then shortly afterward vanished into retirement.[9]

The most famous penetration of the U.S. embassy was the Great Seal bug, also discovered during Kennan's ambassadorship. Having requested a thorough sweep of his residence and the embassy, Kennan was sent a security team from Washington. To check for any voice-activated bugs, one of the technicians asked the ambassador to sit at his desk at Spaso House after hours and go through the motions of dictating a letter to his secretary. Kennan, with a certain touch of humor, chose to read from his 1936 cable in which he did nothing but recycle his predecessor's dispatches from czarist Russia to show that nothing had changed under the Communist regime. Suddenly detecting a UHF signal coming from behind Kennan's desk, the technician began hacking at the wall behind a wooden replica of the Great Seal of the United States that hung there. He then turned his hammer to the seal itself and pulled from behind the carved eagle's beak a three-quarter-inch-diameter diaphragm-covered cylinder, attached to a short rod antenna.[10]

The seal had been presented as a gift from Russian schoolchildren to Ambassador Averell Harriman in 1945 and had hung there ever since. The American engineers who discovered it dubbed it "the Thing." Its principle of operation was ingenious. The Thing was entirely passive, requiring no power supply and giving off no signal itself until it was illuminated by a microwave radio beam aimed from an adjoining building. As the diaphragm vibrated in and out in response to sound waves coming from the room, it minutely changed the shape, and thus the resonant frequency, of the cavity formed by the small cylinder. That slight distuning of a resonant frequency around 1800 MHz caused the strength of one of the harmonics of the incoming illuminating signal to fluctuate, producing a modulated radio signal of the same kind generated by an AM radio transmitter. The resulting signal could be picked up from a nearby location outside the building.[11]

Even more remarkable was the story behind the device. In 1987 the English-language *Moscow News* ran a series of articles about the musical inventor Léon Theremin, revealing him to have been the secret genius behind the Great Seal bug. Theremin, inventor of the eponymous electronic instrument, had come to America in 1927, hailed as "the Soviet Edison." In New York he performed concerts, including two at Carnegie Hall, on his futuristic musical instrument, cutting a dashing appearance as a slim figure in white tie and tails standing intently before the strange device, hands hovering near two antennas and circling in small, dramatically precise motions to vary the pitch and volume of the otherworldly sounds emerging from the theremin. He also invented during the prewar years a remote-control device for aircraft, a wireless intruder alarm (the "radio watchman"), and a prototype television system.

He was also a Soviet spy, having been recruited by the GRU before leaving Russia, and throughout his time in America he supplied reports on aircraft and avionics technology gleaned from his consulting work with Bendix and other U.S. defense contractors. In 1938, Theremin abruptly vanished from New York. Friends were convinced he had been kidnapped by the Soviet authorities. In fact, his business going bad, deeply in debt, and with a messy trail of marriage, divorce, and girlfriends behind him, he had decided to return on his own to Russia. He chose the worst possible time. It was the height of Stalin's purges, and Theremin was immediately arrested and forced to confess to being a "fascist spy." He was sentenced to eight years in the Gulag. But after a few months at a labor camp under brutal conditions, he was transferred to a *sharashka,* a special facility where prisoners with scientific training were put to work on research projects for the state. (Other famous *sharashka* inmates during this time were Andrei Tupolev, the aircraft designer; Sergei Korolev, a major figure in the future Soviet space program; and the writer Aleksandr Solzhenitsyn.) It was there that Theremin designed the Great Seal bug. He later also developed the Buran, an eavesdropping device that reflected a beam of infrared light off the glass of a window to detect vibrations generated by sounds inside a room. Freed in 1947, Theremin was awarded the Stalin Prize for his achievements while a prisoner of the state.[12]*

---

*Theremin continued to suffer from the paranoia of the Soviet regime over the following decades. In 1967 the *New York Times* music critic Harold Schonberg was astonished to encounter Theremin at the Moscow Conservatory; his American acquaintances, who had never heard from him again after his disappearance, were convinced he had died during the war. But after Schonberg's article appeared in the *Times* revealing that the

Despite all of the warnings that U.S. embassy officials had received by the early 1950s, they continued to fall prey to Russian surveillance and bugging that made codes and codebreaking irrelevant. In 1953, when construction of a new American embassy on Moscow's Tchaikovsky Street began, Ambassador Bohlen had guards keep a close eye on the Russian workmen during the day, but blithely assumed there was no need to extend the vigilance to after hours. In April 1956, State Department security technicians examining the U.S. embassy in Prague uncovered a network of microphones with wires snaking into the attic; similar discoveries were made in Budapest and Belgrade. Only in 1964, after a tip from KGB defector Yuri Nosenko, however, were the experts called in to conduct a thorough search of the Tchaikovsky Street building. Two security men spent ten days pulling the top floor apart, tearing out electrical wires and phone jacks, jackhammering the plaster walls (burning out one jackhammer in the process), ripping up the parquet floors, prying loose doorjambs, finding nothing.

Only when they cut the heavy iron radiator from the floor did they discover a tiny hole, three-sixteenths of an inch in diameter, drilled into the wall behind where the radiator stood. Hacking into the plaster, they soon found a microphone. The building's top two floors, which housed the most sensitive parts of the embassy, including the ambassador's office, the CIA station, and the code rooms, were in fact honeycombed with hidden listening devices: eventually fifty-two were found, feeding into a coaxial cable that ran to an antenna in the attic. A search of the new embassy in Warsaw was then ordered; it turned up fifty-four bugs hidden behind radiators.[13]

Interviewing Nikolai Nikolayevich Andreev, the retired head of the KGB's Eighth (Communications) Chief Directorate, in 1996, the historian of cryptology David Kahn concluded that during the Cold War "the Soviet Union seems to have gained most of its communications intelligence not from cryptanalysis, but from bugs and traitors." Although the KGB was able to cryptanalytically solve the Hagelin machines used by the Swiss and Italians plus a few other less secure systems in the postwar period, its success against other countries' communications mostly

---

famous inventor was not only alive and well but continuing to create new electronic musical instruments, he was abruptly fired by nervous conservatory officials.

relied on direct measures, including break-ins of the Japanese embassy and those of a number of Middle Eastern countries to steal code materials, and the planting of ubiquitous bugs. The choice of Andreev to lead the Soviets' communications intelligence effort in the 1960s and 1970s underscored the point: he was not a mathematician, Kahn observed, but a bugging expert.[14]

Frank Rowlett was never a member of the Ralph Canine fan club. Always contemptuous of those less knowledgeable about cryptology who tried to tell him how to do his work, Rowlett was irked from the start by the new director's "hip-shooting" management style, in the words of a declassified NSA history. When Canine in 1952 abruptly directed his top civilian cryptologists to swap positions as part of his pet rotation scheme, ordering Rowlett to become head of communications security and giving up to Abe Sinkov his plum assignment as head of production, Rowlett left in a huff and took a job at CIA. Many of his old colleagues viewed him as a "traitor" for going over to the rival agency, which aside from all of the other sources of tension was even setting up its own small signals intelligence unit to try to chip away at NSA's monopoly in the field. Getting an old hand like Rowlett was a coup for CIA. He was soon head of its clandestine electronic surveillance section, Foreign Intelligence/Staff D.[15]

Rowlett did not think much, either, of all of the outside panels and committees brought in to review the stalled progress on breaking high-level Soviet code systems. When CIA was sent the findings of the Clark committee for review and concurrence, he filed a dismissive rejoinder that began, "This report is dangerous." Rowlett witheringly observed that it had been prepared by "individuals who . . . have no experience or background in the field" of signals intelligence. "Truisms and cliches are profuse throughout the discussions and from them specious conclusions are drawn. Nowhere does the report get down to brass tacks and examine step by step either professionally or expertly the essential elements of COMINT production." There may have been little love lost between Rowlett and his old agency, but he sarcastically pointed out that NSA's director surely did not need a committee to remind him what "everybody in Washington knows": that "as far as cryptanalysis is concerned there is a desperate need for the production" of decrypted high-level Soviet traffic. He suggested in effect that all of the outside "experts" shut up and let NSA get on with the job, which only it knew best how to do. "Most of the people on these panels would not have known a Russian cipher if it hit him on the head," he later said. "Rule by committee is a terrible way to run a spy agency."[16]

The truth of the matter, however, was that it was Rowlett's own growing doubts over NSA's failure to make cryptanalytic progress against the Russian codes that in part led him to make the move to CIA. The shift of Soviet communications from HF airwaves to landlines and line-of-sight UHF and the increasing sophistication of Soviet cryptography helped persuade him that the future of communications intelligence lay more in cloak and dagger than pencil and paper. In 1951, Rowlett confided to Bill Harvey, who had founded CIA's Staff D, his frustrations at the loss of intelligence as a result of the unreadability of Soviet signals. Shortly thereafter, a young civil engineer from the agency's Office of Communications was summoned to a meeting where Rowlett and some officials from the Office of Plans were present. "The only question they asked was whether a tunnel could be dug in secret," the engineer recalled. "My reply was that one could dig a tunnel anywhere, but to build one in secret would depend on its size, take more time, and cost more money."[17]

In December 1952, Harvey arrived to take charge of CIA's Berlin Operations Base with his and Rowlett's plan to make a bold end run around the Soviets' cryptographic security already well under way. Berlin was the hub for telephone circuits running throughout Eastern Europe. With assistance from the British SIS, which had run a similar tunneling operation in Vienna, Operation Regal aimed to beat the Russians at their own game, tapping directly into cables that carried long-distance telephone calls and teleprinter traffic of Soviet military and civilian officials between Moscow and East Germany and other Soviet satellites.[18]

Harvey was far from the obvious man to run the operation. A former FBI agent who joined CIA at its start in 1947, he bristled at the Ivy League types who filled the agency, a distinct contrast from the blue-collar atmosphere of the FBI. His physical appearance announced that he was definitely not a member of the polished East Coast elite: he was short, fat, with a bullet head and a smudgy mustache. His agents called him "the Pear." He kept a large collection of pistols he was always toying with, and took special satisfaction in belligerently parading his midwestern uncouthness before the "nice Yale boys," as he sneeringly called them, salting his conversation with crude observations deliberately intended to discomfit his more refined colleagues.

In Berlin—and now separated from his first wife, whose honor he had defended with a drunken punch at Guy Burgess at Kim Philby's home the year before—Harvey took up residence in a huge, stuccoed mansion with a garden and swimming pool in the American Sector where he hosted alcoholic lunches for the other CIA officers; guests arrived at noon and were served pitcher upon pitcher of martinis until four o'clock, when some food finally appeared. "Trial by firewater," his staff called it. Harvey was more than just a heavy drinker; he would in his declining years suffer the ravages of advanced alcoholism. But there was a fierce intellect under that crude exterior. Harvey had graduated from high school at age fifteen and had earned a law degree from Indiana University; his FBI background check reported that he had been considered "brilliant" by his teachers. He was also a thoroughgoing, street-smart case officer who left nothing to chance.[19]

In January 1953, having recruited several agents inside East Berlin's telephone exchange, the Berlin Operations Base arranged a clandestine sampling of the long-distance cables that its sources had identified as assigned for official use. Late one night, a telephone operator in the main exchange of the East Berlin Post Office surreptitiously patched a connection from a prime target circuit to a line connected to the West Berlin exchange for fifteen minutes; a CIA technician posing as a West German employee of the office was there to record the take.

Over the next six months the sampling continued, usually just a few minutes at a time whenever the operator felt she could safely elude detection. CIA officers never knew ahead of time when that would be and had to maintain a twenty-four-hour watch at the West Berlin Post Office site. By August they had two hours' worth of material, enough to make a persuasive case that a considerable amount of valuable intelligence was to be gleaned from the unguarded telephone conversations passing over official channels.[20] Rowlett made a final visit to Berlin to prepare a formal proposal to the CIA director for their audacious plan to bore a tunnel under the border to tap directly into the East Berlin cables;

Allen Dulles's approval came in January 1954. Construction began the following month.[21]

From a tradecraft standpoint the operation was a technical tour de force. Specially designed tunneling equipment was tested on a 150-foot-long mockup in New Mexico; 125 tons of steel liner were transported on freight trains crossing the East Zone to West Berlin, packed in double-crated boxes as a safeguard against accidental exposure; an entire two-story warehouse was built at the site next to the border chosen for the tunnel's terminus to house receiving equipment and, in its basement, conceal the thirty-one hundred tons of earth removed in excavating the 1,476-foot-long bore over the course of a year; air-conditioning ductwork, later hastily supplemented by a separate line of cold-water chiller tubes when tests proved the first system inadequate, kept the tap chamber cooled so that heat from its vacuum-tube-filled preamps would not create a telltale ring of melted ice on the East Berlin roadway above in winter. From May to August 1955 technicians completed the delicate task of making the actual connections to the three targeted cables. On average 121 voice circuits and 28 telegraphic circuits were in use at any given time. All were collected on what would become a staggering pile of fifty thousand reels of magnetic tape during the time the tap remained in operation. The voice recordings, containing sixty-seven thousand hours of Russian and German conversations, were sent to London for transcribing by a special section staffed by 317 Russian émigrés and German linguists; the teleprinter signals, many of them multiplexed, were also collected on magnetic tape and forwarded to Rowlett's Staff D for processing.[22]

The operation was a brilliant technical success; it also threatened to bring the bureaucratic infighting of the U.S. intelligence services to the breaking point. CIA had not bothered even to inform NSA officials about the tunnel's existence until a month after the first tap went into operation, and even then refused to discuss technical details or allow more than a small number of specially cleared senior NSA analysts access to the material. Now it was CIA's turn to refuse to share credit, and at a time when precious little else was available about Soviet military organization and activities the tunnel taps provided current intelligence "of a kind and quality which had not been available since 1948," a CIA report boasted.

Among other things, the intercepts revealed the location of a hundred Soviet air force installations in Russia, East Germany, and Poland; the names of several thousand high Soviet military officers; the identification of several hundred scientists involved in the Soviet atomic program; a doubling of Soviet bomber strength in Poland and the equipping of the Soviet air army in East Germany with nuclear-capable bombers and twin-jet fighters with airborne intercept radars; and the order of battle for ground force units in the Soviet Union "not previously identified or not located for several years by any other source."[23]

The triumph would prove short-lived. At one of the joint CIA-SIS planning meetings for Operation Regal, held in London on December 15–18, 1953, the British representatives included George Blake, a Dutch-born SIS officer who had just been released from three years' captivity in a North Korean prison. Blake had been a member of the anti-Nazi resistance in the Netherlands as a teenager, escaping to England in 1942 disguised as a monk. A gifted linguist, he was recruited by the British spy agency and later sent to the British embassy in Seoul, where he and the rest of the staff were cut off when the capital was overrun by North Korean troops in the opening days of the war. A month after the London planning meeting, at a carefully prepared rendezvous with a KGB officer atop a double-decker bus, Blake turned over a carbon copy of the meeting minutes.

It was not until 1961, acting on evidence supplied by a Polish defector, that SIS discovered Blake had been a Soviet spy ever since returning from North Korea. On his third day of interrogation he suddenly blurted out a confession, provoked when one of the SIS interrogators mildly suggested that his actions were perfectly unstandable, as he had surely been tortured and brainwashed during his captivity in North Korea.

"Nobody tortured me!" he shouted. "Nobody blackmailed me! I approached the Soviets and offered my services to them of my own accord." He later said that it was the relentless bombing of North Korea by the U.S. Air Force that convinced him he "was on the wrong side."[24]

In 1966, Blake escaped from Wormwood Scrubs prison and made his way to East Germany and then to the Soviet Union, where he lived the rest of his life on a KGB pension, receiving the Order of Friendship from Vladimir Putin on his eighty-fifth birthday in 2007. In 2015, researchers examining the archives of the Stasi, the East German secret police, were able to establish what happened to six of an estimated four hundred Western agents whom Blake betrayed: five were held in Stasi prisons for up to seventeen years; one was taken to Moscow, most likely shot. The hundreds of others no doubt encountered like fates.[25]

For decades it was not known why the Soviets had not acted earlier on the material Blake provided on the Berlin Tunnel; there was even speculation that the KGB engineered a massive deception operation, feeding disinformation to the listening CIA tappers. But information revealed by former KGB officers in the 1990s pointed to a more prosaic explanation. The KGB's own high-level communications went on a separate system of overhead lines that could not be tapped without its being obvious, and, concerned above all with protecting Blake as a valuable source inside SIS and unwilling to share its secrets with rival agencies, the KGB had simply left both the GRU and the Stasi in the dark about the tunnel's existence. Bureaucratic infighting and the red tape of security restrictions were hardly the sole province of the U.S. intelligence agencies. (As General Canine was fond of reminding NSA's production staff in their moments of frustration, "Don't forget, the Russians also have to put on their pants one leg at a time.") And so for nearly a year insecure military communications continued to flow into the CIA tap while the KGB got around to inserting its second leg in its trousers. A small KGB team was formed to secretly locate the tap but did nothing further until a plan was ready to stage the tunnel's discovery in a way that would not implicate Blake. Finally, on the night of April 21, 1956, a Soviet army signal company, ostensibly searching for the cause of a short circuit caused by recent heavy rains and flooding, began digging on the street directly over the tap chamber. The next day the Soviets and East Germans were triumphantly announcing their discovery of the "American spy tunnel."[26]

Plans for Operation Regal called for the United States to issue "a flat denial of any knowledge of the tunnel" if it was discovered: CIA thought the Soviets would rather join a tacit conspiracy of silence than admit that their communications had been successfully breached. But Khrushchev saw another chance to squeeze the Americans in their anatomically vulnerable spot and ordered an all-out propaganda offensive denouncing the "perfidy and treachery" of the United States for abusing its position in the German capital, and insinuating that the West sought to keep

Berlin divided merely to exploit it as a base for such illegal provocations against the German Democratic Republic. The East German authorities offered tours of "the capitalist warmongers' subterranean listening post" and provided a guestbook where visitors could express their "indignation." Khrushchev's only restriction on what should be revealed was that nothing should be mentioned about British involvement in the project: "Despite the fact that the tunnel contains English equipment," the Soviet Foreign Ministry instructed its ambassador in Berlin, "direct all accusations in the press against the Americans only." The Soviet leader was at that very moment on his visit to England and did not want any diplomatic boats rocked.[27]

With the secret in the open, CIA likewise concluded that Operation Regal was now more valuable as a public relations weapon and quietly leaked impressive facts and figures underscoring the technical ingenuity and logistic challenges of the $6.7 million project—and probably coming off the better by at least giving Americans and West Germans a sense of pride that, as a report from the NBC radio correspondent in Berlin put it, "We pulled off an espionage trick on the Reds for a change."[28]

The teleprinter traffic from the tunnel taps included both plaintext and enciphered signals, and Rowlett had hoped that even some of the otherwise indecipherable material might be readable owing to stray emissions that teleprinter devices were known to give off: these ghost signals could in theory be carried along the metallic twisted-pair telephone wires for some distance, revealing the plaintext letters as they were typed into the machine.

Whether this was ever successfully exploited on the Berlin intercepts appears doubtful, but the underlying phenomenon was real enough. During World War II a technician at Bell Telephone doing a routine test on a one-time-tape teleprinter encryption device called SIGTOT noticed that every time the message tape advanced one position, a spike appeared on an oscilloscope on the other side of the room. The Bell researchers quickly traced the signals to the electric relays in the teleprinter that were actuated as each character was read: every time their contacts opened they created a small spark, which induced a radio frequency signal. Moreover, the researchers discovered, each of the characters in the five-bit Baudot teleprinter code had its own distinctive pattern of spikes on the oscilloscope, thereby reducing the cryptanalytic challenge for any comparably equipped eavesdropper from an impossible one-time-pad problem to a simple monoalphabetic substitution cipher. The signals could not only radiate through the air for distances of as much as half a mile, but could induce an electrical impulse that could travel for miles or more in nearby conductors like power lines, telephone wires, even water pipes.[29]

By 1954 the Soviets appeared to be well aware of the problem, too, and in a comprehensive set of standards issued ostensibly to prevent interference to radio broadcasts by electronic equipment, there were curiously stringent shielding requirements for teleprinter devices. As U.S. researchers began delving more deeply into the matter in the early 1950s they found to their dismay that *everything* they tested radiated telltale emissions. Cipher machines whose rotors were operated by electric motors could even be exploited by measuring voltage fluctuations on the power lines they were attached to as they drew varying amounts of current; sounds produced by cipher machines or electric typewriters as their mechanisms operated proved an equal giveaway.

The subsequent discovery of concealed microphones in the code room of the American embassy in Moscow left little doubt in the minds of NSA's experts that the Soviets were well along in exploiting such "side channel" attacks on cryptographic devices. "Most people were concerned about all the conversations that may have been overheard" by the embassy bugs, an NSA security expert later wrote in an article for the agency's in-house technical publication, *Cryptologic Spectrum*. "We were concerned with something else: What could those microphones do to the cryptomachines used there?" The sweep of the embassy also uncovered a large metal grid, embedded six inches deep in the concrete floor of the attic directly over the code room, that apparently was used to collect stray radio frequency emissions from the code machines.

It was all something of a nightmare, and a massive program called TEMPEST was launched to try to develop effective shielding or masking technologies. But many of the problems defied easy solutions. Engineers tried to obscure the stray teleprinter signals by having all five relays that registered the bits of the plaintext letters actuate at once, rather than sequentially, thus producing a single spike instead of five smaller ones,

ARLINGTON HALL, JUNIOR COLLEGE FOR GIRLS
WASHINGTON, D. C.

Arlington Hall, on a prewar postcard. The U.S. Army seized the property for its signals intelligence headquarters in 1942.

The staff of the Army Signal Intelligence Service, 1935. William F. Friedman is standing in the center. Solomon Kullback (second from left) Abraham Sinkov (third from right), and Frank Rowlett (far right) all would go on to hold senior positions at NSA.

"The World's Greatest Cryptanalytic Bowling Team," 1946. Cecil Phillips (standing, far left) and Frank Lewis (below, left) made important contributions to solving the Russian one-time-pad systems.

A German T52 Geheimschreiber teleprinter encryption machine (known to the Allies as "Sturgeon"); the Soviets' Bandwurm machine operated on a similar principle.

U.S. Army divers recovering records of the German high command's cipher bureau, dropped in the Schliersee in twenty-nine sealed boxes.

The Russian version of Boris Hagelin's B-211 cipher machine.

Rear Admiral Joseph N. Wenger, chief of the Navy codebreaking department, Op-20-G.

Brigadier General Carter W. Clarke, who headed the War Department's "Special Branch."

“Goldberg,” one of the leviathan special-purpose electronic comparators built by ERA to attack Soviet ciphers in the late 1940s; it had one of the first magnetic drum memories.

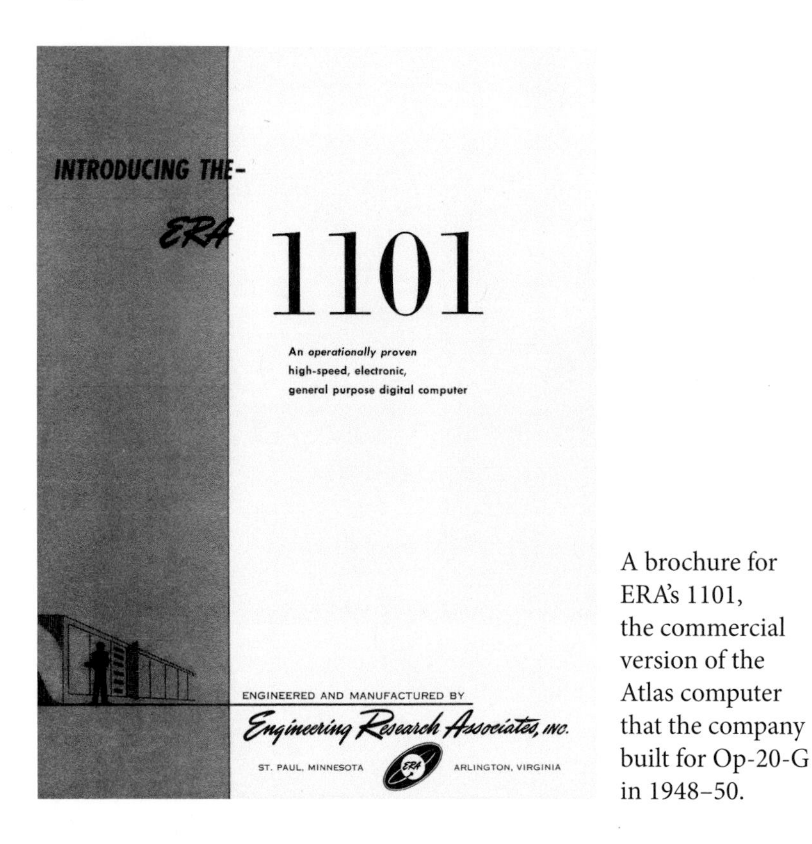

A brochure for ERA’s 1101, the commercial version of the Atlas computer that the company built for Op-20-G in 1948–50.

C-47s lined up at Tempelhof Airport unloading supplies during the Berlin Airlift, 1948.

A segregated unit at Arlington Hall. William Coffee, standing, was the first African American supervisor in the Army's cryptologic organization.

President Harry Truman arriving at Wake Island to confer with General Douglas MacArthur, October 1950.

U.S. Marines at Chosin, cut off by Chinese troops that launched a massive attack in December 1950 against UN forces in Korea.

Ralph J. Canine, NSA's first director, receiving his promotion to major general: pinning on his second stars are his wife and the Army chief of staff, General J. Lawton Collins.

A pointed cartoon about Canine's management style, included in a book presented by NSA employees at his retirement in 1956.

Contestants in the Miss NSA pageant, held in the 1950s and early 1960s.

NATIONAL SECURITY AGENCY Newsletter

VOL. VI NO. 9 Fort Meade, Maryland July 31, 1958

Sept. 12,13 & 14

NSA FALL FESTIVAL

3 BIG GALA DAYS!

HEY MOM! HEY DAD!
HEY KIDS!
HEY EVERYBODY!

CONTESTS
PRIZES

Choral Group

PONY RIDES

HORSE SHOW

HOBBY SHOW

GOOD FOOD

ART EXHIBIT

An NSA newsletter from the 1950s advertising one of the frequent agency events to boost employee morale.

One of the distinctive and unmissable "elephant cage" antenna arrays located at U.S. intercept sites ringing the Soviet Union.

The Russian Fialka cipher machine.

State Department security officials display Léon Theremin's resonant cavity bug found inside the Great Seal replica at the U.S. embassy in Moscow.

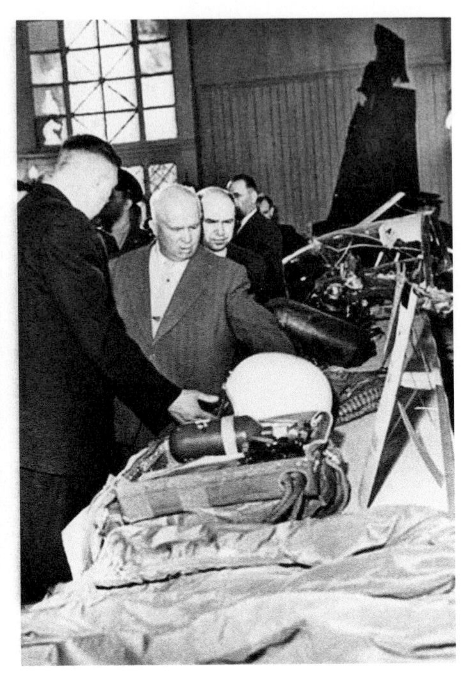

Soviet leader Nikita Khrushchev views items from the wreckage of a U-2 spy plane shot down over the Soviet Union in May 1960.

NSA defectors William Martin (left) and Bernon Mitchell (center), at a news conference in Moscow, September 6, 1960.

A U.S. tank at Checkpoint Charlie during the Berlin crisis, 1961.

The trawler-sized signals collection ship USS *Pueblo,* a few months before its capture by the North Koreans in January 1968.

A team from the Naval Research Laboratory with the first GRAB electronic intelligence satellite before its launch in 1960.

U.S. Army intercept operators taking down manual Morse traffic at Da Nang, South Vietnam.

U.S. fighter-bombers providing very close air support to marines under siege at Khe Sanh during the Tet Offensive, 1968.

The electromechanical KL-7 cipher machine; the larger KL-47, used for the U.S. Navy's Atlantic submarine forces, had the same cipher rotors but incorporated a paper tape reader and punch.

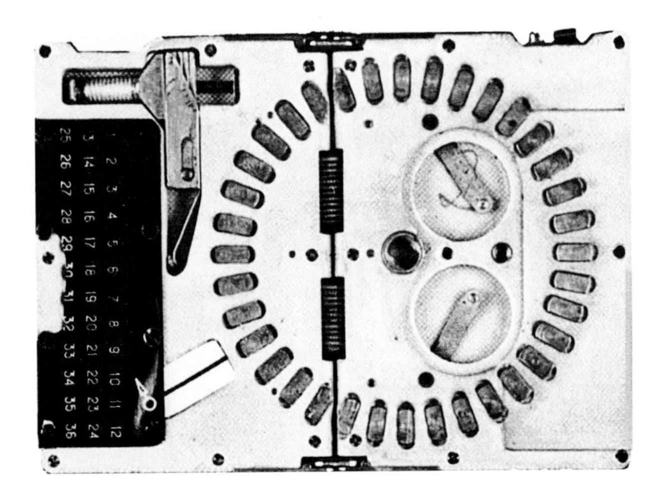

The "rotor reader" supplied to John Walker by the KGB to record the internal wiring of the KL-47 cipher rotors.

NSA's long-serving deputy director, Louis W. Tordella, known as "Dr. No" for his resistance to taking risks.

NSA director Admiral Bobby Inman, with Ann Caracristi, the agency's first woman deputy director, and Frank Rowlett.

Lech Wałęsa in 1980 after successfully registering the Solidarity labor union, a dramatic defiance of Communist Party control in Poland.

A German citizen chips away at the Berlin Wall following the opening of the border in November 1989.

but they found they could still ascertain the letters by measuring the total size of the single spike. Attempts to muffle sound coming from a code room by lining it with acoustical tile actually turned out to make the problem worse, dampening reverberations that obscured the distinctive machine noises. Side channel attacks, and the defenses devised to thwart them, would become an increasingly dominant part of the signals intelligence wars between the United States and the Soviet Union as advances in cryptographic sophistication stymied the traditional tools of the codebreakers.[30]

Making an end run around cryptanalytic solutions seemed to many the only hope in any case. The report of the Baker Panel in the late 1950s pointedly balanced its pessimism over the prospects for ever solving Soviet high-level systems through business-as-usual cryptanalytic attacks with the observation that there is "no limit" to what can be obtained through the CIA's more direct approach: "If machine plans, key usage schedules, and operational information is stolen, any system can be read."

It particularly urged the employment of such second-story cryptanalysis for advancing the so far fruitless efforts to crack any post–Black Friday one-time-pad Soviet traffic: "All permissible efforts should be made to obtain one-time key material."[31] An expanding program of surreptitious entries at foreign embassies around the world, aimed specifically at copying code materials and planting bugs, became CIA's major contribution to the effort after the end of the Berlin Tunnel operation.

Following Canine's retirement, Rowlett returned to NSA as a special adviser to the director. He had always maintained that production of intelligence was not actually the most important aspect of cryptology: far more vital was safeguarding one's own secrets.[32] It was now beginning to look as if that was all cryptology had to offer in any case.

On February 25, 1956, Khrushchev delivered to the 20th Congress of the Soviet Communist Party an astounding indictment of the crimes of his predecessor, cataloging Stalin's purges and "cult of personality." When the Polish Communist Party leader Bolesław Bierut read the text of the speech, he had a heart attack and promptly dropped dead. By June, CIA had obtained a copy of the initially secret document and leaked it to the

New York Times. It was, said Allen Dulles that summer, "the most damning indictment of despotism ever made by a despot."[33]

The new Soviet leader was, in the assessment of John Lewis Gaddis, "genuine—and fundamentally humane—in his determination to return Marxism to its original objective" of offering a better life for people than did capitalism. But Stalin's ghost was not to be "so easily exorcized," for the immediate effect of Khrushchev's call for reform was a wave of rebellion within the Soviets' overextended empire, posing an immediate challenge that seemed to leave little choice between crushing force or compete loss of Russian control.

Taking Khrushchev's speech at its word, the Polish government began freeing political prisoners, including the former Communist Party leader Władysław Gomułka, who had been deposed by Stalin. In October, without seeking Moscow's approval, the Polish Communist Party met in Warsaw and chose Gomułka as Bierut's successor.

Khrushchev's desire to break with his predecessor's brutal methods and to give socialism "a human face" was real enough, but he had not risen through the ranks of the Stalinist regime without learning how to deal ruthlessly with those who threatened his power. He had had Beria shot after Stalin's death, correctly viewing him as a dangerous rival, though ironically on the accusation that Beria was too willing to reach a rapprochement with the West: Beria had proposed accepting a reunified and capitalist Germany if it remained neutral and outside the NATO alliance. (Beria had also nearly dismantled the entire foreign operations of the MGB/MVD in 1953, recalling or dismissing its residents from embassies in every Western country and pulling back seventeen hundred officers from Germany alone. This was likely an effort to remove officials loyal to his rivals, but it gave his enemies a further opportunity to accuse him of being an agent of "English and German intelligence.")[34]

Furious over Poland's independent moves, Khrushchev flew uninvited to Warsaw, threatened an invasion by the Soviet army, and relented only when Gomułka pledged to continue the military and political alliance with the Soviet Union. But that same day, October 24, protests in Hungarian cities against that country's hard-line Communist regime became an outright armed rebellion.[35]

The ongoing impenetrability of Soviet code systems left NSA to rely upon traffic analysis, direction finding, and plain-language communications to follow the chaotic and uncertain events in Hungary. An order sent in the clear to the 2nd Guards Mechanized Division of Soviet Forces in Hungary on the morning of the twenty-fourth was intercepted by an ASA monitoring post at Bad Aibling in Germany: it ordered the unit, normally stationed fifty miles south of the capital, to move without delay to Budapest and use its tanks' cannons against the "rioters." Other signals over the next several days pointed to heavy Soviet casualties and growing hesitation on the part of commanders. After several days of bloody street fighting in Budapest, Soviet army forces that had initially come to the aid of the government pulled back, and it seemed as if Khrushchev was even considering letting Hungary fall out of the Soviet orbit.

On the night of November 2, direction-finding fixes on radio traffic associated with identified Soviet ground force units, however, showed a massive movement of troops within the country and fresh reinforcements pouring across the border, indicating that a final Soviet offensive to crush the uprising was imminent. On the morning of November 4, the Soviet army attacked Budapest and other cities, and the Russian generals who were engaged in negotiations with Hungarian officials to end the Soviet intervention "abandoned the pretense and arrested the conferees," CIA reported. Prime Minister Imre Nagy issued a desperate plea for Western military assistance before he, too, was arrested. He was later shot, adding one more casualty to the twenty thousand Hungarians and fifteen hundred Soviet troops killed.[36]

Amid the Hungarian uprising, Britain, France, and Israel on October 29 launched a secretly planned and ultimately ill-fated military operation in the Sinai, adding another crisis to rapidly escalating international tensions.

The action was ostensibly limited to returning control of the Suez Canal, vital to Britain's and France's oil supplies, to the internationally owned Canal Company following Egyptian president Gamal Abdel Nasser's surprise declaration in July that he was nationalizing the waterway. But it was more broadly intended to deal Nasser a humiliating blow that would lead to his overthrow. Smacking of unreconstructed European colonialism, the Sinai operation infuriated Eisenhower—he had not been told what America's allies were up to—and gave Khrushchev an opportunity to bolster the Soviet position on the world stage by threatening to strike London and Paris with "rocket weapons" if their forces were not withdrawn at once. Behind the scenes, Eisenhower threatened the invaders with economic sanctions, which proved a more believable and effective message. On November 6 the British government, under unprecedented pressure from Washington, agreed to a cease-fire, and by the end of the year the Anglo-French Task Force was withdrawn, replaced by UN peacekeepers.

In his memoirs, Eisenhower insisted that "we were in the dark" about what Britain and France intended to do, that he was unable to "fathom the reason" why Israel was mobilizing its forces, and even after Israeli troops crossed into the Sinai did not believe the invasion was part of a plot involving Britain and France. Secretary of State John Foster Dulles, Allen Dulles's brother, told a press conference in December 1956 that the attack "came as a complete surprise to us." But this was the administration's effort to underscore its dissociation from the action and disapproval of it, rather than an honest reflection of the intelligence that the United States had at its command. Allen Dulles was furious over his brother's remarks, later pointing out to a reporter that U.S. intelligence had accurately foreseen the attack, knew that all three countries were involved, correctly identified the intention, and had reported the day before that it was imminent. As with the Hungarian uprising, most of the intelligence came from NSA, which had followed the Israeli mobilization and the deployment of British troops to Cyprus; during the last two weeks of October, NSA also noted an upsurge in diplomatic communications between Paris and Tel Aviv and between London and Paris, leading it to conclude that France was planning "actions in conjunction with Israel against Egypt."[37]

Strikingly, as much as the event strained U.S.-British diplomatic relations, it had no discernible effect on cooperation between NSA and GCHQ at the working level, which "continued without interruption" throughout the crisis and afterward, according to an NSA internal history—underscoring both the exceptionally strong personal ties that had grown up between the two agencies over the course of a decade and a half of technical collaboration and the unique nature of the UKUSA signals intelligence agreement, which seemed to occupy its own hidden world of international relations completely apart from the normal channels through which the United States conducted its foreign policy.[38]

NSA had acquitted itself reasonably well in providing signals intelligence coverage under far from favorable circumstances. One of its most important findings was a negative one: the absence of any indicators that Soviet forces were mobilizing to intervene in the Suez Crisis, reassuring Eisenhower that Khrushchev's bellicose threats against Britain and France were a bluff.[39]

But trying to put the agency on high alert to respond to the twin crises in the fall of 1956 laid bare just how utterly unprepared NSA was to shift gears in an emergency. The entire organization had been built as a vast assembly line, with the work broken down into individual, specialized tasks; there were few generalists who could jump into a new problem or take over an urgently needed task that demanded extra attention. Emphasizing the factory-like structure, the "production" branch, called PROD, was divided into four sections: GENS, which handled the "general Soviet" problem; ADVA, focused on advanced work on Russian ciphers; ACOM, responsible for Asian Communist countries; and ALLO, which revealingly stood for "All Other."

There was a plan on paper to place NSA on four different "COMINT Alert" levels in the event of an emergency. Alpha was the highest, for an actual war, followed by Bravo, X-ray, and Yankee. A Yankee Alert was intended to increase coverage and reporting when "planned U.S. or Allied activity may stimulate a foreign communications reaction or provoke military or paramilitary action by a foreign nation with respect to the U.S.," and it called for issuing intelligence reports at six-hour intervals at an "intermediate" level of precedence. But even that immediately strained NSA's capabilities beyond its limits when Canine declared a Yankee Alert in the midst of the Hungarian and Suez crises. Canine was flabbergasted to learn it took seven hours for his order even to be distributed: no one had ever worked out the procedure for what to do if an alert was actually declared. Other field units meanwhile kept filing their normal daily reports, clogging the lines and delaying delivery of the alert material, which was supposed to have precedence. The entire communications system that NSA depended on to move intercepts from the field to Washington virtually ground to a halt under the surge of reporting, an inadequacy that a postmortem of the alert said was "appallingly apparent." The agency's internal history of the era bluntly concluded, "As for crisis reporting, all was chaos. The cryptologic com-

Yet Friedman, truth be told, was part of the problem himself: as NSA's chief liaison to the scientific and mathematical community through the 1950s, he repeatedly maneuvered to prevent the agency's top academic consultants—the members of its Special Cryptologic Advisory Group, later the NSA Scientific Advisory Board—from learning much about, much less contributing to, the top-level Soviet problems that had so stubbornly resisted NSA's own efforts at solution. This group of distinguished outside mathematicians and scientists met in Washington only twice a year for two or three days, during which time they were treated to dog and pony shows with little opportunity for making substantive contributions; most of their advice ended up dealing in generalities about developing the agency's scientific and mathematical expertise. Members repeatedly complained about being brought in "to hear a couple of lectures" but not actually working on any specific problems.[44]

Even participants in a more substantive program to bring the nation's top mathematical brainpower to bear on NSA's unsolved problems found themselves getting the runaround. The Special Cryptologic Advisory Math Panel, or SCAMP, was begun in 1952 as an annual summer symposium held on the campus of UCLA. But the agency restricted discussions to the Confidential level, and the initial topic NSA chose for the group to examine, an abstruse area of pure mathematics known as finite projective planes, had at best a "tenuous" connection to cryptology, according to a subsequent assessment. Several years later, when panel members protested that NSA was giving them little more than "sales presentations" while denying them any essential information about important cryptanalytic problems, the Albatross machine in particular, Friedman airily replied that NSA did not believe any outsiders could make a contribution to solving Albatross unless they were willing to devote months of continuous effort, which precluded the use of consultants such as themselves.[45]

"There appears to be a kind of deadlock due to the fact that the Agency is reluctant to clear and brief a consultant for a high-level, compartmented program unless he agreed to come to the Agency and work on the problem for at least three months, and mathematicians are reluctant to commit themselves to at least three months on a problem about which they know almost nothing," complained Charles B. Tompkins, a SCAMP member and PhD mathematician who had also worked at ERA in its earliest years. In putting together SCAMP, Tompkins pointed out, every effort had been made to recruit the strongest possible group of cleared mathematicians, "and it is completely illogical to withhold from them a problem which is largely mathematical in character and which has defied solution for several years." He added, with obvious frustration, "It is not true that mathematicians claim to be superior to cryptanalysts at cryptanalysis; they do claim to be better at mathematics"—precisely the point Friedman refused to acknowledge.[46]

The best case that the NSA cryptanalysts could make for keeping the job within the agency was their faith—and it was becoming little more than faith—that the faster computers the agency was pouring tens of millions of dollars into acquiring would eventually achieve the long-sought breakthrough on the Russian problem. Largely because of NSA's lavish patronage, the capabilities of commercially produced computers were increasing rapidly, with IBM, Sperry Rand, and other new companies introducing substantially more powerful models every year.* The first magnetic core memories, the first high-speed tape drives, the first all-transistor computer, the first desktop-sized computer, and the first remote workstation were all built in response to NSA orders, and the commercial markets they helped spawn stimulated further innovation. NSA took delivery in 1953 of an IBM 701, the world's first commercially available scientific computer, which offered a tenfold jump in processing speed on standard cryptanalytic tasks over Atlas; the IBM 704 and 705 that followed in 1956 increased speed by another factor of five.[47]

None of these general-purpose machines were specifically optimized for high-speed cryptanalytic mass data processing, however. Programming the early machines was a huge challenge. Programs had to be written in machine language, unique to each different computer; the simpler, higher-order programming languages like FORTRAN and BASIC did not yet exist. A slightly sardonic article in NSA's in-house

*ERA, notably, introduced in 1951 a pioneering commercial machine based on Atlas called the ERA 1101. (The name was an inside joke: 1101 is binary for 13, a reference to NSA's Task 13 that began the project.) This was followed, after the company's purchase by Remington Rand and incorporation into its UNIVAC division, with the UNIVAC 1103, a groundbreaking scientific computer that rivaled the IBM 701.

technical journal in October 1956 entitled "Computers—The Wailing Wall," described the yawning culture gaps between the cryptanalyst attempting to devise an innovative and experimental attack on a recalcitrant cipher; the "methods analyst," whose job it was to try to transform "vaguely expressed analytic ideas into exact logical procedures which can be represented on a machine"; and the programmer—who typically "holds a BS in mathematics, has been exposed to eight weeks of cryptanalysis in training school, and has been learning programming for six months"—whose job it was to make it happen.

Programming the early machines involved a host of highly specialized tricks to synchronize computational steps with fetching data from memory and other input-output functions, all of which were opaque to the cryptanalysts, who "come to regard programmers as very obstinate and undependable people who have no grasp of the problem, bother a great deal about petty details, appear crushed by the slightest change of plans, and never get jobs done on time." The programmer in turn found himself bewildered by the cryptanalysts, "notoriously inarticulate even in their own idiom," throwing out "terms like 'stecker,' 'three wheel cycle,' 'isomorphic poker hand,' and 'wheels not on a true base'" that he "has never heard . . . in his life."[48]

Because of the considerable individual differences among wired-rotor cipher machines, each of which presented its own potential weaknesses around which an attack could be fashioned, the programming of research jobs aimed at recovering the basic configuration and setup of rotor machines proved the greatest challenge, even as they potentially made the best use of the digital computer's inherent flexibility. The easiest jobs to program, by contrast, were the massive, highly repetitive streaming runs to match messages in depth, hunt for busts, or laboriously "set" individual messages against known key sequences in cases where the indicator system of an otherwise solved system could not be broken. The teleprinter scramblers, with their often exceptionally long key sequences, typically involved huge data processing runs as well. These types of jobs were also the most processor-intensive: a search devised to locate a bust in messages produced by a Soviet military cipher machine known to NSA as Silver called for more than 10^{17} tests, far beyond the processing capacity of the most advanced computers at that time. Yet with a powerful enough machine it might be possible even to attempt something approaching a brute-force "exhaustive key search," simply trying every single possible key to unlock a message setting.[49]

In the wake of the Clark committee report, Canine had seen an opportunity to make a pitch for a huge project to make an unprecedented advance in computer speeds in a single leap. Given the "very high standards of security techniques and discipline" the Soviets had introduced in their high-level systems, "the single most important project to improve NSA's capability on the Russian high-level problems is the development of super high-speed machinery," argued a May 1956 NSA proposal titled "Recommendations for a Full-Scale Attack on the Russian High-Level Systems." It recommended hiring nine hundred additional staff and buying $16 million in new computers to tackle the "three main high-level problems," which included both machine ciphers and manual one-time-pad systems. (The proposal also recommended providing a subsidy to GCHQ "to derive full advantage of their established technical competence.")[50]

Kullback, Sinkov, Snyder, and other NSA cryptanalytic veterans in the meantime had been pushing a proposal from IBM that promised a hundredfold increase in speed over the IBM 705, which they felt would offer the breakthrough required. Negotiating with IBM was a bit like dealing with the court of Louis XIV. Although the company was determined to become the industry leader in computers, it was determined to do so on its terms. By the late 1950s, half of NSA's computers were being supplied by IBM, which was taking in more than $4 million a year in rental fees from the cryptanalytic agency. The company repeatedly held out the promise that its next computer would be "a machine designed primarily for Agency needs," but invariably once it had a contract in hand the design would drift back to a general-purpose machine that could be sold to other customers, none of which had NSA's unique needs for massive data handling and specialized streaming processing. The IBM 701, which IBM originally called the "Defense Calculator," was much more of a number-cruncher designed to meet the needs of Los Alamos's nuclear weapons designers, meteorologists at the U.S. Weather Bureau, and ballistics engineers at the Army's ordnance labs. The new IBM machine that the company was now proposing was turning into the same bait and switch. In the summer of 1955, NSA agreed to provide IBM the $800,000 in funding it needed to develop the high-speed core memory that was to be the heart of the new "Stretch" computer. But meanwhile IBM also negotiated a deal with the Atomic Energy Commission to supply Los Alamos with a Stretch computer, too, for a fixed price of $4.3 million; then the company's top management began to insist that whatever the final design, it had to be marketable to commercial users as well.[51]

"As usual the agency has a firm hold on the IBM leash and is being dragged down the street," an NSA engineer assigned to keep tabs on the company's work reported as the project progressed. "If you want to control an R/D contract you should pick a company other than IBM. If you pick IBM sit back and wait to get something like the equipment you ordered at a premium price. Don't try to direct, you're only kidding yourself."

By the time the first machine was delivered to NSA in 1962, the price of the project had ballooned to $19 million, which did not include $1 million for supplies such as magnetic tapes and cartridges; $4.2 million for training, additional personnel, and software development; $196,045 for "installation"; and $765,000 a year in rental fees. IBM had resolved the problem of building a computer that could simultaneously serve scientific, cryptanalytic, and commercial customers by designing a flexible central processor, a high-speed arithmetic add-on unit for the AEC, and an add-on streaming unit for NSA, modeled on Abner's "Swish" function. The special NSA add-on was called "Harvest," which eventually became the name of the whole system; its official designation was the IBM 7950.[52]

Canine pushed for an even more ambitious research project in the wake of the Clark report. At an NSA cocktail party the director was talking to several of the agency's senior computer managers about the seeming impossibility of ever building a machine fast enough to get ahead of the relentlessly growing data processing load. Harvest, they noted, was to have a 10-megacycle processor speed. Canine exploded in frustration, "Build me a *thousand* megacycle machine! I'll get the money!"

Canine went to President Eisenhower and did just that, securing for NSA a special exception from the rule that all basic research in the Department of Defense had to be funded through a central agency. The Lightning project, as it was called, was strongly pushed inside the agency by Howard Engstrom, who though in ill health agreed to return from Sperry Rand in 1956, where he was a vice president, to become NSA's deputy director. With a budget of $5 million a year for five years, NSA contracted with IBM, RCA, UNIVAC, Philco, and MIT to carry out basic research on microcircuitry and component fabrication.[53]

By 1960, NSA had spent $100 million on computers and special-purpose analyzers. The basement of the Operations Building at Fort Meade held more than twenty general-purpose machines, one of the largest complexes of computing power in the world. There had been a brief flurry of excitement when two significant busts were found in Silver messages in 1956, leading to the hope that a general solution would soon follow. But in spite of a $20 million crash program to quickly build a series of special-purpose analyzers called Hairline, the Soviet machine cipher resisted regular solution; only about 3 percent of the traffic was exploitable by late 1957. A later agency review, apparently referring to the same machine, noted that even though NSA's cryptanalysts had devised a method for reconstructing the machine's internal configuration "half-way through the typical two-year crypto period, on average," reading individual messages still was limited to instances of operator carelessness or machine malfunction. "While valuable, Silver's messages contained rather low-level information," an agency history of NSA computers acknowledged. It would take more than that to bring back the golden age of World War II codebreaking.[54]

Even with all the computing power in the world, the classical methods of cryptanalysis were also hitting fundamental limits. "To rush a computer to completion by an extravagant expenditure of both money and of our technical resources" was to put brawn over the brains actually required, the Baker Panel critically observed, all the more so since it was clear now that "the goal of the 1,000-megacycle repetition rate can no longer be regarded as a near magic solution to the problem of breaking" Soviet high-level ciphers. The panel pointed out that in the case of the most secure Soviet machine systems, an exhaustive key search was beyond the bounds set by the laws of physics.

"There is not nearly enough energy in the universe to power the computer" that could test every setting of such a rotor machine, which had an effective cryptanalytic keyspace on the order of 10^{44}. Even the "more modest undertaking" of recovering the setting of an individual message enciphered on such a machine whose internal configuration has already been recovered, which would involve testing about 10^{16} possibilities, would cost $2,000,000,000,000,000,000,000 per message for the electricity required to power any known or projected computing devices.[55] (In 1998 a $250,000 machine built with 1,856 custom-made chips successfully carried out an exhaustive key search on the 56-bit key DES encryption system—a keyspace slightly greater than 10^{16}—in two days. But a 128-bit key, with a keyspace of the order 10^{38}, can be shown to resist an exhaustive search even by the most theoretically energy-efficient computer that the laws of physics permit.)

More to the point, the Baker Panel questioned the entire thrust of NSA's reliance on massive data processing, a legacy of a traditional approach to both cryptanalysis and signals intelligence that no longer made sense: "In the past, an overwhelming emphasis has been put on volume and completeness of interception. Today, volume of intercept is out of proportion to the value of its content." Trying to chase the bygone World War II successes of its predecessors, NSA was becoming a "Frankenstein-like monster which amasses constantly greater heaps of material which a dozen or 20 cryptanalysts . . . cannot even lift, let alone survey."

The massive searches for coincidences and depths and letter-frequency counts that computers had been pressed into service to carry out were in fact just mechanized versions of the same old cryptanalytic tricks that had long predated the computer era. Engstrom was one who agreed that these traditional methods of bust searches, brute-force data processing, and simple statistical tests—which had become enshrined as the canonical tools of NSA cryptanalysts, and the entire aim of the agency's multimillion-dollar computer development—were never more than "expedients," which "had to be replaced" by a more mathematically sophisticated approach.[56]

In fact, many of these standard cryptanalytic tests had been developed precisely because they were all that manual methods or punch card machines *could* do: they really were tricks rather than fundamental mathematical solutions to the problem. Looking for double hits in enciphered codes or counting the number of E's in a message were only very crude approximations of the real hypotheses cryptanalysts were seeking to test as they tried to reconstruct an unsolved code or cipher.

The habit of "muddling through, laboriously bludgeoning answers out of a problem with outmoded techniques," in Frank Lewis's words, was inescapably reinforced by NSA's culture as it had evolved from "a small, more personalized, and highly motivated fraternity" to a large, overorganized, and ever more security-conscious bureaucratic giant. Need-to-know and compartmentalization were the very antitheses of the kind of open inquiry, freewheeling exchange of ideas, and give-and-take bull sessions that academic researchers took for granted as a mainspring of creative advances. Lewis regretted the tendency for ever-higher walls to prevent the exchange of technical ideas even within the agency, and his humorous description of the "conservative supervisor who likes to survey a nice quiet roomful of deeply concentrating, strong-but-silent types" captured an all too serious problem in the agency's work culture.[57]

A striking sign of how far NSA's cryptanalysts had fallen behind developments in academic mathematics, the Baker Panel noted, was the complete absence of any research by the agency on communication theory, linguistic structure, or higher-order language statistics. "This is a field of great subtlety and challenge, affording many opportunities," they wrote. "However, NSA appears to have no work going on in this or related information-theoretic directions."

Several seminal papers published in the late 1940s and early 1950s by Claude E. Shannon, a researcher at Baker's own Bell Labs—he was also a member of the Special Cryptologic Advisory Group—had laid the foundation of the new field of information theory and set off an explosion of work of considerable significance to mathematical cryptanalysis. But none of it seemed to have made much of an impression on NSA's business-as-usual approach.

Shannon's key insight was that written language is highly *redundant:* rules of spelling, syntax, and logic all constrain the ways letters are combined into words, words into sentences, sentences into longer passages. The result is that any written text contains far less actual information than a string of characters of equal length in theory could contain. For example, certain letters regularly follow others as a result of particular grammatical constructions or spelling conventions (such as, in English, the combinations TH, ED, LY, EE, IE, OU, ING) while others are rare or impossible (XR, QA, BG, KK).

The uneven frequency with which individual letters appear in normal text, that mainstay of paper-and-pencil cryptanalysts from time immemorial, is merely the simplest, lowest-order manifestation of this redundancy. Shannon calculated that English is in fact about 75 percent redundant: in other words, written language betrays a huge amount of extra clues beyond what is strictly needed to convey meaning. He pointed out that it is often possible to guess the next letter or word in a sentence as a result of this high degree of excess signal to information; one can, for example, eliminate vowels altogether and still usually understand the meaning of a typical sentence (T B R NT T B, THT S TH QSTN for "To be or not to be, that is the question"). Higher-level redundancies in the structure of language likewise make it easy to predict which words follow others, even over long intervals of text; in a phrase such as "There is no way that she possibly could have misunderstood me," many of the words can be omitted with no loss of meaning: "No way she misunderstood me."[58]

The relevance to cryptanalysis was that information theory likewise implied that much of the redundant statistical structure of language must persist in any message encrypted with a finite key, and detecting those ghosts of plaintext can reveal significant information about the cipher machine or its setting that was employed. Traditional cryptanalytic tests such as searching for unrandomly frequent repeats of particular short character strings or code groups at significant intervals were elementary examples of how these persistent statistical structures could be exploited. But they barely scratched the surface of what information theory could do.[59]

Friedman and Kullback, who wrote important papers on the index of coincidence and other fundamental statistical tests of cryptanalysis in the 1930s, were solid statisticians but never at the forefront of the field, and by the late 1950s they were unmistakably behind in embracing the latest research. The entire cryptanalytic leadership of NSA remained in the hands of Friedman's original protégés. Even their admirers had begun to feel that the old guard—Kullback, Sinkov, and others from the World War II generation—had stayed too long, and that NSA had notably failed to bring in truly "high-calibre research mathematicians" or provide the working conditions amenable to the kind of research that the problems demanded, as the agency's Scientific Advisory Board noted in 1957.[60]

The failure to develop methods that put the new insights of information theory to use to exploit the higher-level statistical structure of language in cryptanalysis was the most glaring evidence of this. William Friedman had seen cryptanalysis as a unique profession, demanding a peculiar kind of puzzle-solving mentality combined with patience, a solid but not brilliant mathematical mind, and an apprenticeship in its arcane mysteries. But the Baker Panel pointed out that the world had changed: "The required skills and ability of a new generation of cryptanalysts may . . . be somewhat different from that of the old generation." What it would take to break the most difficult problems now facing NSA were skills and techniques that went beyond the rote application of the same old "bag of tricks" that the "traditional cryptanalyst could rely entirely on."[61]

The Baker Panel's tough words were effectively an indictment of the entire culture of extreme secrecy that had left the agency increasingly isolated and cut off from the advances in higher mathematics that modern cryptanalysis depended on, and whose wartime legacy of individualism and innovation was rapidly succumbing to the paralysis of bureaucratic sclerosis. "Everything secret degenerates," the British statesman Lord Acton had warned a century earlier. Now approaching only its second decade, NSA was encountering the truth of that observation with a vengeance.

8

Days of Crisis

On September 6, 1960, the English-language broadcast of Radio Moscow began with a very Middle American–accented female announcer introducing "the appearance of two employees of the National Security Agency of the United States" at a press conference in the Soviet capital. Through the crackling atmospherics of the shortwave band, William H. Martin and Bernon F. Mitchell explained their reasons for seeking political asylum in the USSR.

A page-one story in the *New York Times* the next day would describe the two men as "appearing in the best of health and spirits," dressed in "neat American suits." Martin, twenty-nine years old, with crew-cut hair and a confident, relaxed manner, did most of the talking, reading a long statement that offered an encyclopedic description of NSA, its functions, the names and responsibilities of its major subdivisions, and a detailed and highly accurate explanation of the intrusive ELINT missions conducted by the United States along the Soviet border, including an account of the flight of a U.S. C-130 ferret aircraft shot down over Soviet Armenia in September 1958.

"We were employees of the highly secret National Security Agency, which gathers communications intelligence from almost all nations of the world for use by the U.S. government," Martin began:

> However, the simple fact that the U.S. government is engaged in delving into the secrets of other nations had little or nothing to do with our decision to defect. Our main dissatisfaction concerns some of the practices the United States uses in gathering intelligence information. We were worried about the U.S. policy of

> deliberately violating the airspace of other nations and the U.S. government's practice of lying about such violations in a manner intended to mislead public opinion. Furthermore, we were disenchanted by the U.S. government's practice of intercepting and deciphering the secret communications of its own allies. Finally, we objected to the fact that the U.S. government was willing to go so far as to recruit agents from among the personnel of its allies.

In response to a question from the *Izvestia* correspondent, Martin stated that the neutral nations whose communications NSA intercepted included "Italy, Turkey, France, Yugoslavia, the United Arab Republic, Indonesia, Uruguay—that's enough to give a general picture, I guess."[1]

The act of intelligence gathering was supposed to support policy, not *be* policy, and it certainly was never supposed to drive world events in a headlong stampede by becoming the news itself. But like the Berlin Tunnel and even more the shootdown on May 1, 1960, of one of the CIA's secret U-2 spy planes over Sverdlovsk—in the very heart of Russia, thirteen hundred miles from the nearest border—Martin and Mitchell's defection was news in anyone's book.

The flip side of secrecy and plausible deniability was that concealment and lies were bound to create a sensation when they became public. The U-2 incident had been the first real test of how the United States could sustain the ethical double standard of engaging in acts of espionage and surveillance that it condemned as illegal and oppressive when performed by totalitarian states like the Soviet Union, and it had dramatically flunked.

Eisenhower—who once admitted that had the Soviets violated American airspace in a similar manner he would have asked Congress for an immediate declaration of war—had been assured by the CIA's director that the U-2 flew so high, seventy thousand feet, that were one ever to be shot down neither the plane nor the pilot would survive to tell any tales. And so the president of the United States proceeded to tell a series of lies about the plane, its mission, and his own involvement in approving the flights until Khrushchev triumphantly revealed that the pilot was alive, well, in Soviet custody, and had already admitted everything.[2]

As with the Berlin Tunnel, the American public seemed to admire CIA technological know-how more than they were troubled by official government mendacity, but coming as it did just two weeks before a summit of British, French, Soviet, and U.S. leaders that was to resolve once and for all the problems of Berlin, the repercussions on world events were stunning. Khrushchev arrived at the Paris summit, delivered a forty-five-minute opening tirade, shouting out his denunciations of the "treachery" and "bandit acts" of the United States, and withdrew his invitation to Eisenhower to visit the Soviet Union following the summit. The Soviet deputy premier, Anastas Mikoyan, was beside himself over Khrushchev's "hysterics"; years later he said that by torpedoing the Paris conference, Khrushchev "was guilty of delaying the onset of detente for fifteen years." KGB chief Alexander Shelepin blandly remarked, "All I know is that there have always been spies and always will be. So there must have been a way for him to find another time and place to tell off Eisenhower."

It was not the flights themselves so much as Eisenhower's deceptions that put the Soviet leader in an impossible situation. "I became more and more convinced," Khrushchev later explained, "that our pride and dignity would be damaged if we went ahead with the conference as if nothing had happened." The Soviet leader had tried to give the president an out by initially suggesting that Eisenhower himself had not known about the U-2 incursions. Eisenhower for his part later acknowledged, "I didn't realize how high a price we were going to have to pay for that lie. And if I had it to do over again, we would have kept our mouths shut."[3]

Martin and Mitchell cited the U-2 flights as one of the more egregious proofs that current U.S. policies had become "dangerous to world peace"—they accurately noted that statements by Vice President Richard Nixon that the flights were aimed only at forestalling a surprise attack by the Soviets was a patently false cover story—but as usual with the labyrinthine psychological story of defection and betrayal the reasons for their jumping sides were not so neat and simple.

The men had left on vacation together on June 24, ostensibly to visit their parents on the West Coast. On Monday, July 25, a week after they were supposed to return, their supervisor tried to reach them at their local addresses. He then phoned the missing employees' parents, who reported they had not seen them at all during their leave. At that point NSA's security office scrambled its entire twenty-two-man staff. By the afternoon of the twenty-eighth they had discovered that Martin and Mitchell had purchased one-way tickets on a flight to Mexico City at Washington's National Airport on June 25. On July 1 they had continued on to Havana, at which point the trail went cold. A key to a safe-deposit box at the State Bank of Laurel had been left in a prominent spot in Mitchell's house, and at NSA's request the Maryland State Police obtained a court order on August 2 to open it. Inside they found a copy of a "Parting Statement to the American People" with a note signed by the men requesting that it be made public.

Over the next six months the NSA security officers spoke to 450 people who had any association with the defectors, hoping to learn what their motivation had been. There was little or no indication that they were ideologically committed Communists. Both had similar backgrounds, raised in upper-middle-class families: Martin's father was the former head of the local Chamber of Commerce, Mitchell's a small-town lawyer. The two men had served together in the Navy in Japan, where they met, and had joined NSA the same day in May 1957 as GS-7 cryptomathematicians.

The bespectacled Mitchell seemed more of a loner, insecure and "extremely unsophisticated and naive," according to the investigators' findings, but both were socially awkward, arrogant, and bitterly resentful that they had not been accorded the recognition they felt was commensurate with their intellectual attainments. Martin's mother told an NSA security officer that "her son was a genius, far superior to ordinary men." His coworkers described him as "an insufferable egotist." Their decision to defect had apparently been an impetuous one. Mitchell left behind a new car and baby grand piano, both fully paid for, and had given his neighboring landlady $100 to pay his rent through August 15. Nor could investigators find any evidence that they had been recruited by the KGB, had tried to enlist other NSA employees to engage in espionage, or had taken any classified documents with them.[4]

Little attention was paid in the ensuing maelstrom to Martin and Mitchell's revelations of NSA spying on U.S. allies or the provocative dangers of the ELINT missions. What made headlines instead was the sensationalistic charge by the chairman of the House Un-American Activities Committee, Representative Francis E. Walter of Pennsylvania, that the defections had exposed NSA as "a nest of sexual deviates."

Newspaper stories alleging that the "two defecting blackmailed homosexual specialists" were part of a "love team" who "recruit other sex deviates for federal jobs" quickly followed in the *Los Angeles Times* and the Hearst papers.

In fact, NSA investigators found that though neighbors considered the pair "odd young men who kept to themselves," there was no reason at all to think that they were homosexual. Their "Predeparture Statement" included a bizarrely incongruous passage praising the Soviet Union's encouragement and utilization of the talents of women, which "we feel . . . makes Soviet women more desirable as mates," an observation that while affirming the picture of the two NSA mathematicians as remarkably naïve and lonely young men did not exactly peg them as "sexual deviates." A female friend of Mitchell's told NSA security officers of having "frequent and normal sexual activity with him during the entire period of their acquaintance," and as for Martin, though he had mentioned to friends certain "sex problems," including sadomasochism, those problems were similarly confined to women. (Among Martin's regular companions was a Baltimore stripper who went under the stage name Lady Zorro, who told investigators that she and Martin had had more than forty "dates," for which Martin paid handsomely in cash.)[5]

None of the facts mattered. A flurry of changes in security procedures that followed went after all the wrong but easy targets. Under pressure from Walter's committee, NSA dismissed twenty-six "sex deviates" whom a reinvestigation of current employees uncovered. The HUAC, in a final report on its own investigation, correctly called attention to the unreliability of the polygraph and the false sense of security it offered, quoting J. Edgar Hoover's strong skepticism about the device. But NSA's reaction was to redouble its reliance on polygraph screening and make it even more intrusive and arbitrary, allowing information obtained during polygraph examination to be turned over to the Office of Security Services for further investigation; previously, NSA applicants had been assured that in return for "voluntarily" submitting to the test, anything they told the examiner would be kept confidential.[6]

Martin and Mitchell almost immediately regretted their decision, discovering that life in the Soviet Union bore little resemblance to the sunny picture painted in *Soviet Life* and other glossy propaganda publications that had shaped their image of the country. Both soon married Russian women, but Martin divorced a few years later; neither ever had children. Over the years a number of prominent visiting Americans, including the clarinetist Benny Goodman and several business executives, were approached by the men asking for help to return to the United States. Both ended their lives as alcoholics, Martin confiding to one associate that he was under constant surveillance and had never been permitted to do any meaningful work at the laser research lab where he was employed by the Russian government.[7]

The focus on rooting out homosexuals and "deviates" from NSA predictably did nothing to prevent the next spy scandal to hit the agency: even red-blooded, heterosexual, married American men with seven children and expensive blond mistresses to support were not immune to temptation. Staff Sergeant Jack E. Dunlap was the holder of a Purple Heart and Bronze Star for "coolness under fire and sincere devotion to duty" in the Korean War. On July 22, 1963, he was found sitting dead in his car at his home near NSA headquarters, a length of radiator hose from the exhaust pipe running through the right front window and the engine idling.

A month later his widow turned over to Army investigators a pile of classified documents from the attic of their home. She said her husband had told her that since mid-1960 he had been meeting a member of the Soviet embassy staff at rendezvous around Washington; in exchange for $40,000 he had supplied documents and hundreds of rolls of film containing pictures he had taken of classified material. Dunlap had worked since 1958 as a driver and courier to the NSA chief of staff, Major General Garrison B. Coverdale, and as an analyst in the Soviet section. Although his evaluation described him as "a traffic analytic assistant of limited ability," he had access to a considerable range of high-level information, having been entrusted with a key to his area and adjoining offices and regularly sent to retrieve materials from the Central Files office.

Dunlap's motive was money pure and simple. He had walked into the Soviet embassy to offer his services, and the air attaché, Mikhail N. Kostyuk, had been all too happy to make the deal on behalf of the GRU. Among the documents in Dunlap's house that he had not yet gotten around to handing to his Soviet contact was an August 17, 1962, report entitled "Use of Radioprinter Scrambler on Soviet Missile Test Ranges," detailing NSA's ability to derive from traffic analysis of this apparently otherwise undecodable material indications of when a Soviet missile test was under way, the approximate time the launch would take place, and whether a successful firing or an in-flight failure occurred.[8]

Three months before his suicide, after applying for conversion to civilian employment at NSA, Dunlap admitted on a polygraph examination to having had "immoral sexual relations" with women and was moved to a "nonsensitive" position. As in the Petersen case, NSA subsequently pointed to this finding as proof that the polygraph "works" and should be mandatory for all NSA employees, military and civilian alike. They played down the embarrassing fact that had anyone been using their eyes instead of pseudoscientific voodoo machines and obsessive fixations on "sex deviates" to identify security risks, they would have noticed long before that Dunlap had left a blindingly obvious trail. On an Army sergeant's salary of $100 a week, he owned two Cadillacs, a baby-blue Jaguar sports car, a thirty-foot cabin cruiser, and a world-class racing hydroplane; he told coworkers a series of contradictory and patently fantastic stories to account for his sudden wealth, including that his father owned a large plantation in Louisiana, that he had made a successful investment in filling stations, that he owned land containing a valuable mineral used to make cosmetics, and that he had won the money as prizes in boat races. Nor did it exactly require a polygraph examination to uncover the fact that a married NSA employee who had begun dating an NSA secretary was possibly engaging in "immoral sexual relations."[9]

In the course of HUAC's investigation, other embarrassing facts had come out about NSA's security program and more generally its cozy self-assurance that it was accountable neither to the public nor even to Congress. Several former NSA employees told the committee investigators that NSA's director of personnel, a former Army major named Maurice H. Klein, wielded such power that even though they had left the agency they feared telling what they knew about problems there because Klein would see to it that their clearances were revoked and they would never be able to work in any classified programs again. Further digging revealed that Klein, on joining AFSA in 1949, had originally lied on his application—he thought that Harvard sounded better than the New

Jersey School of Law as the source of his law degree—and NSA's security director, a former FBI agent named S. Wesley Reynolds, subsequently covered up for his colleague when he turned up the inconsistency. Klein had then slipped a new copy of his application into his personnel file, revised to reflect the correct information about his educational credentials. The only trouble was that the substitute form had been printed by the Government Printing Office after the date of his original application. When HUAC investigators requested a copy of Klein's personnel file, he realized that his attempt to cover his tracks might now be exposed, and pulled yet another switch, locating a blank copy of the older version of the form and filling that out with the true information—but typing it on an IBM electric typewriter that hadn't existed at the time. All of it came out, and both men were forced to resign.[10]

That two senior government officials had succeeded for so long in such a cover-up was a stark demonstration of the impunity that secrecy conferred. During his tenure as director, General Canine had played Congress like an accordion, telling congressmen who tried to ask any substantive questions about NSA's work, "I don't think you would want to be burdened with the responsibility of that information," and hinting that the nation's very security would be gravely endangered if the agency did not immediately receive the extra tens of millions of dollars he always seemed to be asking for.[11] During the HUAC investigation, for the first time, thirty-four senior NSA officials were summoned to testify before Congress, albeit behind closed doors. It would be another decade before Congress or the courts would at last begin to exercise any regular oversight of NSA's programs. But already it was clear that the blissful era of invisibility of all things cryptologic that a generation of America's codebreakers had assumed as a virtual birthright was coming to an end.

Not that they didn't keep trying. A postmortem of Martin and Mitchell's defection concluded that the damage they had done was minimal. Nonetheless, in an exercise of doubtful utility, NSA undertook to rename every section of the agency revealed by the pair at their press conference and all of the "Codewords, Nicknames, Pseudonyms, Short Names, and Mythological Designators" they might ever have seen.[12] GENS and ADVA, the Soviet-bloc sections of PROD, were merged and renamed

A Group; ACOM, dealing with Communist countries outside Moscow's immediate orbit, which included China, North Korea, North Vietnam, and Cuba, became B Group; ALLCOM, responsible for the entire rest of the world, was now G Group. All other operating and administrative sections of the agency received similar new single-letter designations.

The Soviets had long since understood the vulnerabilities of cryptographic systems to an attack by a sophisticated opponent like NSA, but there were still reasons for hope that as long as the wall of secrecy remained intact, less sophisticated targets, including many smaller nations, might continue to be unaware of just how readable their codes could be. The East German police, remarkably, continued to use the Enigma for manual Morse code messages as late as 1956; NSA retained one of its more than one hundred World War II–era bombes in Building 4 of the Navy's Nebraska Avenue installation just to break this traffic, using bombe menus couriered over from Arlington Hall each day. (The police signals were mostly mundane reports of insignificant arrests and fires, but NSA appears to have kept tabs on the traffic in support of the Berlin Tunnel operation, watching for indications that the East Berlin police had noticed the tunnel's construction or were about to take action to expose it.)[13]

Although some of this reliance on old machines may indeed have been cryptographic innocence of the kind that NSA wished to keep blissfully undisturbed, the force of sheer bureaucratic inertia was not to be discounted as an explanation either. During World War II, for example, the Germans had been confronted with repeated evidence that the Enigma was insecure but steadfastly refused to believe it possible. It had similarly taken months of maddening bureaucratic wrestling within the U.S. Navy in the spring of 1943 for Op-20-G to convince top naval commanders that the Germans must have broken the U.S.-British convoy codes, which contained information about the timing and routes of merchant ships crossing the Atlantic. First they had to overcome security restrictions that prohibited the American cryptanalysts from looking at the Allies' own signals. But when they finally were permitted to do so it was unmistakable that Enigma messages sent to Nazi U-boats, containing exact information on "expected convoys" with latitudes and longitudes specified to one degree and speeds to a tenth of a knot, coincided precisely with identical information transmitted just before in the

Allied convoy code.[14] Knowing that the cost and difficulty of ordering a wholesale replacement of a code system was a huge deterrent to any cryptological department venturing on such a decision, all the more so when dealing with the militaries and foreign offices of totalitarian states, run by powerful men who did not look kindly on technical experts who presumed to meddle in their business. It was always easier to take refuge in the belief that the existing codes were good enough.

Cryptographic innocence and bureaucratic inertia both played a part in a shadowier effort by NSA from the late 1950s on to ensure that the Hagelin machines used by dozens of smaller countries remained readable. After the war, Boris Hagelin had moved his development work to Switzerland to escape a Swedish law that allowed the government to appropriate inventions deemed necessary for national defense, establishing the Hagelin Laboratory in Zug, a beautiful lakeside town in central Switzerland. (The entire firm subsequently moved there in 1959, reincorporated as Crypto AG.)

Hagelin had also become a close personal friend of William Friedman's, going back to before the war when Hagelin first tried to interest the U.S. Army in his B-211 and M-209 machines; the Army's subsequent contract to build under license 140,000 M-209s (which were still being used for tactical low-level field communications into the early 1960s) earned Hagelin's gratitude, and made him a millionaire.

The two friends shared a certain familiar Russian gloominess—Hagelin was born in Baku in 1892, one year after Friedman's birth in Kishinev—which they instantly recognized in each other. "We were both born in Russia. But more important was that we were neurotics," Hagelin once observed, not entirely joking. "We both suffered from depressions, but never at the same time—so we could help one another."[15] The men exchanged dozens of letters over the following years, sharing news of their children, their ailments real and imagined, their vacations and travels, often enclosing jokey or sentimental little gifts.

As early as 1950, Friedman was undertaking informal discussions with his friend Boris about AFSA's and CIA's concerns that Hagelin's new postwar models might "cause some previously readable sources to disappear." A much-improved machine called the CX-52 that Hagelin was about to put on the market employed a bank of rotors that moved in a far more complex fashion than the predictable stepping of the M-209;

on some versions of the machine the user could rearrange at will the letters of the alphabet on the input and type wheels, adding an extra substitution (much like the plugboard of the Enigma) that utterly defeated the standard cryptanalytic attack of subtracting two messages in depth. Hagelin had also devised a drop-in module for the CX-52 that replaced the rotor bank with a random-tape reader to allow the device to be used for one-time encipherment. His new line of "Telecrypto" machines similarly used either a standard Hagelin rotor assembly or a prepared key tape to encrypt a standard teleprinter's five-bit output.[16]

In July 1954, Canine asked Friedman to see if he could find out what countries had placed orders for the CX-52. Then in December 1954, Canine sought USCIB's approval of a Top Secret plan to send Friedman to Zug to present a "proposal to Mr. Hagelin."[17] The details remained classified six decades later, but reports from Friedman and other NSA representatives who followed up on the matter noted the "unexpectedly successful" outcome of their discussions. Friedman wrote Boris in November 1956, "I was particularly glad to learn that everything is going along as planned, and that you remain content with the arrangement which was initiated several years ago."[18]

From partially declassified documents released in 2015, it was clear that the core of the deal was NSA's promise that it would certify the CX-52 for purchase by NATO countries, throwing a substantial amount of new business Hagelin's way. Friedman reported after his first visit that Hagelin, who had accepted the proposal on the spot "without any reservations," was especially grateful for NSA's understanding that he wanted to avoid "any relationship" in which direct payment "would play a prominent part." (An earlier proposal, apparently pushed by CIA, to pay Hagelin $700,000 outright apparently fell through, and Friedman subsequently warned Canine on several occasions that "Hagelin could not be bought with money alone.")[19]

In return, Hagelin apparently undertook not to sell his most secure devices to non-NATO countries. As Hagelin wrote Friedman in August 1956 when he learned that a rival firm, the German company Siemens, was planning to sell a one-time-tape device to Egypt, "We have warned Siemens now that this will surely be frowned on by the NATO; but I doubt very much that our word shall have much weight with them. But for my own business, even if I am quite satisfied to keep my end of the 'gentlemens agreement,' from purely sentimental reasons, it is no fun to lose business on this basis."[20]

Friedman formally retired from NSA in August 1955, but in 1957 and 1958 again took extended trips to Europe on NSA official business that included visits to Boris in Zug, where they had further discussions about their "gentleman's agreement." At one of those meetings company officials revealed another favor they were doing NSA's codebreakers: the company had decided it would produce the Telecrypto machines in three different versions. Outwardly they would appear more or less identical, but in fact they would incorporate varying levels of security. Category 1, the most secure, was reserved for NATO countries; Category 2 was for "friendly neutral countries"; and Category 3 for countries of "doubtful orientation" or "leaning toward the USSR," such as India, Egypt, and Indonesia. The operating instructions for each version were to be distinguishable only by "secret marks" printed on them.[21]

The most extraordinary part of the "gentleman's agreement" involved Hagelin's allowing NSA to secretly write the operating manual for NATO customers of the CX-52; even Hagelin himself would not be permitted to see the resulting instructions for "proper usage" of his own machines. The intent seems to have been mainly to improve the security of NATO communications, but there is at least a hint in the documentary record that NSA may have been out to ensure that it retained its own ability to read the traffic of its allies as well.[22]

On August 8, 1958, Friedman wrote to Howard Engstrom, who had just left the agency as deputy director, with a circumspect allusion to the current state of the "Boris deal":

> I have also to report to you, either with tears or laughter, I don't know which, that Sammy [NSA director John A. Samford] made it crystal clear to me, in words of one syllable, that he did not want me to write any more to our friend Boris except on social matters. The thing is now in the hands of you-know-who and he thinks that we (including, especially, myself) should have absolutely nothing to do with it any longer. I am beginning to wonder, in connection with this project, whose ox is being gored? To whose interest is it that the project go through successfully?

Engstrom replied, "I am very anxious to find out how the Boris deal is coming and hope it doesn't die after all the effort you expended on it." The writer James Bamford speculated that "you-know-who" was CIA, whose Staff D was definitely seeking ways to defeat target cryptosystems by means other than cryptanalytic.[23]

The trail disappeared there but popped up two decades later in 1976, when the British writer Ronald W. Clark was completing his biography of Friedman, *The Man Who Broke Purple*. As he recounted in his introduction to the book, NSA began to show "nervous interest" in his manuscript when it learned that he made reference to Friedman's 1957 and 1958 European trips: the agency, he said, expressed "serious concern" that details about these visits not be made public.[24]

Another two decades later the trail again resurfaced, this time following the arrest of a Crypto AG salesman, Hans Buehler, in Iran in 1992. Buehler was on his twenty-fifth visit to the country when he was thrown in jail, accused of being a spy for the United States and Germany, and held in solitary confinement and threatened with torture unless he admitted that Crypto AG was "a spy center." After nine months, Iran released him on $1 million "bail." The ransom was paid by his employer—which then promptly turned around and fired him and demanded he repay the full amount.

Incensed, Buehler went public with the accusation that the company had in fact inserted known weaknesses in the equipment it sold. The company then sued him for defamation. But just days before a Swiss judge was to begin hearing testimony, including from several former Crypto AG engineers who were prepared to testify that the company had indeed rigged its machines, it agreed to settle the case.

Two *Baltimore Sun* reporters subsequently interviewed the engineers Buehler had enlisted to testify, one of whom recounted Boris Hagelin's admitting to having deliberately weakened the machines sold to Third World countries: "He said different countries need different levels of security." Records of one meeting at the company in the 1970s, at which the design of a new model was being discussed, included on the list of attendees a woman whom the reporters were able to identify as a cryptologist employed by NSA at the time. According to one account, the back door in some of the later Crypto AG devices operated by surreptitiously inserting into the cipher text stream a key that disclosed—to those in the know—the setting of the machine used to encrypt the message.[25]

Planting ingeniously designed bugs or executing artful break-ins of embassies to overcome the increasingly evident limitations of classical cryptanalytic attacks was one thing, but seeking to undermine the cipher security of commercial systems across the board was a much more dangerous game. Anything that made it easier for NSA to read traffic also made it easier for others to do the same, endangering the security of at least some commercial and government communications that the United States and its allies had an interest in protecting. And as Buehler's fate underscored, involving private companies in such deceptions raised serious ethical concerns when their employees were unwittingly placed in compromising and even perilous situations.

Friedman's own growing doubts about the ethics of such activities—and indeed about the entire field of signals intelligence, which he, more than anyone, had created—cast an ever-darkening shadow on his last years. He was in declining health, having been through several heart attacks and hospitalizations for severe depression, and his anger over NSA's attempts to muzzle him with pettifogging security regulations and their ham-fisted seizure of dubiously classified materials from his home had by 1961 sent his relations with the agency "over the 'brink of the precipice,'" he confided to a friend. (The agency would spend four decades after his death in 1969 trying to make up for its shoddy treatment, naming a building and auditorium after him at NSA headquarters, installing a bust of him at the agency museum with a plaque hailing him as "The Dean of American Cryptology," and publishing his six lectures in cryptology, which it had previously removed from his home and reclassified.)

Some of Friedman's bitterness was no doubt that of a man seeing his position as a pioneer displaced by a younger generation and new methods he no longer related to; he was scornful of computers and made sardonic comments about being an "old fogey" whom no one listened to anymore. But his worries about the deeper value to mankind of the arcane field he had devoted his life to were clearly thoughts that had been gestating for decades, not merely the irritated effusions of old age and illness. In 1962 he annoyed NSA again by giving a public presentation in which he had the "temerity" to mention cryptology, presenting a scholarly paper at a meeting of the American Philosophical Society about Shakespeare in which, as he put it, he committed the sin of mentioning "certain shady practices engaged in by the British Post Office from the early years of the Tudor Period." Entitled "Shakespeare, Secret Intelligence, and Statecraft," Friedman's paper took as its point of departure a line from *Henry V*: "The King hath note of all that they intend / By interception that they dream not of." Friedman since his retirement had made the study of Shakespeare a serious scholarly hobby, moving to Capitol Hill largely to be near the Folger Shakespeare Library and Library of Congress collections, and he had once in the early 1950s twitted the security rules at NSA by substituting a picture of Shakespeare for himself on his Top Secret access badge for several days. (None of the security guards noticed.)[26]

It did not require much of a leap of imagination to see that he was invoking the Bard of Avon as his alter ego in the meditative questions he posed at the end of the paper he presented to the learned society founded in Philadelphia by Benjamin Franklin twenty-two years before the birth of the country that, more than any other, had stood for the democratic ideal:

> Did Shakespeare have any private views concerning the ethics of interception, the collection of secret intelligence, and its use in the conduct of public business? I wonder. Did he recognize that it is difficult to reconcile such activities with the democratic ideals of a free and open society that would prefer its government to conduct all its internal or domestic affairs openly, so far as possible, and also to conduct all its external or foreign affairs in the same manner? How far is open conduct of public affairs compatible with the national security of a democracy? I wonder what Shakespeare's answers to questions such as these might be?[27]

It was evident what William Friedman thought, even if William Shakespeare's views on the matter were ever unknowable.

Seven weeks after the U-2 shootdown a Thor-Able-Star rocket blasted into space from Florida's Cape Canaveral in the early morning hours of June 22, 1960. A *Washington Post* story the next day quoted officials as saying that the "spectacular 'double-header' launching" of two satellites on a single vehicle was proof that America was "moving into space for real."[28] Three years after the launch of *Sputnik* the United States was still smarting from the humiliation of being beaten to space by the Soviets. *Sputnik* did little more than send out a stream of beeps for the few weeks its transmitter lasted before its batteries died, but the Soviets' achievement had let loose an orgy of national agonizing over what had gone wrong with America: scientific leaders blamed the nation's failure to invest in basic research and the national habit of deriding scientists as impractical eggheads while the Soviets treated their scientists as national heroes; critics of the educational system derided the lax standards of American high schools, the practice of allowing students to choose most of their own classes, and the glorification of sports over intellectual achievement; newspaper editorialists and politicians bemoaned the postwar loss of national purpose and a rising malaise of small-mindedness in a nation that had always thought big and prided itself on an ability to get things done.[29]

A string of embarrassing launch-pad explosions and failures only added to the anxiety that, even if the Soviets had not scored a genuine technological victory over the United States in the space race, they had scored a psychological victory on the stage of world opinion. Every launch from Cape Canaveral became an event of public importance that vastly outweighed the relatively minor purposes of the early small satellites, limited to low earth orbits, a short lifetime, and payloads of a couple of hundred pounds. So did their price tags: each Thor-Able-Star launch cost at least $3.5 million just to get off the ground.

The larger of the two satellites carried on the "double-header" launch in June 1960 was a naval navigation system, Transit II-A. The smaller, which was separated from its mate into orbit by a spring-activated pogo-stick-like device, was named GRAB, which stood for Galactic Radiation and Background. Its ostensible purpose was to study cosmic radiation from space for basic astronomical research. Its real purpose was something far more revolutionary, which would transform SIGINT operations as nothing else in the postwar era. The satellite, built by the Naval Research Laboratory, was the first attempt to take signals interception into space. Its highly secret mission was to collect radar signals from two

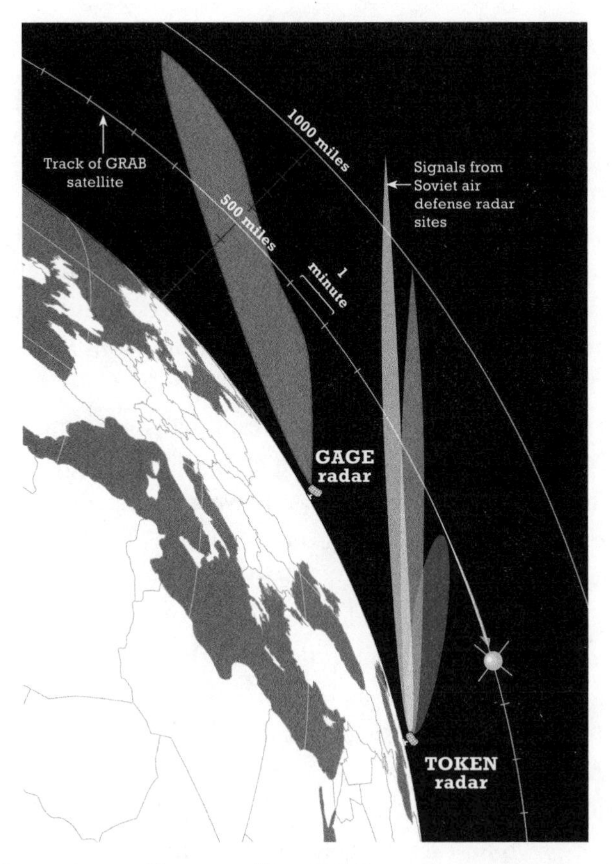

The first electronic surveillance satellite, GRAB, was launched in June 1960 in an over-the-pole orbit to collect signals from key Soviet air defense radars.

Soviet air defense systems, code-named Gage and Token. An NRL engineer, Reid D. Mayo, had come up with the idea in March 1958 when, stranded in a snowstorm in a restaurant in Pennsylvania, he whiled away the time working out some calculations on the paper placemat to see if a radar detector he had developed for submarine periscopes could be packed into a twenty-inch Vanguard satellite and pick up Soviet air defense radars from the much more distant vantage of low earth orbit.[30]

Flying in a polar orbit just five hundred miles high, GRAB swept a continually shifting swath of territory as the earth turned below during the hour and a half it took to make each pass; at that low altitude it flew through the cone of microwave signals coming from a targeted Soviet radar installation for only a minute or less at a time. A tape recorder collected the signals and then played them back as the satellite passed over Hawaii. The first ground stations were literally no more than mobile two-man huts; a hand wheel affixed to a metal shaft running down from the ceiling between the two operator stations permitted the antenna on the roof to be steered in the right direction.[31]

The first GRAB satellite had a life of only three months, but from this modest beginning a new age of SIGINT collection had begun that would eventually make most of NSA's traditional ground intercept stations, as well as the extremely dangerous ferret missions, obsolete. Satellites afforded unimpeded line of sight, at least when they happened to be in the right spot overhead, and were especially well suited to pick up radar and other microwave signals, which (unlike HF signals, which are deflected by the ionosphere) travel relatively unimpeded through the atmosphere and remain readable even thousands of miles into space.

Getting to that promising new age, though, required traversing not only a host of technological challenges but the usual rough bureaucratic road crowded with other agencies all working at cross-purposes. The Air Force was racing ahead with its own program of ELINT space probes designed to piggyback, on a space-available basis, with its Samos photoreconnaissance satellites intended to pick up where the now-canceled U-2 overflights had left off mapping Soviet ICBM sites. The F-series ELINT probes would each be able to cover the entire territory of the Soviet bloc about every five days. Three F-1 satellites were planned to be launched along with E-series photo satellites in 1960 and 1961, but only one made it into orbit.[32]

Overall responsibility for ELINT meanwhile remained a bureaucratic free-for-all. Halfhearted attempts to give NSA the job had been largely shrugged off by the Navy and Air Force, which since 1952 had been operating their own joint ELINT collection and analysis program. With LeMay's single-minded focus on SAC's ability to "execute the SIOP"—to deliver his "Sunday punch" of nuclear bombs in an all-out attack on the Soviet Union, as prescribed in the Single Integrated Operations Plan—the Air Force commander had insisted that as far as he was concerned the only purpose of ELINT collection was to locate the targets for his bombers and missiles and the Soviet air defenses that stood in the way of their carrying out the mission. In laying out plans for the Samos program, the Air Force gave top priority to collecting radar information on the antiballistic-missile system the Soviets were developing to intercept American ICBMs. Next on the list were conventional air defense radars in remote locations: Gage and Token, which the GRAB satellite was sent to map and probe, were search and surveillance radars associated with the SA-1 surface-to-air missile batteries, and their deployment in the north of the Soviet Union along the polar route that U.S. bombers would fly to carry out a nuclear strike placed them out of reach of the border-hugging ferret planes and ground-based intercept stations. At the very bottom of the Air Force's intelligence requirements for Samos, at "Priority IV," was "COMINT—to the extent development proves feasible." The military men were concerned about radars, not chitchat.[33]

Meanwhile CIA was staking out for itself the lead role in running the entire satellite program. CIA already had its own ELINT office, and by virtue of having been in charge of the U-2 program had built up a large photointerpretation staff, housed in the former Steuart Motor Company Building at 6th and K Streets in downtown Washington, which became the obvious choice to handle photographs taken by the new satellites as well. Just before leaving office in January 1961, Eisenhower gave CIA overall supervision of overhead reconnaissance; under a special access program code-named Talent-Keyhole, CIA controlled who was permitted to have any information about the program or its products, and the agency denied NSA all but a handful of the limited number of T-K clearances. Although NSA worked out a separate deal with the Air Force for sharing ELINT data, it still was shut out from decisions on the design or tasking of any of the signals-collecting satellites.[34]

With his zeal for organizational efficiency and cost control honed at Harvard Business School and the Ford Motor Company, the new defense secretary, Robert McNamara, took office determined to impose some order on the sprawling defense establishment, and he moved quickly to centralize the satellite programs in a newly created Defense Department organization, the National Reconnaissance Office, that brought together the separate CIA, Air Force, and Navy projects. NSA's new director, Vice Admiral Laurence Hugh Frost, protested that NSA had the authority under the National Security Council's directives to develop its own signals intelligence satellites; McNamara's staff brushed off Frost's objections, pointing out that the secretary of defense was the "executive agent" for carrying out NSC decisions and he had all the authority he needed to lay down the law and stop the squabbling. McNamara told the admiral his decision was final: NSA would receive the product from the satellite intercepts, but NRO would run the program.[35]

Frost, in office only a year and a half, was not a brisk executive in the mold of McNamara's "whiz kids." In meetings with senior Pentagon and White House officials he habitually spoke in so low a voice, literally mumbling his words, that no one could understand what he was saying. Although an experienced naval intelligence officer, he had little experience with civilians. Bewildered by the large staff at NSA that did not always follow orders as he was accustomed to having them followed in his thirty-five years in the Navy, he surrounded himself with fellow Navy men, which further isolated him from the agency's rank and file, who considered him ineffectual, uncommunicative, and given to occasional unpredictable outbursts in which he vented his frustrations on subordinates, usually picking the nearest available target, deserved or not. He once humiliatingly dressed down Frank Raven, a top NSA civilian cryptanalyst, doing the whole "finger-on-the-chest bit," as one embarrassed witness to the scene described it.

When Frost threatened to go to the White House to have McNamara's decision on the signals satellites overturned, the impatient defense secretary seized the opportunity to fire him; he was gone by July 1962. Frost ended his naval career with an ignominious tour commanding the Potomac River Naval Command.[36]

Yet another spoil fought over in the unending interagency wars for control of new types of signals intelligence and new methods of interception was the telemetry data radioed back to earth from Soviet missiles and space missions. These mostly VHF-band signals posed a special collection challenge since they required line of sight to a missile's path as it rose from Soviet test ranges and launching sites in Kazakhstan, particularly the Baikonur Cosmodrome and the antiballistic-missile test facility at Sary Shagan. The data was encoded using what was called pulse position modulation, which represented information such as acceleration, fuel pressure, and temperature by the time interval between a steadily repeated reference pulse and another pulse that stood for each variable. NSA intercept stations recorded the data on special one-inch-wide magnetic tape on twelve-inch reels. "We'd get them from field military posts in strange places," recalled John O'Hara, who worked on Soviet telemetry at NSA in the early 1960s, "and a lot of them came in in very bad shape." Although the Soviets sometimes tried to hide the signals by using frequencies very close to TV broadcast channels, the data itself was unencrypted. U.S. posts in northern Turkey, Peshawar in Pakistan, the northern island of Hokkaido in Japan, and central Norway were the prime ground-intercept locations. Back at Fort Meade, the tapes of the intercepted radio pulses would be transcribed by chart recorders as traces on four-hundred-foot-long rolls of emulsion paper for analysis.[37]

Just as with ELINT, CIA and the Air Force also each laid claim to telemetry intelligence. Satellites were tailor-made for this mission, giving direct coverage of even the remotest locations; a satellite in geostationary orbit, about twenty-two thousand miles above the earth, could even remain permanently "parked" over Baikonur and Sary Shagan, orbiting at the same rate the earth turns below to keep a constant watch on the missile sites. CIA proceeded to develop a series of signals intelligence satellites code-named Spook Bird and later Rhyolite to do just that—again without clearing anyone at NSA to know about the project.

NSA had meanwhile been developing its own plans for a communications intercept satellite in geostationary orbit over the Soviet Union and China that could pick up telephone signals from microwave repeater towers that escaped into space due to the curvature of the earth as they were beamed from one link to another along the relay chain. Even at VHF and UHF frequencies huge high-gain antennas were required to capture a detectable signal at such extreme distances, and the satellites presented an unprecedented feat of space engineering involving unfurling a seventy-five-foot-diameter umbrella-like array. The Defense Department tried to kill the program, but at the end of 1965 the rival spy agencies arranged an uneasy truce; CIA agreed to include an NSA COMINT package within their Rhyolite mission and to establish a joint processing center, with NSA supplying half of the telemetry staff and all of the COMINT staff.[38] The cost of maintaining this overhead presence would grow from tens of millions of dollars per satellite in the 1960s to a billion dollars apiece by the 1990s.

As the race to the moon increasingly began to dominate the U.S.-Soviet competition, NSA spent tens of millions of additional dollars setting up deep-space-tracking stations with huge parabolic dish antennas, spaced out along the equator at three locations (Asmara, Ethiopia, was one), specifically to grab data from Soviet lunar and Mars probes and rush to NASA the findings, which were expected to be "of great value" to the U.S. Apollo manned lunar landing program.[39] If space was where the money was, then the U.S. spy agencies were going to be there, too.

Spy scandals, unprecedented congressional scrutiny, new ethical questions, revolutionary technologies, and undiminished interagency rivalries were more than enough to keep NSA's managers busy dealing with internal crises at the start of the new decade of the 1960s, but world events were about to intrude with their own more urgent reality.

Nikita Khrushchev's warmth and enthusiasm for the Communist regime of Fidel Castro's Cuba were animated in part by the simple nostalgic feelings of an old Marxist-Leninist. "I felt as though I had returned to my childhood!" exclaimed Khrushchev's deputy Anastas Mikoyan, a veteran of the Bolshevik revolution, upon meeting the Cuban leader for the first time. Here was a youthful, energetic, and committed apostle of Communism who had secured a beachhead for the revolutionary movement in Latin America; to Khrushchev, always agonizingly sensitive to the perceived stature of the Soviet Union on the world stage, the fate of Castro's regime was a test of Soviet power and importance. The botched attempt by fourteen hundred Cuban exiles to overthrow the Cuban Communists in April 1961, the fiasco known to history as the Bay of Pigs Invasion, had been a galling embarrassment for the new Kennedy administration (whose denial of U.S. involvement had quickly proved anything but "plausible"), but to Khrushchev it added another reason to deploy the full military might of the USSR to defend Cuba—and to maintain "Soviet prestige in that part of the world," as he later explained.[40]

Kennedy had run for the presidency decrying the "missile gap" between the United States and the Soviet Union. It had not taken McNamara more than a few weeks as secretary of defense to discover that all of Khrushchev's boasts of Soviet missile strength—he had once declared that Russia was turning out missiles "like sausages"—was a bluff. Photographs taken by the new U.S. reconnaissance satellites over the previous six months showed that the Soviets had ten operational ICBMs, compared to fifty-seven in the U.S. force, and the disparity was rising rapidly. The Soviets' SS-6 and SS-7 missiles moreover took hours to fuel and had to have their unstable liquid propellant drained every thirty days to prevent them from blowing up on the launch pad; the new U.S. Minuteman missile, entering final testing, was powered by solid propellant and could be launched in minutes.

McNamara had been at his job three weeks when the Defense Department's public affairs officer came to him and said that as he had not yet met the Pentagon press corps, he ought to hold a press conference. McNamara tried to demur, noting that he was still learning the ropes, but finally agreed.

The first question was, "Mr. Secretary, what have you learned about the missile gap?"

McNamara innocently replied, "Oh, I've learned there isn't any, or if there is, it's in our favor." He was astonished when a few seconds later the room emptied in a mad rush as the reporters raced to the door to break the sensational story the new defense secretary had just handed them.[41]

The Soviets did, however, have an abundance of medium-range missiles capable of delivering nuclear weapons; similar U.S. weapons based in Britain, Italy, and Turkey were already targeted at Russia. In August 1961, Khrushchev had given another yank on the West's anatomy in Berlin with the sudden erection of a wall dividing the East and West halves of the city. By the next year he had decided on a more provocative demonstration of Soviet force with a move that would bring the two superpowers to the brink of nuclear war. Sending nuclear-armed medium-range ballistic missiles to Cuba would not only defend Castro's revolution and "throw a hedgehog down Uncle Sam's pants," as Khrushchev put it, but would restore the credibility of the Soviet nuclear threat that McNamara's announcement had so dramatically deflated. It would be no more than giving the Americans "a little of their own medicine," he insisted in his memoirs, to have dozens of missiles pointing at them.[42]

Throughout the Soviet military buildup in Cuba the Soviets insisted on absolute secrecy. Several times Castro and his aides questioned the wisdom of that policy, arguing that if they indeed had the legal and moral right to match U.S. nuclear forces with their own, why engage in duplicity and not just announce the Cuban-Soviet military alliance openly? It was never clear how much Khrushchev thought through the entire gambit: years later his speechwriter, at an extraordinary conference of American and Soviet officials who met to reexamine the crisis, said there had always been "irrational reasons" behind Khrushchev's decision. In any case, the Soviet leader waved away Castro's concerns, confidently asserting that if the Americans were presented with a sudden fait accompli they would have no choice but to accept the presence of Soviet missiles ninety miles from their shores, just as the Soviets had swallowed U.S. missiles based right over their border in Turkey.[43]

At NSA, Cuba was the responsibility of the B1 office, since July 1961 under the direction of Juanita Morris Moody, who had gotten her start at Arlington Hall during World War II and since risen through the ranks to become one of the most senior women at the agency. Juanita Morris, like Cecil Phillips, was the quintessential accidental cryptanalyst: she was also from western North Carolina, had also left college after a year, having attended Western Carolina College from 1942 to 1943, and also happened to be living where Army recruiters from Arlington Hall were scouring for GS-2 clerks to hire during the war. While waiting for her background check to be completed she was assigned to an unclassified course in elementary cryptanalysis and was crestfallen when her clearance came through and she had to drop what she already found a fascinating study and go to work sorting messages and punch cards.

One problem that Arlington Hall's cryptanalytic branch had put aside during the war was the German diplomatic code called GEE, a one-time-pad system used for the highest-level traffic. Morris started working on it on her own time after her day's shift was done and soon was one of the key members of a small after-hours group that tried to find a way into the system. They succeeded in January 1945 when, with the extensive help of IBM sorts, they discovered an underlying pattern in the one-time-pad sheets, the product of a mechanical printing system the Germans had used to generate the sheets automatically. It was one of Arlington Hall's most brilliant feats of pure cryptanalysis, and helped to inspire and encourage the effort to attack the Soviet one-time-pad messages.[44]

In 1961, disappointed at being moved from A Group, where she had been assigned for much of the 1950s, Moody was reassured by Louis

Tordella, who had become NSA's deputy director after Engstrom retired in 1958, that she was going to have the agency's "most interesting job for the next five years." The B1 office had begun to pick up a few wispy trails of the growing Soviet military relationship with Cuba the previous year, when monitoring of the Czechoslovak air force's ground control voice circuits picked up Spanish-speaking pilots flying piston-engined trainers at Czech airfields.[45] NSA had begun conducting "hearability" tests to see what Cuban communications could be intercepted from the closest U.S. listening posts, a Navy installation at Puerto Rico and the Army's large World War II–era station at Vint Hill, Virginia. The answer was little. But a Navy destroyer, the USS *Massey,* was able to pick up signals from Cuba's microwave-based telephone network when it circled the island in July 1960.[46] A conversation between operators at two Cuban airfields intercepted the following May offered another clue of the deepening Soviet military relationship with the Castro regime:

Operator 1: Did you know that I've . . . we have to learn Russian.
Operator 2: Who?
Operator 1: Everybody here.
Operator 2: Everybody?
Operator 1: Yes.[47]

Concerned for some time over a lack of coverage of sub-Saharan Africa and South America and the growing political difficulties in establishing American intercept bases in those regions, NSA had asked the Naval Security Group to study the idea of having a dedicated naval vessel do the job. In 1960 the Navy began building the first of its "Technical Research Ships": the USS *Oxford,* a World War II Liberty ship taken out of mothballs and equipped with two dozen radio intercept positions and a bristling array of antennas. Two smaller, equally overage transport vessels were leased by NSA from the Military Sea Transportation Service and hastily converted to seaborne listening posts as well. On its shakedown cruise in September 1961 the *Oxford* was able to intercept both military and commercial microwave telephone communications from Cuba with considerable success as it slowly trolled along the coast just outside the twelve-mile international limit.[48]

Throughout the first half of 1962, monitoring of plain-language communications from Russian merchant vessels had tracked a surge of mysterious shipments to Cuba. On May 1, a U.S. ferret aircraft flying near the westernmost tip of the island collected signals identified as coming from a Soviet Scan Odd radar, carried on MiG-17 and MiG-19 jet fighters and used for airborne interception. On July 24, NSA reported "an unusual number of Soviet passenger ships" en route to Cuba, carrying as many as 3,335 Russians.[49]

The crisis began in earnest on August 29, when a U-2 flying over the island returned with photographs showing the construction of eight SA-2 missile sites. The SA-2 was the antiaircraft missile that had shot down the U-2 over Sverdlovsk in 1960: it was the most advanced surface-to-air missile in the Soviet military, and it was expensive. Soviet officials had repeatedly insisted that none of the weapons it was sending to Cuba were offensive. But CIA director John McCone—at this point virtually a minority of one among the president's advisers—concluded that the only logical purpose of the SA-2s was to defend something of exceptional importance, and the only thing he could think of that fit that bill was launch sites for ballistic missiles that could strike the United States. On September 19, NSA detected signals from a radar known as Spoon Rest that was part of the SA-2 system, indicating that the air defense sites were operational. Then, on October 15, a U-2 overflight photographed unmistakable evidence of SS-4 ballistic missile sites under construction.[50]

In the frantic fourteen days that followed, NSA moved one hundred analysts and linguists from A Group to Moody's area in B1. A makeshift command post was set up across the hall from the A Group front office. Admiral Frost's successor as NSA director, Air Force lieutenant general Gordon A. Blake, had been in the job only three months but had already shown himself to be a master of tact both in dealing with the civilian staff and in handling the agency's external relations around official Washington. Juanita Moody remembered the director dropping by at one point in the midst of the crisis to ask if there was anything he could do to help; Moody replied that she could use some additional staff to deal with a sudden problem that had arisen. The next thing she heard was the three-star general speaking on a phone at a nearby desk: "This is Gordon Blake calling for Mrs. Moody. Could you come in to work now?"[51]

Fearing that public disclosure of the existence of Soviet ballistic missiles in Cuba would force his hand by creating a domestic political furor, leave the Soviets no diplomatic option to back down, and jeopardize any hopes for a peaceful resolution, Kennedy tried to cut off all access to intelligence on the matter within the government; McCone countered that withholding evidence—Kennedy was even insisting that top intelligence officials be kept in the dark—was "extremely dangerous," but finally settled on an arrangement in which distribution would be strictly limited. Only members of the U.S. Intelligence Board and "their personal offices" were cleared to receive the material, code-named Funnel. That included the NSA director, but it cut off all the normal flows of information through the U.S. intelligence community: even Moody was not cleared.[52]

NSA's crucial reporting came from ELINT, plain-language intercepts, and traffic analysis. On Monday, October 22, Kennedy announced in a televised address the presence of the Soviet missile sites and his imposition of a naval "quarantine" of Cuba: the word was chosen to sidestep the fact that a blockade by any name was an act of war.

Early the following morning NSA intercepted a series of HF Morse code signals from the main radio station of the Soviet merchant marine, located in the Black Sea port of Odessa, to each of the twenty-two Soviet vessels en route to Cuba, ordering them to stand by for an extremely urgent cipher message. An hour later the coded message, with a preamble marking it as the highest priority, came through. Its contents were unreadable by the U.S. analysts. But by noon, direction-finding fixes on the Soviet freighters began to show that some had stopped dead in the water, while others had actually turned around and were heading back. It was the first hint that the Soviets might be prepared to back down.[53]

The report went to the director of naval intelligence, one of the select few cleared to receive Funnel material; he in turn informed the CIA that evening only that he had "unconfirmed" information about the Soviet ship movements. It was not until noon the next day, after ONI had corroborated the reports with visual reconnaissance, that the Navy passed to McNamara and the White House the fact that all of the twenty-two merchant ships had either halted or turned back. McNamara blew a gasket, demanding to know how ONI could have sat on such important information for more than twelve hours in the midst of a crisis.[54]

On Friday, October 26, with the largest assemblage of U.S. military forces since the Korean War gathering in Florida, the Caribbean, and the southeastern United States in preparation for airstrikes by 579 combat aircraft followed by an invasion of the island by forty thousand marines and one hundred thousand Army troops including two airborne divisions, Khrushchev dictated a long, rambling, emotional letter to the White House seeming to propose a deal to remove the missiles if the United States promised not to invade Cuba. The next day, as an encouraging reply was being drafted at the White House, Radio Moscow broadcast a second Khrushchev letter, attaching new and unacceptable demands, including the withdrawal of U.S. missiles from Turkey. An American U-2 was shot down over Cuba later that day, the pilot killed.

Returning to the Pentagon that evening, McNamara looked up at the sky and wondered aloud to an aide how many more sunsets he might see.[55]

The breakthrough came late that night, when the president's brother, Attorney General Robert Kennedy, held a long meeting with the Soviet ambassador, Anatoly Dobrynin, at the Department of Justice in which he warned that time was running out. It was not an "ultimatum," Kennedy carefully said, but the U-2 flights were going to continue and the next time one was shot at, U.S. forces would answer by striking the SA-2 site that had fired on it, and the result would be an uncontrollable escalation that could lead to a nuclear war neither of them wanted: if the only issue was Turkey, that could be dealt with quietly. (In fact the United States had decided months earlier to withdraw its missiles from Turkey, a move delayed only because the Turks objected that it would be seen as undermining U.S. support for their country.)

The next day, Sunday, October 28, Khrushchev ended the crisis with a letter to President Kennedy read over Radio Moscow declaring his intention to accept the American terms, without mentioning the behind-the-scenes understanding on Turkey, and stating that the Soviet government had issued "a new order to dismantle the arms which you described as offensive, and crate and return them to the USSR."[56]

NSA made an important contribution to the management of the greatest crisis of the Cold War, but signally failed to offer either advance warning of Soviet intentions or evidence of the arrival of the SS-4 missiles that were the crux of the entire matter. The only tangential indicator that signals intelligence provided came from monitoring plaintext telegrams sent by Soviet military officers to their wives and families back home telling them of their safe arrival in Cuba—"love and kisses messages," NSA analysts called them—and the discovery that a number of the officers were known to be associated with the Soviet rocket forces. In the assessment of NSA's declassified history of the period, the Cuban Missile Crisis "marked the most significant failure of SIGINT to warn national leaders" since the Japanese attack on Pearl Harbor.[57] The inability to decipher any of the high-level cryptographic systems of the Soviet government or military was continuing to take its toll.

Just as in the Hungary and Suez crises of 1956, the Cuban crisis of 1962 jammed and overloaded NSA's global communications systems. In the aftermath of the earlier crises, NSA had pressed a plan for a new communications network called Criticomm that would directly link two hundred NSA field sites to a central hub at Fort Meade; the aim was to be able to notify the president and National Security Council of any critical signals intelligence information—such reports would be labeled with the designator "Critic"—within ten minutes of interception. Louis Tordella, only a few days into his job as NSA deputy director, was summoned to the White House in August 1958 to brief the president, who approved it on the spot.

The heart of the system would be a new hundred-word-per-minute online encryption system that NSA had developed for use aboard Navy ships and at Army and Air Force tactical field stations. The KW-26, mounted in a six-foot-high equipment rack, used vacuum-tube shift registers in place of wired rotors, and NSA already had an order out for the first fifteen hundred units. But Criticomm would never have more than a limited capacity to handle only the most urgent messages. The overwhelming majority of intercepts still arrived either after hours of travel through a series of manual teleprinter relay stations or on the tens of thousands of reels of magnetic tapes sent to Fort Meade each month by air, or even by ship. A seemingly endless jurisdictional battle with other Defense Department agencies delayed plans for a more comprehensive overhaul of the communications network. More than one abortive attempt to develop a so-called automatic switch that could eliminate the mountains of paper tape that had to be printed out and repunched at each relay point compounded the bureaucratic gridlock with a series of embarrassing technical missteps.[58]

Following the Cuban crisis, things finally began to move. The mechanical demodulators with rotating distributors and printers that were still being used to separate out intercepted teleprinter traffic at field sites—their design was literally unchanged from the 1940s devices that had been copied from captured German gear at the end of World War II—at last began to be replaced with digital equipment that recorded the signals directly onto magnetic tape, with the first test models arriving at field sites in January 1963. Manual Morse traffic, which was still being transcribed by operators on typewriters, the copy then retyped on teleprinter machines for transmittal, was semiautomated with new terminals that could directly punch special eight-bit paper tape for transmission over a dedicated NSA communications network called Strawhat; the eight-bit format included special symbols that flagged header information such as call signs and times of transmission for automatic sorting and filing. By the mid-1960s NSA was developing a system using a Honeywell 316 computer at each field site that would accept manual Morse data from up to 128 operator terminals, format it directly onto magnetic tape for storage, then transmit the accumulated messages every six hours over a dedicated data link to Fort Meade. There, other computers repacked the data and passed it on to a large mainframe such as an IBM 360.

By the end of the decade, NSA's efforts to digitize data transmission and handling had led directly to the development of technologies that would make their way with revolutionary consequences into the commercial computer world in the ensuing two decades, among them optical character readers, 2400- and then 9600-baud data links, an embryonic system of e-mail to allow analysts to exchange messages, and the sharing of data from a central server to individual workstations to create an all-electronic "paperless environment," allowing analysts to call up intercepts, reports, and databases on a screen without the need for cumbersome printouts.[59]

The Cuban crisis also dealt a final blow to the last remnants of NSA's nine-to-five work culture. The events had hit home the reality that it no longer demanded a particular military or diplomatic confrontation to generate a crisis: a world with two nuclear-armed superpowers possessed of ICBMs and long-range bombers was by definition on the brink of nuclear war all the time, a state of constant crisis that demanded keeping an around-the-clock watch. General Blake had set up a makeshift command post at NSA during the Cuban crisis, but afterward ordered the creation of a permanent command center, which in 1969 evolved into the Current SIGINT Operations Center and then in 1973 the National SIGINT Operations Center, or NSOC (pronounced EN-sock), a mission control–like facility with maps, big screens, colored telephones, and secure voice links to CIA, the White House, and other key Washington offices. The NSA command center maintained a twenty-four-hour-a-day watch to monitor incoming reports and could communicate directly to field sites over a teleprinter network called Opscomm, to order alerts or quickly shift the focus of their monitoring in response to developing events. The Criticomm system was reserved for formal reporting of urgent Critic messages, but by 1964 a system was devised for sending "tip-offs" to alert the operations center of Soviet activities that bore closer watching, such as movements of Soviet nuclear-capable Tu-16 jet bombers in Eastern Europe; a field site could fire off over Opscomm a short formatted report known as a Bullmoose (plural: Bullmeese) that would arrive in a matter of minutes at Fort Meade and the U.S. Air Force's center at Zweibrücken, Germany, and from there automatically be repeated over the teleprinter network to all of the other NSA field sites monitoring Soviet air activity west of the Urals.

"All of us felt we were on the parapets watching for the start of World War III," recalled one NSA official from that era. The transformation of NSA from an academic "Sleepy Hollow," in the words of the agency's internal history, to a sentry post on the front lines of the Cold War was complete; more than 80 percent of the total "intelligence product" of the entire U.S. government was by the mid-1960s coming from SIGINT.[60]

But there was still a bottleneck when it came to NSA talking to the White House. Kennedy's national security adviser, McGeorge Bundy, shared McNamara's exasperation over the delay in receiving timely signals intelligence reports during the Cuban crisis, and wanted NSA to have a direct feed to the White House Situation Room from now on. CIA, however, ran the Sit Room—and lodged the familiar protest that NSA had no charter to perform intelligence analysis, or even send its results directly to the White House; that was the job of CIA, State, and the military intelligence offices. (In preparing NSA's products for wider distribution, CIA also rigorously enforced the security rule that anything that identified it as coming from intercepted communications had to be stripped off, which not only overstated the role of its agents but often added a spy-novel absurdity. Admiral Bobby Inman, who would serve as NSA's director in the late 1970s, would forever remember one CIA-sanitized COMINT report he read that attributed the source to "an elderly Tibetan horsekeeper.") During the Cuban crisis NSA had already overstepped the letter of its supposed charter by issuing six-hourly wrap-ups, and then continued afterward to issue a daily "SIGINT Summary" (nicknamed the Green Hornet for the color of its cover wrapper), to the extreme irritation of CIA and the military intelligence agencies, but with the approbation of the White House—which Deputy Director Tordella and other top NSA officials rightly concluded was more important as far as the political fortunes of their agency were concerned. By the start of America's involvement in the Vietnam War, NSA had a permanent liaison in the White House and a direct circuit running from Fort Meade, not to mention a soaring reputation as the one unfailingly reliable source of intelligence that an increasingly secretive and embattled president was willing to trust.[61]

9

Reinventing the Wheel

Lyndon Johnson was fascinated by signals intelligence. Like no world leader since Winston Churchill, Johnson constantly demanded to see the actual translations of individual intercepted messages. Like Churchill, too, who shortly after becoming prime minister in 1940 ordered that "all the Enigma messages" be sent to him daily, not imagining how many there could be or how meaningless most of them were without interpretation and context, Johnson told NSA officials that he wanted to be called personally whenever a Critic message came in. Given the number of Critics triggered by routine movements of Soviet bombers over the Arctic, NSA took it upon itself to quietly ignore the request. But Tordella made sure that his agency's most important customer was kept well supplied with its best products. Johnson, the politician par excellence, especially wanted to see any intercepts in which foreign officials or leaders mentioned or quoted him by name, and Tordella was too good a politician himself not to seize such an opportunity. He established a special courier service from Fort Meade to place this especially sensitive material directly in the president's hands. To handle the regular supply of other raw signals intelligence coming into the White House, NSA now had a permanent representative stationed in the Situation Room to brief White House staff and try to add at least some of the context still almost completely absent from NSA's reports and translations.[1]

Like Harry Truman—another self-made man who assumed the presidency upon the death of an immensely popular, eastern establishment, Harvard-educated predecessor—the Texas-born Johnson had an inexhaustible energy for studying documents and absorbing facts. He once confided to Hugh Sidey of *Time* magazine, "I don't believe that

I'll ever get credit for anything I do in foreign affairs, no matter how successful it is, because I didn't go to Harvard." Keenly aware of the contempt of intellectuals for his folksy political persona, his unsophisticated upbringing, and his Machiavellian mastery of the Washington insider game, LBJ compensated by seemingly trying to know everything, and through sheer force of formidable natural intelligence often succeeded.[2] The appeal of NSA's reports was irresistible to a man of Johnson's sharp mind and sharper political instincts: signals intelligence promised an unfiltered source of information that gave him the ultimate insider's advantage over both enemies abroad and political rivals at home.

Johnson once called Vietnam "a raggedy-ass little fourth-rate country." But in 1964 it was rapidly becoming a foreign policy problem that threatened to overwhelm his entire presidency, much less earn him any credit.

America's involvement in Southeast Asia had been marked by secrecy and deception from the start. Following the Communist leader Ho Chi Minh's successful armed rebellion against French colonial rule after the end of World War II, the United States had taken over military assistance to the non-Communist South Vietnamese government created by the 1954 Geneva settlement ending the war and the French presence; because the agreement strictly limited outside military forces, the American troops sent to the region were called "advisers," even when their numbers grew to thousands by 1962, under the command of a U.S. Army four-star general.

By March 1964, when McNamara warned that a renewed Communist insurgency in the South was "a test case of U.S. capacity to help a nation meet a Communist 'war of national liberation,'" Johnson was sliding toward a policy that took concealment to a previously inconceivable height; he was in effect proposing to fight a major war while keeping it a secret from the American people as long as possible, apparently in the hope that it would all be over before the truth became known. When at the end of the 1964 he authorized a substantial military escalation, Johnson instructed his aides, "I consider it a matter of the highest importance that the substance of this position not become public except as I specifically direct."[3]

There were alarming signs from the start, as well, that this was not a winnable war. The South Vietnamese government was led by a corrupt

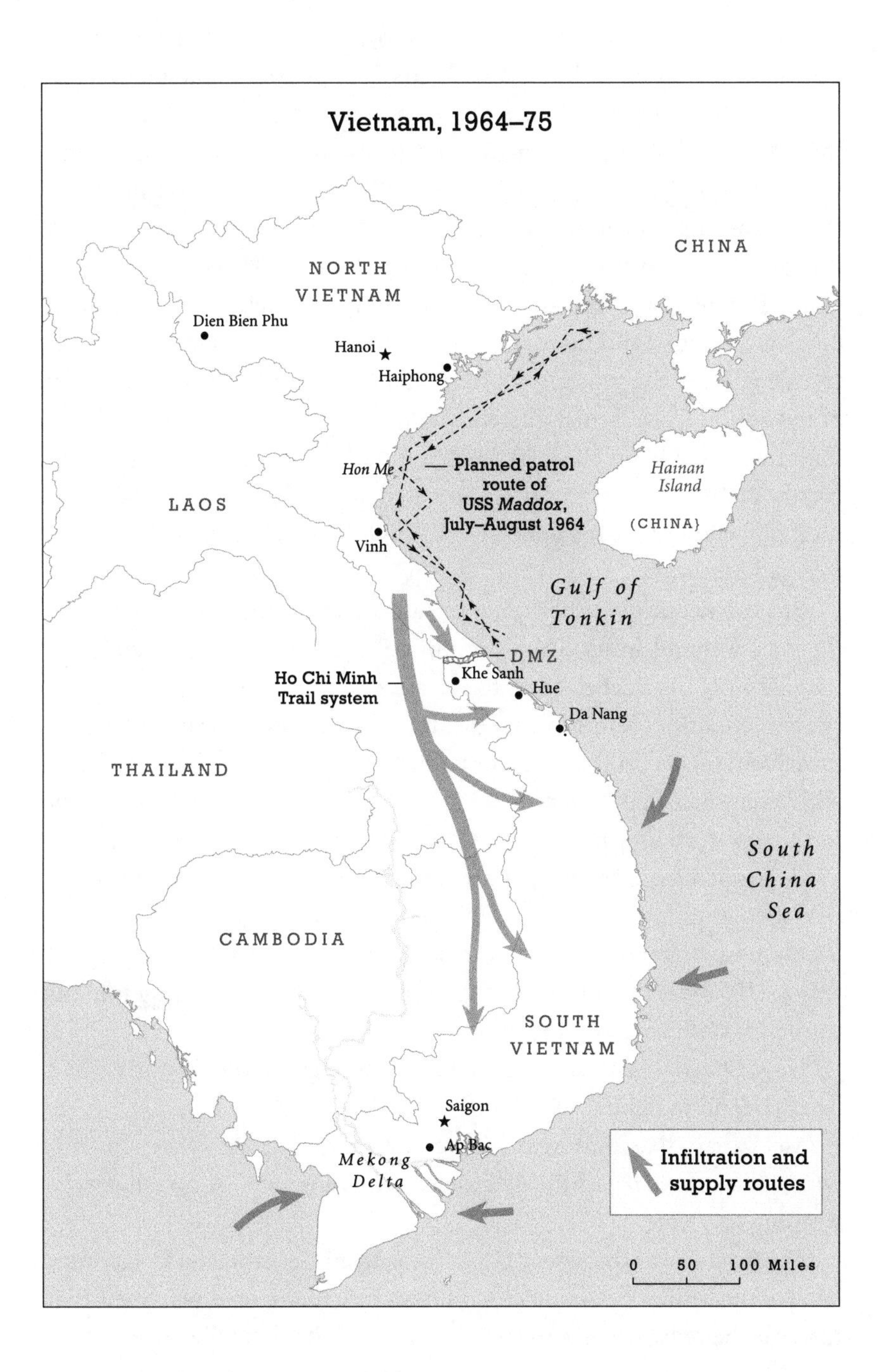
Vietnam, 1964–75
CHINA
NORTH VIETNAM
Dien Bien Phu
Hanoi
Haiphong
Hon Me
Planned patrol route of USS *Maddox*, July–August 1964
Hainan Island
(CHINA)
LAOS
Vinh
Gulf of Tonkin
DMZ
Khe Sanh
Ho Chi Minh Trail system
Hue
Da Nang
THAILAND
South China Sea
CAMBODIA
SOUTH VIETNAM
Saigon
Ap Bac
Mekong Delta
Infiltration and supply routes
0 50 100 Miles

regime that refused to hold elections and was made up largely of refugees from the North who had fled Ho's Democratic Republic of Vietnam; nearly all were Catholics and former soldiers or police officers of the French colonial government, and to many of the indigenous and primarily Buddhist South Vietnamese, they represented nothing more than a continuation of the hated colonial rule. American reporters noted that South Vietnamese villagers did not welcome Americans the way South Koreans had, and that U.S. and South Vietnamese government officials in Saigon did not dare venture even a few miles into the countryside without the protection of a military convoy. Across vast stretches of the South, Communist guerrilla fighters—which intelligence reports found were made up 80 to 90 percent of locally recruited peasants, not North Vietnamese infiltrators—operated with impunity, assassinating local officials and setting ambushes by night, then melting back into the rice paddies by day.[4]

But the necessity of believing in quick American success started at the very top and infused the thinking of Pentagon officials and commanders on the scene; even those closer to the fighting, who daily encountered uncomfortable truths, found that it did not pay to report unwelcome facts up the chain of command. McNamara had an apparently boundless faith in his ability to size up a situation by going straight to the raw data and numbers; on his first visit to the country, in May 1962, he sped from briefing to briefing, filling his notebook with statistics, then announced to reporters that the United States and the South Vietnamese had the Communist insurgents on the run. Neil Sheehan, a young UPI reporter, intercepted the secretary as he was getting into his car on his dash back to the airport.

How, Sheehan asked, could he be so confident about "a war we had barely begun to fight"?

McNamara fixed him with his trademark steely gaze. "Every quantitative measure we have," the defense secretary intoned, "shows that we're winning this war."

The trouble was that every U.S. official, from General Paul D. Harkins, head of the Military Assistance Command Vietnam, on down knew that that was the only message the bosses wanted to hear. Intelligence reports from the field were rigorously edited by Harkins's operations staff, who systematically reduced the number of enemy forces, inflated the number of enemy casualties, and swept under the rug the pathetic performance of South Vietnamese army troops. The weekly report Harkins sent back to the Pentagon was titled the "Headway Report."[5]

In June 1964, William P. Bundy, assistant secretary of defense, had concluded that some form of congressional authorization would be required for the expanding American military operations in Vietnam; seeking to avoid a formal declaration of war, he drafted a joint resolution of Congress authorizing the president to use U.S. forces to protect any Southeast Asian nation threatened by Communism. But the feeling was that some provocation would be needed to justify the step. The situation had all of the ingredients for one of the worst intelligence fiascoes in American history: extreme secrecy, political pressure, an intelligence agency short on analytic experience but ambitious of White House entrée, an overconfident reliance on signals intelligence as an "unimpeachable" source, and a president and secretary of defense who insisted on seeing everything for themselves and were supremely confident of their ability to act as their own intelligence officer at a moment of crisis.

Since 1962 the U.S. Navy had been running patrols along the Chinese coast to assert "freedom of the seas." That was a traditional mission of the U.S. Navy going back to the Great White Fleet, if not the War of 1812. But the destroyers on these "Desoto patrols" also carried fastened to their decks a Top Secret ten-ton van that concealed a self-contained radio intercept post, operated by a Naval Security Group detachment.

On July 28, 1964, USS *Maddox,* a World War II–era destroyer, steamed out of the Taiwanese port of Keelung with orders to conduct a Desoto patrol along the North Vietnamese coast. Its mission was to locate and identify all North Vietnamese coastal radar transmitters, to assess whether the junk fleet was supplying Communist forces in the South, and to "stimulate and record" the North Vietnamese reaction to its presence. The *Maddox* was ordered to stay eight miles off the coast, outside the three-mile territorial limit Vietnam claimed, but to go provocatively near—as little as four miles—several islands in the Gulf of Tonkin where the North Vietnamese had small coastal defense installations and a patrol boat base.[6]

The *Maddox* arrived on station July 31, just hours after a raid by

South Vietnamese commandos on radar stations at Hon Me and Hon Ngu islands. The attacks were part of a highly secret U.S. Navy operation called Oplan 34A that optimistically hoped to put pressure on the North Vietnamese government to halt its support of the insurgency in the South. The U.S. Navy trained and provided logistical support for the South Vietnamese special forces that carried out the attacks. The *Maddox*'s captain, John Herrick, had been briefed only in general terms about Oplan 34A and was not told anything of the operations taking place in the area during his assigned patrol. At NSA, the analysts in the B2 office, responsible for North Vietnamese communications, were equally in the dark. "None of us had been cleared for 34A, and we did not know that there were actions under way," recalled Milt Zaslow, who headed the office.[7]

On the morning of August 1, ASA intercept station USM-626J at Phu Bai, near the city of Hue, on the coast of South Vietnam near the DMZ, intercepted a signal between two North Vietnamese patrol boats, sent in Morse code using a "low grade cipher system," passing on tracking information that correlated with the *Maddox*'s assigned course near Hon Me Island.[8] Shortly before midnight, Navy intercept station USN-27 in the Philippines read a message from the North Vietnamese naval base at Ben Thuy:

> DECIDED TO FIGHT THE ENEMY TONIGHT [one group unreadable] WHEN YOU RECEIVE DIRECTING ORDERS.[9]

A report was sent as a flash message to the *Maddox*, and Herrick at once ordered her east out of the patrol area at ten knots. Throughout the night intercepted messages showed the North Vietnamese radar stations following the U.S. destroyer, but by the next morning, with no more overtly hostile action having occurred, Herrick resumed his patrol.

Shortly before noon on August 2, Phu Bai read a message stating that a planned attack by high-speed torpedo boats was under way. Phu Bai issued a Critic that reached the *Maddox* at 2:15 p.m. A few minutes later the *Maddox*'s radar picked up three North Vietnamese boats approaching at thirty knots from Hon Me. The *Maddox* turned east and increased speed to twenty-five knots. By three o'clock the torpedo boats were within five miles of the U.S. warship, closing rapidly at their top speed of fifty knots. Herrick sent an urgent message requesting air support from the carrier USS *Ticonderoga* and stating his intention to open fire on the pursuing boats. In the ensuing brief engagement the North Vietnamese boats launched their torpedoes, all of them missing; the *Maddox* returned fire with 250 five- and three-inch shells, hitting one of the boats and leaving four dead and six wounded. A few minutes later the *Ticonderoga* air patrol arrived and drove off the attackers. Aboard the *Maddox* there were no injuries; the ship was found to have sustained a single machine-gun bullet hole.[10]

It was 5 a.m. in Washington when news of the maritime skirmish reached the White House; by the time Johnson met with his advisers at 11:30 a.m., senior NSA officials who were asked to attend the briefing had additional information suggesting that the entire action might have been the result of "miscalculation or an impulsive act of a local commander," as McNamara concluded: an order issued by the North Vietnamese naval headquarters in Haiphong two hours before the attack but not transmitted until it was actually under way instructed all boats to return to shore and to relay specifically to the torpedo boat squadron the order NOT TO MAKE WAR TODAY. It was a sign of NSA's growing influence that CIA director John McCone was not invited to attend the White House meeting. Johnson ordered that U.S. plans for retaliatory airstrikes be placed on hold, but that the maritime patrol be reinforced; he called reporters into the Oval Office to warn that any further North Vietnamese interference with U.S. ships in international waters would have "dire consequences."[11]

Captain Herrick proposed cutting short his patrol as an unacceptable risk, earning him a swift rebuke from the commander in chief of the Pacific Fleet, Admiral Thomas Moorer: "Termination of Desoto patrol after two days of patrol ops subsequent to Maddox incident . . . does not in my view adequately demonstrate United States resolve." Then, on the night of August 4, eighteen hours after the initial skirmish—"the darkest night I'd ever seen at sea," in the words of one of the *Maddox*'s radar operators, in rough seas with a heavy chop, with a low overcast sky—the *Maddox* and a second destroyer, the *Turner Joy*, fired hundreds of rounds in a wild, four-hour-long zigzagging encounter in which their crews claimed to have seen gun flashes, searchlights, torpedo wakes, and radar and sonar contacts indicating attacks by multiple enemy boats that fired twenty-six torpedoes.

A welter of confusing and contradictory evidence in the ensuing few hours cast doubt on the whole incident. For one thing, the entire known North Vietnamese force of twelve torpedo boats could have fired at most twenty-four torpedoes. The *Turner Joy*'s far more experienced sonarman had detected no torpedo contacts. Neither ship had suffered any visible damage. The radar contacts had appeared and disappeared at all points of the compass; not a single continuous track was followed. The white streaks in the water that some crewmen reported, Herrick quickly determined, had been nothing but the churning created by the American ships' own wild evasive maneuvers, dodging nonexistent torpedoes. Air patrols reported they had not seen *any* enemy vessels or wakes. "Review of action makes many recorded contacts and torpedoes fired appear doubtful," Herrick reported. "Freak weather effects and overeager sonarmen may have accounted for many reports. No actual visual sightings by Maddox. Suggest complete evaluation before any further actions."[12]

It was at that moment, with orders for the retaliatory airstrikes pending, that McNamara decided to become his own intelligence analyst in earnest, seizing on two signals intelligence reports that had just come in: one was a Critic from Phu Bai issued the night of August 4 reporting POSS DRV NAVAL OPERATION PLANNED AGAINST THE DESOTO PATROL TONITE 04 AUG. The second, which arrived at the White House just two hours after Herrick's message casting doubt on the whole business, appeared to be an after-action report from an unidentified North Vietnamese naval authority: SHOT DOWN TWO PLANES IN THE BATTLE AREA. WE HAD SACRIFICED TWO SHIPS AND ALL THE REST ARE OKAY. THE ENEMY SHIP COULD ALSO HAVE BEEN DAMAGED.

To McNamara this was decisive; he would ever afterward maintain that proof of the North Vietnamese attack on August 4 had come from "intelligence reports of a highly classified and unimpeachable nature." At 10:30 that night in Washington, the president of the United States went on television and announced, "Air action is now in execution against gunboats and certain supporting facilities in North Vietnam." Three days later Congress, by a vote of 88 to 2 in the Senate and unanimously in the House, approved the resolution William Bundy had prepared months earlier for just such an occasion.[13]

The information NSA provided on the August 2 attack had shown the agency at its nimble best: it had decoded messages in virtual real time, flashed an alert to the commander on the scene in time to give him tactical warning, and had sent the White House within hours crucial additional evidence that the attack might have been an unauthorized adventure by an overly aggressive North Vietnamese patrol.

Its reporting on the August 4 phantom attack that precipitated America's large-scale military intervention in Vietnam was another matter. McNamara undeniably seized and ran with the evidence he wanted to believe, but NSA's inexperience in intelligence analysis and frantic efforts to supply the White House with information in the heat of crisis was what allowed him to do so. "Everybody was demanding the SIGINT; they wanted it quick, they didn't want anybody to take any time to analyze it," said Ray Cline, the CIA deputy director at the time.[14]

In fact, it had been a leap of complete guesswork on the part of the analyst at Phu Bai who issued the Critic on August 4 that a new attack on the Desoto patrol was about to take place: the actual intercepted North Vietnamese message, which McNamara did not see, referred only to unspecified "operations" by patrol boats that night. And as for the second message, the seemingly even more decisive after-action report, analysts at the NSA watch center later acknowledged that there had been a difference of opinion whether this referred to the earlier August 2 attack or a new incident. But under the pressures of the moment it had been sent out as evidence of a second attack.[15]

NSA's subsequent efforts to cover up its mistake turned its sin from venal to mortal; what began as an innocent lapse became an act of deliberate falsification as the agency systematically concealed the truth, issuing a series of summary reports over the following days that backed with obedient certainty the administration's position even as the evidence pointed completely the other way.

Within days NSA analysts were privately convinced that no second attack had occurred. The evidence was overwhelming: unlike on August 2, there had been no tracking reports transmitted by any of the North Vietnamese coastal radar stations on the night of August 4. At the very time the August 4 "attack" message was intercepted, other messages from North Vietnamese boats repeated orders to steer clear of the Desoto patrol altogether and left little doubt that the only "operation" taking place that night was a salvage operation to recover two boats damaged in the August 2 skirmish. And as for the August 4 "after-action" report, that had been transmitted while the *Maddox* and *Turner Joy* were in the midst of their four-hour phantom engagement and was merely a propagandistic recapitulation of the earlier action, not a formal after-action report at all.[16]

A classified, searingly honest accounting by NSA historian Robert J. Hanyok in 2001 found that in bolstering the administration's version of events, NSA summary reports made use of only 15 of the relevant intercepts in its files, suppressing 122 others that *all* flatly contradicted the now "official" version of the August 4 events. Translations were altered; in one case two unrelated messages were combined to make them appear to have been from the same message; one of the NSA summary reports that did include a mention of signals relating to a North Vietnamese salvage operation obfuscated the timing to hide the fact that one of the recovered boats was being taken under tow at the very instant it was supposedly attacking the *Maddox* and *Turner Joy.* The original Vietnamese-language version of the August 4 attack message that had triggered the Critic alert meanwhile mysteriously vanished from NSA's files.[17]

Although there was never any evidence of direct orders from the White House to NSA to supply the confirmation it was looking for, there was no need: the doubters among NSA's analysts kept silent while NSA's middle and upper managers gave the administration what it wanted. Within a few days the agency was in too deep to admit it had been wrong, even if it had the inclination to do so. In 2005, after a story about the distortion of the Tonkin Gulf intelligence appeared in the *New York Times,* NSA finally agreed to partially declassify Hanyok's study and 140 supporting documents. The *Times* noted that NSA's top managers had resisted the move for several years, "fearful that it might prompt uncomfortable comparisons with the flawed intelligence used to justify the war in Iraq" launched by President George W. Bush in 2003.[18]

Among the first American "advisers" to arrive in Vietnam in 1961 was a contingent of the Army Security Agency, dressed in civilian clothes, who set up a makeshift headquarters in an empty hangar at Tan Son Nhut Air Base outside Saigon. To create workspaces they piled up boxes of C-rations to form seven-foot-high partitions and cobbled together desks from sheets of plywood and scrap lumber.[19] It had been a decade since NSA or the service cryptologic agencies had operated in a war zone, and it was quickly apparent that all of the lessons and skills learned in Korea had been forgotten. Manning the electronic parapets of spy satellites in space and permanent listening posts at large bases in Germany to watch for the start of World War III was different in every way from lurking on the edge of a jungle trying to direct fire on a shifting guerrilla force a few miles off.

The initial idea was that the main challenge facing the South Vietnamese army in fighting the Communist insurgents on its own territory was simply finding the enemy: it was, ASA analysts thought, a problem much like the Navy faced in locating an enemy naval force amid the vast expanses of the ocean, and the solution was thus the same as well. Direction-finding (DF) units would generate a fix on enemy units when they used their radios; South Vietnamese forces would then be dispatched to root them out.

They had not, however, reckoned with what it was like to operate DF equipment in a heavy jungle. The moist air and heavy foliage so attenuated HF radio signals that the ground wave, traveling directly from a transmitter, could not be detected more than ten or fifteen miles away. Some of the Communist transmitters used as little as 1 watt of power, making detection even more difficult. The sky waves produced when HF signals bounced off the atmosphere traveled longer distances, but came down at such a sharp angle that they could not be detected with the old equipment the South Vietnamese army had inherited from the French. In any case, there was a ninety-mile-wide "skip zone" where neither the direct ground wave nor the sky wave could be heard at all, effectively blacking out most of South Vietnam from any listening post based in Saigon. The only solution was to put DF equipment in mobile units, consisting of a three-quarter-ton truck and two jeeps.

The result, in the words of an internal NSA history of Vietnam, was "sheer chaos." The field missions had to get dangerously close to their targets and even then could detect only about 5 percent of enemy transmitters. The danger was made all too clear when a DF unit was ambushed ten miles from Saigon while returning from a mission on the south coast on December 22, 1961; an ASA soldier, Specialist James T. Davis, and nine South Vietnamese were killed. Davis was the first American fatality of the Vietnam War.[20]

The ambush hastened a project that the Army already had under way which it thought could solve both the technical and logistical problems of obtaining radio fixes on the Communist forces: namely, to mount DF equipment on light aircraft. By March 1962, Army engineers had overcome a puzzling technical challenge to HF airborne direction finding and had a system flying in Vietnam on a single-engine plane, the U-6A Beaver, and working well. (The system basically used the plane's metal skin as an antenna, turning what had seemed an insoluble interference problem into a virtue; instead of rotating a movable loop antenna to determine at what direction the signal was the strongest, the pilot yawed the plane itself in a fishtailing course until a pair of fixed antennas mounted on each wing registered the same intensity. He would then fly to two different spots and obtain additional bearings to complete the triangulation of the target.)[21]

The American SIGINT specialists had not reckoned, however, with how South Vietnamese or American military commanders would use the information obtained by such technological innovation. Under orders from South Vietnam's unelected strongman, Ngo Dinh Diem, who feared that his troops would not be available to protect him from coups in Saigon, his army was forbidden to engage in actions that might result in casualties to themselves; when they were alerted to the presence of Communist forces, the South Vietnamese moved slowly or not at all, inching forward in American-supplied armored personnel carriers only after bombing and strafing indiscriminately with close support aircraft, thereby invariably giving the insurgents time to get away unscathed.

A failed South Vietnamese army assault at Ap Bac, south of Saigon, in January 1963, launched on the basis of an accurate DF fix on a Communist transmitter, was a more serious debacle: the Communist forces shot down five helicopters, damaged nine others, and inflicted two hundred South Vietnamese casualties for the loss of a handful of their own fighters. General Harkins hailed the capture of the village as a decisive victory. But Ap Bac would prove a turning point in American military involvement in Vietnam, convincing U.S. commanders that American ground troops would ultimately have to carry the burden of the fight.[22]

Yet throughout the war U.S. troops were often no better at employing DF fixes. The Communists quickly learned to place their radio huts some distance from command posts, and not infrequently U.S. ground commanders called in strikes to "blast a patch of jungle just because a transmitter had been heard there," as NSA's declassified history related. Fundamentally, the problem was that a generation of U.S. Army commanders had grown up without learning to put signals intelligence to practical use on the battlefield the way their World War II predecessors all had. "Very few commanders had any training in SIGINT. In the 1950s it had been kept closeted, a strategic resource suitable only for following such esoteric problems as Soviet nuclear weapons development." Later in the war when airborne intercept posts on RC-130 aircraft were able to instantly spot and report activations of SA-2 surface-to-air missile radars in North Vietnam, the Air Force refused to act on the information because SAC procedures called for photoreconnaissance confirmation before carrying out any strikes on air defense sites, another instance of Cold War doctrine trumping battlefield realities.[23]

Adding to the difficulties, all of the old fights over control of signals intelligence in the field resurfaced. The hard-won lessons from previous wars of the importance of centralization seemed to have been utterly forgotten; it was as if Korea or World War II had never happened. The Army and the Military Assistance Command Vietnam furiously opposed NSA's move to take charge of ASA's field center, which had relocated from Tan Son Nhut to Phu Bai in 1962, protesting any attempt at "removing these SIGINT resources from the control of military commanders in the area." NSA's director, Gordon Blake, subsequently worked out a deal directly with the Joint Chiefs of Staff under which NSA would run all fixed sites within the country while tactfully agreeing to "delegate" its authority to ASA to command "direct support units" that operated with U.S. troops, effectively placing them under the control of local Army commanders.

The result, as usual of such compromises, tried to satisfy everyone while leaving a mess that only exacerbated suspicions and left NSA and the service cryptologic agencies duplicating and tripping over one another. The Air Force, which wanted its own airborne DF capacity, tried and failed to develop a workable system early on but in 1966 was back with four dozen specially equipped C-47s—and insisting that because they were really doing ELINT, not COMINT, the operation came under Air Force rather than NSA control.[24] NSA's unilateral move to shift signals intelligence processing from the field to a center at Clark Air Base in the Philippines, and later Okinawa and then finally Fort Meade, aroused suspicions of local commanders that NSA was not giving its full support. So did NSA's abrupt decision, following a series of communications changes by Communist forces in April 1962, to essentially abandon efforts at cryptanalysis of high-level systems altogether and focus on DF and traffic analysis instead.

The enemy code changes—which NSA officials more out of reflexive habit than sense immediately blamed on a recent leak by South Vietnamese government sources of U.S. cryptanalytic success—were "catastrophic for the American SIGINT effort," in the words of NSA's Vietnam history. In fact, the move had clearly been in the works for some time, and could not have been a reaction to recent events, as it involved not only new procedures for frequent and regular changes in call signs, schedules, and radio frequencies but also entirely new codebooks and additive pads, which would have taken months to prepare and distribute. Throughout the war U.S. cryptologists would successfully read low- and medium-grade codes employed at the tactical level by enemy units at the level of regiment and below (so much so that by 1968 the processing and decryption of this traffic would have to be automated to handle the considerable volume), but tackling Hanoi's higher-level codes was considered unlikely to be worth the investment of time NSA thought they required.[25]

One of the other forgotten lessons about signals intelligence in a real war would prove even more costly to U.S. forces. In 1965, NSA analysts examining traffic on a network used by Chinese air forces in Vietnam noticed that some of the messages began with an unusual character, a "barred E," . .—. . in Morse code. Recalling that this same character appeared as a prefix on urgent messages transmitted by German U-boats in World War II to report convoy sightings, the chief of B21 division, E. Leigh Sawyer, suggested on a hunch that they compare the timing of these messages with the launch of U.S. bombing missions against North

Vietnam. (Named Operation Rolling Thunder, the airstrikes had begun in March 1965 in an effort to halt the movement of supplies to the South and to increase political pressure on North Vietnam.) The correlation was perfect: barred E messages had been sent ahead of 90 percent of Rolling Thunder strikes that targeted the northeast quadrant of the country. The warnings were giving North Vietnamese MiG pilots time to scramble and be waiting—and add to the toll of more than nine hundred U.S. aircraft shot down during the three years of Rolling Thunder.[26]

Under a program called Purple Dragon, NSA teams were sent to look for leaks in U.S. communications that might be tipping off the enemy. They found a torrent. Tactical commanders had received as little instruction in communications security as they had in the use of signals intelligence. American radio operators frequently made up their own amateurish codes rather than follow required procedures. Strict rules required commanders to account for all code materials and return them for destruction at regular intervals; rather than risk getting a black mark for having to account for lost items, the NSA investigators found, many simply locked the codes up in their safe rather than distribute them for use in the field. As a test, NSA experts playing the part of an enemy analyst tried to see how much they could learn from intercepted U.S. signals. From traffic analysis of a single encrypted voice channel between two air bases, they were able to successfully predict eighteen of twenty-four actual air missions over the North.

The North Vietnamese army, which had a signals intelligence staff of five thousand, proved as adept at exploiting traffic analysis as NSA was. Every U.S. bombing mission was preceded by an upsurge of traffic involving logistics, ordnance loading, weather flights, and aerial refueling tankers, and even if none of the content of the signals was readable, the pattern was a dead giveaway.[27]

The U.S. Air Force generally spared the North Vietnamese even that trouble. Reflecting bureaucratic inertia, American overconfidence, and more than a little disdain for the intelligence capabilities of the enemy, nearly all radio communications of the U.S. air operations used unencrypted tactical voice. NSA's efforts to have the Air Force install voice encryption equipment on aircraft had gone nowhere for years: as one NSA official involved in the program remarked in frustration, the Air Force would accept such a device "only if it had no weight, occupied no space, was free, and added lift to the aircraft." The Air Force insisted that air operations moved too quickly to require such security measures anyway.

But a trove of North Vietnamese signals intelligence documents subsequently captured revealed that U.S. Air Force plain-language voice had been the North's major source of advance warning of U.S. airstrikes throughout the war. Between January and September 1966, 228 aircraft were shot down over North Vietnam, and the captured documents showed that the North Vietnamese had at least thirty to forty-five minutes' warning of 80 to 90 percent of Rolling Thunder missions. Despite NSA's occasional success in tightening up particularly leaky communication practices, the problems continued throughout the war. SAC, which flew B-52 bombers from Guam to strike Communist forces in the South starting in June 1965 in an operation called Arc Light, was by far the worst offender, giving the North Vietnamese as much as eight hours' warning and often revealing exact launch times and likely targets.[28]

It had long been an article of unshakable NSA doctrine that the least disclosure of success in exploiting enemy communications would instantly cause the source to be lost. But the serene indifference of U.S. forces seemed to be proving exactly the opposite proposition: no matter how glaring the evidence that one's communications systems had been compromised, it was impossible to get anyone to do anything about it.

Even as McNamara began to doubt by 1967 that airstrikes could force North Vietnam to halt its operations against the South and accept a negotiated end to the war—the longest sustained bombing campaign in history, Rolling Thunder would deliver 643,000 tons of bombs by the time Johnson ordered a halt in 1968—the defense secretary retained an unwavering confidence that high-tech wizardry held the ultimate solution to defeating an elusive and resourceful enemy in a jungle counterinsurgency campaign.

By the end of 1967 construction began of an elaborate networked system of ground sensors to locate and target infiltrators and supplies moving into the South through rainforest trails in Laos and South Vietnam. McNamara's Wall was the U.S. answer to North Vietnam's Ho Chi Minh Trail. It cost billions of dollars by the time it was done. Dropped from aircraft and camouflaged to resemble vegetation, twenty thousand electronic sensors picked up sounds, vibrations of footsteps, magnetic fields from a passing soldier's rifle, electrical emissions from a nearby truck engine, even human urine. Aircraft orbiting overhead relayed radio signals from the sensors to a data center in Thailand, where IBM 360/65 computers correlated all the information and controllers called in airstrikes on identified targets. NSA contributed to the effort with a series of wiretapping devices for intercepting landlines along the trail routes: early models required a man to stay behind, hidden in the bush, ready to detach the tap if a Communist patrol approached to inspect the line, but later models were designed to be indistinguishable from standard Vietnamese insulators. NSA engineers subsequently devised a helicopter-dropped pole that could pick up signals from a nearby landline without any direct connection at all.[29]

The other high-tech, computer-based intelligence-processing marvel of the war was an NSA and Air Force system that re-created the tactical warning system for fighter aircraft that had proved so effective in Korea. It was slow in coming: once again the lessons of an earlier war had been largely forgotten and had to be reinvented from scratch. An intercept site at Da Nang Air Base on the coast near the DMZ, was able to pick up the highly stereotyped radar tracking plots sent in manual Morse by North Vietnamese radar sites to air defense headquarters; these consisted of a series of numbers or letters giving the altitude, speed, direction, and identity of aircraft, both friendly and enemy. As in Korea, this was an irreplaceable source of distant warning of incoming MiGs that threatened U.S. aircraft. But the security rules that required concealing the source of signals intelligence proved a maddening obstacle to putting it to use: the intercept stations were forbidden to pass the information on to the Seventh Air Force until the enemy aircraft were within range of a U.S. radar station that could plausibly have been the source of the warning.

Even then, the Air Force insisted on a convoluted procedure intended to ensure that its own commanders called the shots but which added deadly delays. The intercept site at Da Nang passed the information to the Air Force Tactical Air Control Center at Tan Son Nhut Air Base in Saigon, which "validated" the tracks and relayed the information to an Air Force radar center on Monkey Mountain, back near Da Nang, which could then warn aircraft. The whole process took twelve to thirty minutes. Similarly, RC-130 aircraft flying over the Gulf of Tonkin that were able to monitor North Vietnamese air-ground VHF voice channels were not allowed to talk directly to pilots at all. In April 1965 two F-105s were shot down by MiGs even though an RC-130 had information that could have warned them in time of the approaching threat.[30]

The system that NSA developed to replace this jury-rigged arrangement in 1967 used a computer link that automatically translated into a geographical coordinate the North Vietnamese tracking data entered by a U.S. operator at a manual Morse intercept terminal; those coordinates were then fed directly into the main air defense computer at TACC and were automatically integrated with the plots that showed up on U.S. Air Force controllers' radar screens. The system, called Iron Horse, reduced the time from intercept to warning to less than three minutes, and sometimes as little as eight seconds. A later system called Teaball, used to vector F-4s to intercept MiGs that threatened B-52s striking targets in North Vietnam, took voice intercepts of North Vietnamese ground controllers collected by a SIGINT U-2 and passed the information to U.S. air controllers; it was credited with thirteen of nineteen MiGs shot down during its two months of operation during Operation Linebacker, the American bombing of Hanoi and Haiphong ordered by President Richard Nixon in 1972 to try to force the North Vietnamese to the negotiating table.[31]

The scientific aura that always imbued signals intelligence cut two ways. The flip side of a failure to understand and effectively use SIGINT by commanders who were mystified by or distrustful of its esoteric complexities was an overreliance on SIGINT by others who were dazzled by its apparent infallibility. By 1966, NSA analysts had worked out a series of "SIGINT indicators" that reliably indicated a pending Communist military operation; these included new call signs and other unscheduled changes in radio procedures, the activation of networks to communicate with forward command and observation posts, the sudden movement of known transmitters by ten kilometers or more, and an increasing tempo of signaling leading up to the actual moment of attack. But it all depended heavily on interpretation and inference. As accurate as the results obtained through traffic analysis and airborne DF proved to be as the ground war intensified throughout 1966 and 1967, there was no guarantee that enemy intentions were always going to be what analysts had assumed them to be: in the absence of decipherment of high-level messages, it was impossible to know for sure *why* an enemy unit had moved. And even as larger, regular North Vietnamese and southern Communist units increasingly began to enter the fighting in the South, no action as of late 1967 had ever involved the movement of more than one division at a time, which minimized having to deal with the greatly complicating intelligence problem of identifying diversions staged as part of a coordinated attack on multiple objectives.[32]

That changed suddenly in November 1967, when intercepts showed two North Vietnamese divisions on the move simultaneously, both crossing into the South in the area of Khe Sanh, where the U.S. Marines held a key base near the DMZ. Khe Sanh bore an uncomfortable resemblance—at least in the minds of senior American officials—to Dien Bien Phu, where the French army had made its last stand in 1954: it was at the bottom of a bowl surrounded by a ring of hills and guarded routes passing through the central highlands of South Vietnam from Laos and the North. The commander of U.S. forces, General William C. Westmoreland, however, saw a chance for a decisive battle and moved U.S. forces north, even as NSA reports beginning on January 25, 1968, called attention to an "accumulation of SIGINT data" pointing to a "coordinated offensive" throughout South Vietnam, to commence on a simultaneous "D-Day":

> During the past week, SIGINT has provided evidence of a coordinated attack to occur in the near future in several areas of South Vietnam. While the bulk of SIGINT evidence indicates the most critical areas to be in the northern half of the country, there is some additional evidence that Communist units in Nam Bo may also be involved. The major areas of enemy offensive operations include the Western Highlands, the coastal provinces of Military Region 6, and the Khe Sanh and Hue areas.[33]

Nam Bo was the region immediately surrounding Saigon. But Westmoreland, increasingly certain that Khe Sanh and the northern province areas were where the real attack was going to fall—and sure that the Communist military commanders were incapable of such coordinated operations—easily convinced himself that the other movements were diversions. NSA's reporting quickly followed suit. After that one passing reference to Nam Bo, NSA analysts at Fort Meade never even mentioned the southern provinces again in the subsequent twenty-six reports they issued before the Communist attacks began a week later. In the days immediately leading up to what would become known as the Tet Offensive, only eight provinces in the north and the west central highlands were identified by NSA as places where attacks would occur.[34]

A subsequent NSA history tried to blame the lapse on a lack of intercept sites in the southern provinces: "SIGINT cannot report what it does not hear."[35] In fact, ASA had five intercept stations in the immediate Saigon area, supplemented by an airborne DF aviation company, and following that initial January 25 report those stations continued to detect and report indications of possible Communist military operations in the southern area. None of their warnings, however, made it into the NSA reports that went to Westmoreland and the White House.

It was probably not so much a case of NSA skewing its findings to tell the commanders what they wanted to hear as a matter of focusing on what they knew they were most interested in, Westmoreland having committed himself to a major American military operation in the northern areas. But the effect was the same. The SIGINT reporting reinforced itself; intelligence to the contrary was brushed aside. Captured documents and prisoner interrogations that strongly suggested a major offensive was being planned for the Saigon area and southward were likewise given short shrift in the face of what now appeared to be decisive signals intelligence evidence. NSA's practice of issuing a blizzard of individual SIGINT reports rather than attempting to synthesize the entire picture further blurred indicators that pointed to a general offensive in the works.

"Such nuanced indicators as highly unusual long-range moves by [Communist] formations, new command relationships, the extensive references to security concerns, morale and propaganda messages, and the concentration of combat units lost their significance in the welter of other information contained in the reports," concluded NSA's historian Robert Hanyok in a critical reappraisal. In a pattern familiar from the Battle of the Bulge in World War II and other similar episodes that had caught by surprise commanders accustomed to assuming that they knew everything the enemy was up to, SIGINT had become "a victim of its own success."[36]

Intelligence historians even had a name for that: "the Ultra Syndrome." It was exacerbated by the fact that inevitably NSA had edged more and more into the role of performing intelligence analysis even as it had to pretend it was not. The very act of selecting, collating, and reporting translations inherently emphasized certain conclusions and interpretations over others; in many ways it was the worst of both worlds, analysis that concealed the analysis that had taken place; analysis performed by analysts who, as CIA correctly complained, often lacked the experience and means to synthesize all available sources into a complete picture.

On January 31, 1968, the first day of Vietnam's celebration of Tet, the lunar new year, eighty-four thousand Communist troops launched attacks against virtually every major population center in South Vietnam. Sixteen provincial capitals in the Mekong Delta south of Saigon were hit, along with nearly one hundred other cities and towns in thirty-eight of the country's forty-four provinces. Power plants, radio stations, army bases, police offices, and government buildings across the country were struck. In Saigon, five battalions of Communist troops that had slipped into the city singly or in small groups, disguised as peasants or South Vietnamese army soldiers, attacked the presidential palace and the American embassy; it took embassy security forces six and a half hours to regain control of the building.

The siege of Khe Sanh was real enough, but in fact *it* was the diversion; it would take seventy-six days, one hundred thousand tons of bombs from B-52s and other U.S. warplanes, and a relief force of thirty thousand troops to lift the siege. Meanwhile, U.S. troops fought desperate battles to retake most of the Mekong Delta and dozens of cities. In the old imperial capital of Hue, American marines endured weeks of fighting straight out of the assaults on the Japanese island redoubts of World War II, employing flamethrowers, grenades, and bayonets in an inch-by-inch struggle. Seventy percent of the city's homes were in ruins when it was over.[37]

That the offensive had failed in purely military terms, with U.S. and South Vietnamese forces losing four thousand killed to considerably higher Communist casualties, and that the mass Communist uprising it was supposed to trigger failed to materialize, was small consolation. The assault on the American embassy in Saigon—which a tone-deaf U.S. spokesman tried to dismiss as "a piddling platoon action"—made a mockery of Westmoreland's confident assertion just two months earlier that the Communist forces were being whittled down and the war might soon be over.

By the spring of 1968 the United States had 549,000 troops in Vietnam, and Johnson's military advisers now were telling him that as many as half a million more would be needed to prevent an American defeat. The war was costing two thousand American lives and $2 billion a month, but it had also taken yet another irreversible toll on America's credibility abroad, and that of its military and political leaders at home. The utter futility of the entire effort, and the sense that the generals leading the war had no real plan for winning it, seemed to be underscored when after the months of bloody fighting to successfully lift the siege of Khe Sanh, Westmoreland issued a matter of fact announcement that the Marine base there was being "inactivated."[38]

On March 31, 1968, Johnson announced a halt to U.S. bombing above the 20th parallel, the opening of peace negotiations with North Vietnam, and his decision not to seek reelection as president. The war would drag on for another five years until a peace treaty signed on January 27, 1973, gave President Nixon what he declared to be "peace with honor," but in fact offered little more than North Vietnam's face-saving concession that it would, in effect, slightly delay its conquest of the South while American troops withdrew. On April 30, 1975, the last South Vietnamese troops defending Saigon surrendered. Late the previous night the last NSA officer still in the country, the Saigon station chief, Tom Glenn, was lifted out of the U.S. embassy compound on one of the helicopters that shuttled back and forth to a fleet of Navy ships assembled just offshore to evacuate the last remaining Americans.[39]

U.S. signals intelligence failures in Vietnam would remain unknown to the public for decades, but the growing disarray of NSA's operations burst into public view in a disastrous and humiliating incident that riveted world attention for almost a year in the midst of the fighting in

Southeast Asia. On January 16, 1968, two weeks before the start of the Tet Offensive, a small U.S. Navy vessel fighting rough winter seas, bitterly cold weather that each night would leave the deck coated with two inches of solid ice to be laboriously chipped off by the crew the next day, an antiquated steering mechanism of uncertain reliability, a failing heating system, and a generator that had blown up on the voyage in, arrived at the 42nd parallel in the Sea of Japan, just south of Vladivostok and the North Korean–Soviet border.

The USS *Pueblo* was neither a happy nor a modern ship. A World War II light transport originally built to ferry supplies between the Pacific islands for the U.S. Army, the *Pueblo* was one of three similar cargo ships pulled out of mothballs beginning in 1964 and hastily reconverted as floating electronic surveillance platforms. Less than half the length and one-tenth the displacement of the eleven-thousand-ton *Oxford*-class "Technical Research Ships," the *Pueblo* and her sister ships were intended to be a cheap and quick answer to the Russians' ubiquitous fleet of forty-eight antenna-laden "fishing trawlers" that prowled along American coasts and overseas bases and shadowed U.S. fleet operations: NSA originally hoped to have twenty-five of the small ships available for deployment against targets of interest throughout the world.[40]

Even under the best of circumstances, life aboard the 177-foot-long *Pueblo* could hardly have been very comfortable for its seventy-six men, but adding to the physical discomforts were a captain sourly resentful of his assignment, personal enmities among the officers that left several barely on speaking terms, and a raw and unconfident crew. Lieutenant Commander Lloyd Mark Bucher had spent eleven years in the Navy in single-minded pursuit of his goal of commanding a submarine; being handed the *Pueblo* was the Navy's way of telling him he was never going to make it, and that his days as a submariner were over. "The orders came as a painful turning point in my career," Bucher later frankly admitted, and "dashed the last of my hopes."

To his further irritation, he learned that the Naval Security Group contingent that made up nearly half of his crew were not to be under his direct command but rather that of the detachment's chief, Lieutenant Stephen R. Harris, a Russian linguist and Harvard grad—from whom Bucher had to request assistance, not give orders, whenever he wanted some of Harris's men to stand watches or make up damage control or firefighting teams. From the start Bucher and his executive officer, Lieutenant Edward R. Murphy, developed a mutual dislike that destroyed any semblance of normal command arrangements and communication, all the more important on a small ship. Bucher sneered at Murphy's strait-laced "pristine perfection" and abstinence from alcohol, tobacco, and even "the stimulant of strong navy coffee," while Murphy was offended by the captain's casual inattention to the proprieties of naval dress and professionalism and his attempt to cultivate popularity with subordinates by playing favorites.[41]

Though briefed in general terms about the goings-on in the equipment-jammed Special Operations Department, or "SOD hut," Bucher had no direct involvement in its "mysterious" operations. Only two of the twenty-nine members of the NSG detachment had ever been to sea before. But even their training in their supposed specialty and preparation for the signals-collecting mission was surprisingly deficient. The SOD hut contained multiple intercept positions and equipment racks filled with gear to receive and record radar ELINT, telemetry, teleprinter, manual Morse, and radiotelephone signals, but several of the men had never used the equipment required to search for and detect teleprinter signals; the documentation they had been provided listing North Korean manual Morse call signs and frequencies turned out to be out of date and useless; and the two marine sergeants assigned to the unit as Korean linguists—who boarded the ship just as it left Yokosuka, six days before the start of the mission—made no secret of the fact that they barely knew any Korean at all and were completely unable to translate or even transcribe the small number of intercepted North Korean voice communications that the detachment was able to collect.[42]

Bucher had expressed some concerns when the ship was being fitted out about the need to ensure a means for emergency destruction of all the documents and gear in the SOD hut, but he never pressed the point, not wanting to make waves in his first command. He also never bothered to exercise his crew in emergency destruction procedures or the standard "repel boarders" maneuver. On earlier patrols, Russian and Chinese warships had harassed and intimidated the American SIGINT ships; twice they had hoisted a signal with the order "Heave to or I will open fire," but in the end had let the U.S. ships depart. A Pentagon review considered the risk of the *Pueblo*'s mission to be minimal, but at the last minute the chief of naval operations ordered that the ship be fitted with two .50-caliber machine guns; installation was completed the day before the *Pueblo* departed Yokosuka. Disregarding Murphy's suggestion that the guns be mounted on the port and starboard sections of the superstructure where they were better protected, Bucher ordered them installed on the forward and aft of the main deck. Only one of the crew, a former Army soldier, had fired one before, and the guns themselves proved to be temperamental, requiring ten minutes of fiddling adjustments every time they were used.[43]

On the passage north Bucher had kept his ship forty miles off the coast and the SIGINT unit had been able to copy little but a few HF Morse signals. For the return trip, the plan was to steam as close as thirteen miles to pick up VHF and UHF signals, particularly those associated with coastal radars and shore defenses, as well as North Korean air force and naval voice communications. On January 22, 1968, the *Pueblo* spent most of the day dead in the water fifteen miles east of Wosan, flying a flag declaring that it was conducting "hydrographic operations." That afternoon two North Korean trawlers made several close passes, once coming as near as thirty yards. Bucher called his Korean "linguists" to come up to the bridge and translate the names of the ships that were painted on their hulls in Korean characters. With the help of a dictionary they finally were able to do so: that was Bucher's first inkling of how little the two marines actually knew of the language.[44]

After standing out to sea for the night, the *Pueblo* resumed its station off Wosan the next morning. A little before noon a North Korean subchaser was seen coming up fast; it circled the *Pueblo* and raised a flag signal asking the ship's nationality, to which Bucher replied with the American ensign. That was when things started to change in a hurry. The North Korean ship signaled with the hoist "Heave to or I will open fire," and shortly afterward three torpedo boats came speeding toward the *Pueblo*, a pair of MiGs made a pass at four thousand feet and began to circle the area, then a fourth torpedo boat joined the small armada that was clearly intending to surround and board the American ship: one of the torpedo boats was rigged with fenders, and an armed party crowded in the bows.

Bucher ordered his ship to make for open sea, which momentarily caught his pursuers off guard, but the PT boats quickly began trying to chivvy him back toward the Korean coast, crisscrossing his bows at twenty-five to thirty-five knots, cutting as close as ten yards. Hoping to avoid giving the North Koreans any excuse to open fire, Bucher ordered the crew not to man the guns or go to general quarters. Harris asked for permission to begin emergency destruction; Bucher denied that request, too, though the crew in the SOD hut by this point just ignored the captain and began to start trying to destroy its gear and papers as best they could.[45]

Had the ship's Korean linguists known Korean, they might well have had an additional eighty minutes to get going on the job, since the subchaser after its first contact had radioed a message in plain language identifying the *Pueblo* as an American surveillance ship, and even as Bucher was still hoping the North Koreans were bluffing, a radio message from the subchaser was asking for permission to open fire.[46] The *Pueblo* had only one small paper incinerator and two shredders that could each handle six sheets of paper at a time. The SOD hut was equipped with axes and sledgehammers, but the crew found they did not have enough room to swing the tools in the confined space. Most of the blows they did land just seemed to bounce off the equipment, a testament to rugged American engineering. Bucher would later say he had no idea how much material there was to try to dispose of: "There was a just fantastic amount of paper, almost I would say ten times what I would have expected that we would have had on board."

At about 1:20 p.m. the subchaser opened fire with its 57mm battery, joined by machine guns from the PT boats. It was quickly apparent that the North Koreans wanted to seize the ship intact, not sink her: they were aiming high, for the bridge, while avoiding the waterline.

Bucher gave the order now for emergency destruction, but the more important reality was that he was losing control of the entire situation. As the fire from the 57mm guns continued with the subchaser now moving up to point-blank range, the *Pueblo*'s engineering officer, Chief Warrant Officer Gene Lacy, turned to Bucher and said, "Are you going to stop this goddam ship before we're all killed?" When Bucher did not reply, Lacy rang the order himself over the annunciator: All stop.

After sending a final coded Critic message reporting their capture, the radio technicians in the SOD hut managed to smash the wired rotors of their cipher machines using a chipping hammer on the steel deck, just before the North Koreans boarded. Some of the material that they had been unable to burn had been shoved into lead-weighted canvas destruction bags, mattress covers, and laundry sacks on deck, but Bucher—incorrectly—understood that Navy regulations required that they be in water at least a hundred fathoms deep before jettisoning classified material, and the bags remained on deck when the ship was forced to turn toward land after failing to reach open sea.[47]

The crew was held for nearly a year by the North Koreans, who beat, tortured, and threatened to kill them to extort a series of highly publicized "confessions." Insisting that the crew would be released only if the United States issued a formal apology for its "crimes," the North Koreans finally accepted an absurd charade in which the American military representative, meeting with his counterpart at Panmunjon in the Korean DMZ in December 1968, would sign an acknowledgment that the ship had violated North Korean territory, but first make an oral statement repudiating the concession ("I will sign the document to free the crew and only to free the crew"). The *Pueblo* remained in North Korea, later becoming a museum and tourist attraction.[48]

A subsequent damage assessment by NSA upon debriefing the returned crew confirmed its worst fears. The destruction of classified material had been "highly disorganized" and was "accomplished in almost total confusion." Some 80 percent of the documents on board were compromised, and only about 5 percent of the equipment had been destroyed beyond repair or usefulness. But even that was optimistic, as a large collection of maintenance manuals and spare parts gave away nearly as much as the intact machines would have. The 397 SIGINT documents on board included encyclopedic reference material on North Korean communications and a list of 126 "Specific Intelligence Collection Requirements" detailing the current state of U.S. knowledge of North Korean, Soviet, Chinese, and other target systems and the gaps NSA was seeking to fill. Combined with what their captors had learned from interrogating the crew, the loss of the *Pueblo*'s material had revealed "the full extent of U.S. SIGINT information on North Korean armed forces communications activities and U.S. successes in the techniques of collection, analysis, exploitation, and reporting applied to this target," NSA found. Although not definitively linked to the *Pueblo*'s capture, changes in Soviet, Chinese, and North Korean communications security procedures in the following months may have been a result of the information revealed in the compromised material about U.S. successes. The overall loss to U.S. SIGINT capabilities, NSA's "worst case" damage assessment concluded, had been "very severe."[49]

In a final assessment, NSA deputy director Tordella was able to take solace on one point: the loss of the cipher machines had certainly resulted only in "minimal" damage to U.S. communications security. On the one hand, the KL-47 cipher machine and other U.S. encryption equipment had been a particular focus of the North Koreans in their relentless and at times brutal interrogations of the crewmen they identified as most knowledgeable:

> Selected qualified cryptographic technicians of the USS PUEBLO were extensively interrogated by special and apparently highly competent North Korean experts on cryptographic principles, operating procedures, and the relationship of keying materials. The materials were neither displayed to the crew nor used for propaganda purposes. This is not the case for a variety of other significant intelligence materials. It appears that the North Koreans thoroughly understood the significance of these materials and the concealment of the details of the acquisition from the United States.

But, Tordella confidently continued, "All U.S. machines have been specifically designed to withstand attacks" based on recovering an intact machine or detailed information on how it works: only if an adversary also has the specific instructions for setting the machine for each key period would it be possible to actually break and read the traffic. Tordella would have been far less sanguine had he known that a few months before, a U.S. Navy communications technician named John Walker, serving aboard a nuclear submarine, had walked into a Soviet embassy abroad to offer his services, or that the North Koreans had swiftly passed on to the USSR the fruits of the *Pueblo* capture. For the next eighteen years the Soviets would have both the knowledge of how the KL-47 worked and a regular supply of the key lists needed to decode every month's traffic.[50]

. . .

The U.S. Navy reacted to the first capture of an American commissioned vessel on the high seas since the War of 1812 by cutting its losses. All of the mistakes that had led to the loss of the *Pueblo* and its classified cargo were eminently correctable: there was no excuse whatever for a ship engaged in the *Pueblo*'s secret activities even to be carrying hundreds of extremely classified documents; her commander's failure ever to conduct an emergency destruction drill was incomprehensible; the lack of a contingency plan for air support or a destroyer escort to come to the aid of a ship sent into such a dangerous spot was criminal.

And the concept of seaborne collection platforms was still a good one: it offered the best chance to carry out the close-in monitoring that was increasingly necessary given changes in radio frequencies taking place throughout the world and the growing limitations and difficulties of operating fixed land-based intercept sites. But in fact the Navy had never really been more than halfhearted about the mission, as was evident in its haphazard approach to training the ships' crews and assigning officers. NSA had overall responsibility for "technical direction" of the SIGINT operations but delegated that authority to the Naval Security Group, and in another of those bizarre bureaucratic compromises that do nothing but sow confusion, NSA and the Navy agreed to alternate authority for "tasking" the ships from one mission to the next. The astonishingly poor training of the *Pueblo*'s SOD hut detachment and the rather contemptuous attitude of the regular Navy captain of the ship toward its "mysterious" doings reflected both the fact that no one was really in charge and that the Navy considered SIGINT missions on sardine-can ships a rather annoying distraction to its much more glamorous job of operating aircraft carriers and nuclear submarines. William O. Baker, who was serving on the President's Foreign Intelligence Advisory Board (PFIAB), did not hesitate to blame the fiasco on the long-running inability of NSA to assert its statutory authority over the SIGINT operations of the military services. "It appears there was no real evaluation of the mission, certainly no suitable preparation for it, and of course no coordination with the experts and responsible managers in the field," Baker wrote presidential adviser Clark Clifford, who chaired the PFIAB. "As we have recorded on many prior occasions . . .

there will be a sequence of such episodes (prior ones have been similar aircraft ventures) uncoordinated and unjustified by authentic technical and operational requirements of SIGINT, until appropriate central leadership and authority are asserted."[51]

Rather than try to find and correct the mistakes, the Navy simply pulled the plug on the whole shipboard collection program. The secretary of the Navy, declaring that the *Pueblo*'s crew had "suffered enough," cut short an inquiry that sought to hold the ship's officers or higher commanders accountable; facing budget cuts ordered by the new Republican administration, he then proposed eliminating the intelligence ships altogether rather than reduce the Navy's combatant fleet. By the end of 1969 the last of the seaborne collection platforms was deactivated.[52]

The cutback was part of a larger retrenchment that all the services were facing owing to competing demands for money and personnel to fight the Vietnam War. NSA still depended heavily on the service cryptologic agencies to operate its field collection sites around the world, but the proliferation of microwave telephone links, radars, telemetry, VHF and UHF tactical voice, and radio teleprinter traffic to be covered was running headlong into the increasing difficulties the military services were facing in training and retaining capable intercept operators.

In 1965, NSA decided to try manning an entire foreign intercept site with its own civilian employees, many of them former military enlisted men who had previously served with one of the service cryptologic agencies. In July 1966 an advance party from Fort Meade arrived to take over the U.S. Air Force station at Harrogate in northern England.[53] Harrogate was a faded Victorian spa town not without its charms but abundantly steeped in the shabby postwar austerity that still hung over the country. Most of the new arrivals ended up in the inaptly named Grand Hotel, a decaying relic from the turn of the century. Its halls reeked of leaking gas and its creaky elevator regularly trapped newcomers who had not yet learned to take the stairs. One Sunday the dining room was hastily evacuated after a twenty-square-foot chunk of plaster came crashing down from the ceiling, missing by a few seconds a family that had started to sit down at the table directly underneath, then chose a nearby table instead.[54]

To update the dimly lit and badly subdivided operations building on base, thirty volunteers from the NSA team, headed by its chief, Hugh

Erskine, spent one Friday night wielding sledgehammers to knock out 130 feet of reinforced concrete and cinderblock walls and 3,000 square feet of ceiling. For security reasons the station staff was required to clean the floors itself, so another do-it-yourself-minded NSA civilian came up with a plan to install a central vacuum system, and with some helpers drilled three-inch holes all over the floor, connected by a network of pipes running below. Only when the vacuum unit arrived after months of delay did they discover that there was not sufficient electric power to run the motor. By then an enterprising golfer on the staff had fashioned little numbered pennants for each hole to create a putting course.[55]

Despite the initial obstacles, the "civilianization" of Harrogate worked well. The larger problem, however, was the cost and political vulnerability of NSA's sprawling global collection network, which by 1970 had ninety-one fixed field sites operating in seventeen foreign countries and the United States; in West Germany alone, the services and NSA operated fourteen different facilities. The HF sites were among the most manpower-intensive and expensive to operate; they were also the oldest, and their basic receiving equipment was still of World War II vintage and becoming expensive and difficult to maintain, with vacuum tube receivers that could no longer be accurately calibrated, manual scanners, and rhombic antenna fields that were a generation behind the circular arrays used at the newer elephant-cage installations.[56]

A 1968 study recommended consolidating and closing many sites on efficiency grounds, but also warned that rising nationalism in the Third World meant that the days were numbered for many of the bases NSA had long counted on. Within two years, U.S. intercept sites in Morocco and Pakistan were pulled out in the face of a wave of anti-American feeling; over the following decade NSA would lose its bases in Ethiopia, Vietnam, Iran, Taiwan, Thailand, and the Philippines as well.[57] The cumulative toll that fighting the Cold War had taken on American credibility and standing in the world was for the first time imposing substantial limits on the covert means that from the start had been the chief U.S. way of countering the Soviet threat. The loss of trust in government at home, to which the Vietnam War added immeasurably, was about to intrude even more dramatically on the free hand NSA had always been able to count on to conduct its intelligence operations, free from public scrutiny or even political restraint.

10

Brute Force and Legerdemain

Though NSA may not have equaled General Motors (695,000 employees), AT&T (773,000), ITT (392,000), or America's other largest employers in the year 1970, with a total workforce of 93,067 and a budget of more than $1 billion the agency was quite large enough to be suffering from all the recognized signs of organizational sclerosis. By its third decade of existence NSA had developed its own insular culture and accustomed ways of doing things and a vast multilayered bureaucracy made all the more resistant to change by its hermetically sealed secrecy. In the name of security, the agency had obtained exemptions from many government contracting rules and civil service hiring procedures, but that did not make NSA's business and personnel offices more efficient or streamlined; it merely freed the most bureaucratic spies in the world to generate endless streams of their own red tape, unchecked by any outside scrutiny. An internal investigation a few years later found that it took NSA's supply office ten days to deliver a box of paper clips to someone who had ordered one, which sounded like a joke but was not; it took thirteen days to send a letter from the time it was written, due to all the reviews and approvals required as it made its way through multiple layers of management; it took six months to issue a purchase order after the agency had received bids from all interested contractors.[1]

The NSA in-house newsletter of the 1950s, which read like a church bulletin or a Rotary Club events calendar, was superseded in 1974 by a staff-written Top Secret publication called *Cryptolog*. Visually it resembled a high school newspaper, with amateurish line drawings and clip art and a cacophony of typefaces, and its pages were filled with poems, recipes for spanakopita, humorous articles, and word games.

But what was most striking were all of the insider allusions to the various cliques of the SIGINT world, each with their own fixed and dug-in attitudes, and to the manifest inability of management to make decisions, implement new technologies, or even get the 19,290 workers at NSA headquarters to talk to one another very much. Along with lots of cybernetic, linguistic, and cryptanalytic geekiness on display were endless careerist laments ("Let's Give Linguists a Bigger Piece of Pie!" "Let's Not Lose Our TA Skills"); a series of almost desperate pleas to readers from the agency's reference and support branches to make use of their services ("Have you ever visited the NSA Cryptologic Collection? Have you ever *heard* of it?" "Next time, instead of 15 phone calls, make just one—to the Central Reference Service, x3258s. Central Information maintains an extensive collection of specialized dictionaries, reference books, journals, and documents. We have people trained in the research techniques necessary to make effective use of these sources"); and accounts of hilariously bungled attempts by management to "improve" working conditions. An article titled "The 2,000-Year-Old Transcriber" offered an NSA insider's version of the Mel Brooks "2,000 Year Old Man" routine, describing the primitive tape recorders and painfully uncomfortable headphones they struggled with in the early years (not designed for repeated replaying of passages, the rewind function on the machines would almost always give out, forcing the transcribers to use a pencil stuck into the spindle to manually turn the tape back), and then the agency's elaborately mismanaged program to develop custom-made "Transcribers Consoles" specifically tailored to NSA's needs. They "looked like somebody's idea of a Cape Canaveral space console," but, as the stereotypical product of a committee, got almost everything wrong, with insufficient drawer space, tape recorders placed where they could not be reached while sitting, and metallic hollow spaces that resonated like a struck bell at every clack of the foot pedals or keystroke of the typewriter.[2]

More serious and revealing was a long and thoughtful series of articles by "Anne Exinterne" about the "unhealthy" atmosphere resulting from the large number of new hires who had been "wooed" to make NSA their career with early promotions; by the time they realized that they were in fact ill-suited to work that required not just the language or math aptitude that led to their being recruited but also an extraordinary tolerance for sheer drudgery, such as listening to tapes eight hours a day, many found themselves trapped by a high salary that they could not equal on the outside and "seven blank years on a resume form" that did little to interest a prospective employer in any case. With scant opportunity for recognition in their field and compartmentalization that often prevented employees from having a sense of how their work even served the agency's mission, "morale is likely to plunge" and promotions—the only tangible measure of accomplishment in this closed world—"can become an obsession." A remarkable number of the bright new college graduates accepted into the agency's fast-track intern program, initiated in 1965 to "assure that high quality pre-professionals" were given a broad training in the agency's mission, ended up asking to be transferred to the personnel office within a few years after becoming "disenchanted with [their] assigned field."[3]

Louis Tordella, who retired in 1974 after fifteen years as deputy director, was widely admired within the ranks for his technical competence and for bringing much-needed stability to the organization. But the flip side of stability was the near-total power he exercised, a situation that NSA's extreme secrecy abetted. Even before officially assuming the deputy directorship in 1958, Tordella was already regarded as the most powerful person at NSA: his nickname was "Dr. No" for the decision-making authority he wielded to approve—or more often disapprove—new projects. "Everyone was kind of afraid of him," recalled one long-term NSA official. Throughout his tenure, Tordella was the only executive aware of all of NSA's most important programs and the single official who managed the relationship with CIA; his detractors within the agency found him an "ultrabureaucratic conspiratorialist" who kept information to himself and often used it to deflect outside criticism of, and pressure on, the agency.[4]

(Underscoring his unique position, a week after Tordella's death in 1996, NSA historian Tom Johnson was delivered sixteen boxes of material from a safe that NSA employees found in Tordella's office as they were cleaning it out. Tordella had kept the office while he continued to work as a consultant to the agency after his retirement, and the documents turned out to be a compendium of every single one of NSA's most highly classified, compartmented programs of the post–World War II era. "Everyone knew he must have some such safe, but no one knew where it was," Johnson remarked.)[5]

The President's Foreign Intelligence Advisory Board had warned as early as the mid-1960s of NSA's tendency to become "ingrown and defensive"; a more comprehensive review by the consulting firm Arthur D. Little in the late 1970s found that NSA's management had become "paranoid," "untrustworthy," and "uncooperative," divided into fiefdoms under the control of entrenched senior managers each defending their own institutional interests rather than responding to the mission of the agency as a whole.[6]

The revolving door of military flag officers serving as director for a four-year term certainly had its own problems; all were outsiders to the agency, and the kind of background, personality, and temperament that it took to make it to three-star general or admiral did not always have much to do with the technical, management, and Washington political skills required to run an exceedingly complex organization like NSA. Tordella's firm grip of the technical management of the agency was both a buffer against martinets like Admiral Frost, who tried to run NSA like a carrier battle group, and a welcome relief to technically unsavvy officeholders like Noel Gayler, a vice admiral who served as director from 1969 to 1972 and often seemed frankly bewildered why he had been selected for the job and was happy to leave the running of the place to his civilian deputies. But either way it was a fundamentally dysfunctional situation. The PFIAB meanwhile complained repeatedly about the organizational "anarchy" arising out of the divided loyalties of NSA's military directors, who were often reluctant to challenge the wishes of the Joint Chiefs, notably when it came to asserting NSA control of the ELINT mission. Despite a series of clear directives giving NSA the authority to assume command of the fragmented and highly duplicative ELINT programs of the services, no NSA director ever seemed willing to rock that boat.[7]

Of course, it was hardly a new phenomenon in the annals of government to find a bureaucracy run by the bureaucrats rather than by the political appointee nominally in charge, but NSA's growing inside political influence and privileged relationship with the White House further insulated the agency from even the normal checks of management accountability and government oversight. NSA had always shrugged off efforts to make it more accountable to its governing board or the director of central intelligence, who by statute had responsibility for coordinating the work of all government intelligence agencies, including setting their budgets. A 1958 directive by the National Security Council, approved by President Eisenhower over NSA's objections, had abolished the U.S. Communications Intelligence Board, declared NSA to be a part of the intelligence community, and placed it under a new, unified U.S. Intelligence Board (USIB), headed by the CIA director. But that had had no discernible effect on NSA's increasing independence.[8] A blistering CIA memo in the mid-1970s on the "CIA/NSA Relationship" pointed out just how much its rival agency had been able to become answerable to no one but itself as a result of its favored status and a budget that had "doubled, tripled and quadrupled":

> An organization which began with a serious inferiority complex gradually developed a feeling that it has "a corner on the market" in terms of intelligence information fit to print.
>
> This new feeling of importance by NSA manifested itself in various ways such as the installation of a direct communications link over CIA objections between Ft. Meade and the White House and the issuance of the SIGINT Summary, a SIGINT current intelligence publication designed to compete with the then Central Intelligence Bulletin. CIA also objected, to no avail, to the SIGINT Summary because it contained then as now gists and summaries of what NSA analysts consider to be "hot" items of information which were in the process of being published in individual translation or report form, but for which NSA wanted to get credit in the eyes of top level intelligence recipients. . . . NSA reps let it be known in numerous ways that there was little or no need for "middlemen" such as CIA, DIA, etc. to chew, digest and regurgitate perfectly good SIGINT data and provide it to the real intelligence consumers such as the President, the Secretary of State and the NSC Staff. . . .
>
> The increasingly aggressive, determined and sometimes overbearing policy on NSA's part . . . have resulted almost by default in the emergence of NSA in a Community role in which the tail too often wags the dog.[9]

Hoarding information was also a venerable element of the bureaucratic power game, but NSA could trump even CIA in that contest, given the special handling that SIGINT had always been accorded, and increasingly NSA withheld details in its reports that it considered "technical information" on sources, intercept locations, and cryptographic systems. A later CIA director, Admiral Stansfield Turner, who headed that agency from 1977 to 1981, complained in his memoirs that aside from dangerously skewing the intelligence picture by overemphasizing SIGINT at the cost of other sources, NSA was mainly trying to "get credit for the scoop. . . . There is a fine line here, but there is no question in my mind that the NSA regularly and deliberately draws that line to make itself look good rather than to protect secrets."[10]

The chief of the Situation Room when Nixon entered the White House was David McManis, the first time an NSA official had held the position, and a majority of the intelligence coming in was from the SIGINT system. Continuing where Lyndon Johnson left off, Nixon's national security adviser, Henry Kissinger, ordered that any intercepts mentioning the president or himself be delivered to him and to no one else in government: all were to be marked NODIS—Not for Distribution. Gayler and Tordella were only too happy to oblige, seeing a further opportunity to bolster NSA's standing.[11]

More cravenly, or venally, the two top NSA officials also backed an extremely secret, and undoubtedly illegal, plan drawn up by the Nixon White House to authorize a vastly expanded monitoring of Nixon's domestic critics by the intelligence agencies. The Huston plan, as it was known, was eventually withdrawn in the face of strenuous opposition by FBI director J. Edgar Hoover (who had no moral objections to spying on left-wing and antiwar groups, but insisted that it was the FBI's job). But even after the dropping of the Huston plan, NSA obligingly added hundreds of names submitted by Nixon administration officials to its secret "watch list" of domestic surveillance targets.[12]

Nixon and Kissinger's efforts to turn the intelligence agencies into a tool of the White House greatly exacerbated the dysfunctional relationships throughout the intelligence system, forcing military and diplomatic officials to engage in palace intrigues of their own just to maintain their routine access to NSA intelligence reports. Kissinger tried to cut out Secretary of Defense Melvin Laird and Secretary of State William P.

Rogers from receiving anything but the most "innocuous" intercepts, as Laird's military assistant described the material that arrived in the secretary's office via official channels. Laird shrewdly countered by ensuring his own back channel. As he later told the journalist Seymour Hersh, he called Vice Admiral Gayler and the Army three-star officer he named to direct the Defense Intelligence Agency, Lieutenant General Donald V. Bennett, into his office: "I told them they'd better be loyal to me. If they were, they'd get four stars after four years. And goddamn it, they were loyal." Laird often met privately with Gayler two or three times a week to find out what NSA was giving the White House. Both Gayler and Bennett got their fourth stars.[13]

NSA's extremely cozy relationship with the Nixon White House and the end runs that it forced the secretary of defense and others to make to maintain the normal flow of intelligence was the result of NSA's own assiduous efforts to promote its standing in the corridors of Washington power. "It was not good for SIGINT," acknowledged an NSA internal history, "and it was deadly for the presidency." But while it lasted it remained extremely good for NSA.[14]

For a while it looked as if NSA might even weather the post-Watergate storms that ripped away the covers which had long kept American intelligence operations out of public sight and free from constitutional scrutiny. "The CIA had always operated under minimal Congressional oversight," observed John Lewis Gaddis. "The assumption had been that the nation's representatives neither needed nor wanted to know what the Agency was doing. That attitude had survived the U-2 and Bay of Pigs incidents, the onset and escalation of the Vietnam War . . . but it did not survive Watergate."[15]

The revelation that Nixon had ordered CIA to tap phones and open mail of "left-wing radicals" and antiwar protestors and that former CIA officers were part of the White House "Plumbers Unit," which undertook a series of burglaries, surveillance operations, and wiretaps against domestic "enemies" on Nixon's direct orders, brought CIA under unprecedented scrutiny in the press, and then in Congress, after the Plumbers were caught breaking into Democratic National Committee headquarters in the Watergate office building in Washington during the 1972 presidential election campaign and the whole story started to unravel. By law CIA, like NSA, was restricted to foreign intelligence activities and could not legally operate within the United States. Nixon had swept those concerns aside (insisting to interviewer David Frost even after his resignation, "Well, when the president does it, that means it is not illegal"). In 1973, during the Watergate investigations that would lead to Nixon's resignation the following year in the face of a certain House vote for impeachment, the CIA's director of operations, William Colby, ordered a complete internal review of the agency's past illegal activities. Parts of the resulting 693-page report—known a bit flippantly within the agency as "the family jewels"—were leaked to Hersh, who wrote a long article in the *New York Times* in December 1974 revealing CIA spying on Nixon's domestic critics. In an effort to make a clean sweep, Colby, who became the agency's director in the ensuing shakeup, immediately confirmed the accuracy of the article and pledged full cooperation with Senate, House, and presidential commissions that were quickly appointed to conduct a thorough, and unprecedentedly public, investigation of U.S. intelligence operations.[16]

NSA had largely succeeded in keeping its fingerprints off the Nixon White House's more serious illegal operations; and the special status signals intelligence had long enjoyed as a secret within a secret in the U.S. intelligence system stymied the initial efforts by congressional investigators to learn much of anything about its activities, legal or otherwise. L. Britt Snider was a thirty-year-old lawyer for the Senate Select Committee on Intelligence, chaired by Democrat Frank Church of Idaho, and along with another staff member was given the job to tackle NSA. "They must have done *something*," the committee's staff director kept telling them. Snider began by asking the Congressional Research Service for everything in the public record referring to NSA. The researchers came up with a one-paragraph description of the agency from the *Government Organization Manual* and "a patently erroneous piece from *Rolling Stone* magazine."

Hoping to locate some insider whistle-blowers, Snider and his colleague, Peter Fenn, tracked down and went to see a handful of NSA retirees living in the Washington area; the most egregious abuses they had to report "were complaints about how NSA allocated its parking spaces among employees and a few cases of time and attendance fraud."

But it was clear from the interviews that they were up against another impenetrable barrier, the strict compartmentalization that limited the flow of information within NSA: even if NSA had been involved in improper activities, knowledge of the fact was sure to be restricted to a tiny circle.

Trying the direct approach, Snider and Fenn then decided to simply ask for a meeting with the director. That elicited NSA's trademark dazzle-the-natives treatment: VIP visitor parking spots at the front door, "broadly smiling handlers" waiting to whisk them to the director's top-floor suite, and an earnest promise from Lieutenant General Lew Allen Jr. himself, the agency's director since 1973, of full cooperation—but nothing, even after the weeks of personal briefings by NSA officials that followed, that identified "a single avenue that appeared promising from an investigative standpoint," Snider wrote.

Their first real break came in May 1975 when the committee received a copy of the CIA family jewels report, which on close reading contained two small references to NSA. One alluded to a CIA-supplied office in New York City used by NSA to copy telegrams. The other mentioned NSA's involvement in monitoring the communications of members of the antiwar movement. But it was only three months later, after another *New York Times* story appeared under the headline "National Security Agency Reported Eavesdropping on Most Private Cables" that NSA agreed to provide a full account to the congressional investigators of its long-running arrangement with the cable companies to copy every international telegram. Pressed by Snider on how the program began and who had approved it, the NSA briefer said the only person alive who knew the whole story was "Dr. Tordella," recently retired as deputy director.

Once he saw how much Snider already knew, Tordella proved surprisingly cooperative when Snider showed up at his home in suburban Maryland. A single manager, who reported directly to Tordella, was responsible for the program, code-named Shamrock. A courier would each day collect microfilm copies or punched paper tape of international telegrams that passed through the offices of the three major cable companies in New York, Washington, and San Francisco; in the early 1960s, when the companies switched to magnetic tape, NSA set up an office in New York to duplicate the tapes and keep a copy for itself. Tordella said the collection program "just ran on" ever since its beginnings in World War II "without a great deal of attention from anyone," and actually "wasn't producing very much of value."[17]

That was probably true enough, and it appeared that for the most part the only cables normally retained for further study were those that an initial automatic processing identified as having been sent by foreign embassies or that appeared to be enciphered. Still, the existence of such a "vacuum cleaner" surveillance program, as the press quickly called it, was a grave embarrassment, almost certainly a violation of the law, and NSA vociferously objected to the committee's exposing any further details about it even though Allen assured the investigators he had ordered the program ended that May. The Church Committee nonetheless released in November 1975 a full report on NSA's bulk cable collection practices. Given that the agency since the 1950s had been running a "New Shamrock" program that directly tapped teleprinter and other communication links of sixty to seventy foreign embassies to obtain the same information with far less trouble, the major damage done by the loss and exposure of the old Shamrock was to the agency's reputation.[18]

Far more troubling was the other program that Snider and Fenn stumbled on, code-named Minaret. What had begun in 1962 as a so-called watch list of Americans traveling to Cuba whom NSA was asked to monitor was expanded in the mid-1960s to include suspected narcotics traffickers, and then in 1967, on Lyndon Johnson's orders, members of the antiwar movement, who Johnson believed were being aided by foreign powers. Under the increasingly paranoid Nixon administration the watch list exploded, and NSA was soon up to its elbows spying directly for the White House on the communications of more than sixteen hundred American citizens who had done nothing more than arouse Nixon's or Kissinger's suspicions or dislike. Among them were the civil rights leaders Martin Luther King Jr. and Whitney Young, the heavyweight boxing champion and Vietnam War opponent Muhammad Ali, the *New York Times* columnist Tom Wicker and the humorist Art Buchwald, U.S. senators Frank Church and Howard Baker, and a few names of NSA's own choosing thrown in for good measure, including the author and historian David Kahn, whose groundbreaking 1967 book on the history of cryptology, *The Codebreakers*, had caused the agency considerable panic. (Lieutenant General Marshall "Pat" Carter, the NSA director at the time, seriously put forth a number of other suggestions to stop publication of the book or discredit Kahn, including breaking into his home to steal the manuscript, planting disparaging reviews, or hiring him for a government job so that he would be liable to criminal penalties under the Espionage Act if he revealed classified information. Although NSA officials and the USIB spent "innumerable hours" considering those ideas, none in the end were carried out, other than placing Kahn on the watch list.)[19]

As an NSA lawyer who later examined the program concluded, NSA staff responsible for distributing the Minaret reports were clearly aware that the entire operation was "disreputable if not outright illegal," and took extraordinary pains to conceal what they were doing. The reports bore no trace of which agency they had come from, contained none of the usual sequential serial numbers assigned to NSA reporting, and were disguised to look like agent reports rather than communications intercepts.

None of the names of Americans on the Minaret watch list came out during the investigations, however, and NSA got off rather easy compared to CIA, which had to endure a litany of sordid revelations about foreign assassinations, coup plots, drug experiments on unwitting subjects, and other illegal and immoral activities. General Allen made an unprecedented appearance as an NSA director testifying in public before Congress, explaining the origins of Minaret and the difficulties of drawing an absolute line between foreign intelligence and domestic law enforcement, and revealing that he had personally shut down the program two years earlier. He was widely praised for his performance, which clearly had done the agency more good than harm. Allen, a man of brilliant intellect—he had a PhD in physics, had done nuclear weapons research at Los Alamos in the 1950s, and would go on after his term as NSA director to become Air Force chief of staff and then director of the Caltech Jet Propulsion Laboratory—was also, in Snider's words, "a man of impeccable integrity," and he steered NSA through a storm that might have sunk it.

A 1977 Justice Department review concluded that while both Shamrock and Minaret involved criminal violations of wiretap laws, a prosecution was unlikely to succeed and recommended against taking legal action against NSA officials.* The ultimate problem was the "ill-defined power" NSA and CIA had been granted from their inception: "If the intelligence agencies possessed too much discretionary authority with too little accountability, that would seem to be a 35-year failing of Presidents and the Congress rather than the agencies or their personnel," the Justice Department concluded. Nonetheless, one of Allen's decisive actions following Tordella's retirement was to ensure that no NSA deputy director ever again wielded such untrammeled power.[20]

But nothing could heal the gaping distrust that now yawned where secrecy and plausible deniability once held fast. On October 25, 1978, President Jimmy Carter signed into law the Foreign Intelligence Surveillance Act, which for the first time imposed statutory limits on NSA's activities. The act required the approval of the attorney general and a new Foreign Intelligence Surveillance Court for NSA to conduct electronic surveillance against foreign targets within the United States or against U.S. citizens and permanent residents anywhere in the world; it set up a permanent system of congressional oversight of NSA activities; and it specifically outlawed the kinds of dragnet collection operations within the United States that the Shamrock program had entailed, regardless of whatever subsequent measures the agency made to minimize the reading of communications of U.S. persons that it snared in the process. The most salient passage made it a crime, punishable by fine and imprisonment, to engage in "electronic surveillance under color of law" except as specifically authorized by statute. It was the first recognition that for covert intelligence operations to be accepted as legitimate and justifiable in a democratic society, they also had to be legal.[21]

At the same time, however, the very act of bringing covert operations under a legal framework marked another downward step in the erosion of traditional American values: it was a formal acquiescence in the kind of moral ambiguity in foreign affairs that after decades of Cold War

---

*In a 1967 case, *Katz v. United States,* the Supreme Court reversed its 1928 precedent in *Olmstead* and ruled that private electronic communications were covered by the protections of the Fourth Amendment against unreasonable search and seizure and that a warrant was generally required for a domestic wiretap.

conflict had come to seem normal, no longer a temporary or deplorable necessity but enshrined in permanent and legally chartered institutions of government.

At the bottom of it all was, of course, the superpower nuclear standoff, which, based as it was on the constant threat of mass annihilation of entire civilian populations, had a way of making a mockery of any traditional notions of ethics and principle in the conduct of national policy: when the alternative was cataclysm, many compromises seemed acceptable, even desirable, by comparison. The policy of détente championed by Kissinger during the Nixon and Ford administrations aimed to reduce the risk of nuclear war by stabilizing the relationship between the United States and the Soviet Union, but in seeking to lock in the status quo of the superpower rivalry, détente also further froze in place the Cold War institutions that had grown up along with it. Mutually assured nuclear destruction, espionage and surveillance, and all of the other moral compromises of the Cold War once seen as aberrations were instead becoming parts of the world order.

Détente might reduce U.S.-Soviet tensions, but it was not therefore going to do anything about reducing the now multibillion-dollar-a-year global espionage enterprise. The efforts of each side to learn the negotiating positions of their adversary in talks that began in the mid-1970s to freeze or reduce nuclear arms and stabilize the political status quo in Europe if anything intensified the targeting of government communications links in Moscow and Washington by the rival spy agencies. The need for each side to verify compliance with arms treaties and reassure itself that in stepping away from the brink of nuclear war it had not left itself vulnerable to military surprises or technological breakthroughs in weapons systems by the other placed even more demands on intelligence.

In Washington, a thicket of antennas began sprouting on the rooftop of the Soviet embassy on 16th Street, just a few blocks from the White House. In August 1974, acting on a warning from NSA, Kissinger informed President Ford, "It is very probable that the Soviets are intercepting out-of-city telephone conversations of key Washington officials, since such calls are usually on radio links which can be intercepted with rather simple and commercially available equipment." A report "identified 10,000 leased government circuits terminating in the Washington area for which protection seemed prudent. About 4,000 of these circuits are now on microwaves and exploitable, and the remaining 6,000 are already on cable but must be tagged to see that they remain there."[22]

A Soviet defector, Arkady Shevchenko, subsequently confirmed that other well-situated facilities used by Soviet diplomats in the United States—an eighteen-story residential complex in Riverdale in the Bronx, located on one of the highest spots in New York City; a Long Island mansion on thirty-seven acres in Glen Cove that served as a recreational retreat; and another recreational facility on the Eastern Shore of Maryland—were used to eavesdrop on microwave towers that served not only the major long-distance telephone links to Washington and New York, but also Andrews Air Force Base, where the president's Air Force One was based, and Norfolk, Virginia, home of the U.S. Atlantic Fleet. "It is most unlikely that these sites were selected for any other reason than microwave interception," an AT&T analysis concluded. The Soviets' Washington-area intercepts of unencrypted phone calls by U.S. officials and defense contractors were, according to KGB defector Vasili Mitrokhin, Moscow's "most important source of intelligence on the foreign and defense policies of the Ford and Carter administrations."[23]

Kissinger lodged a protest about the Soviet antennas with Ambassador Dobrynin, but the fact was that the United States was over a barrel in pressing the matter. By January 1977 the White House national security adviser, Brent Scowcroft, was able to report that "government communications in the Washington area have been rerouted from microwave to cable, and government communications in New York and San Francisco are in the process of being moved to cable," but cautioned that implementing more robust protections—such as providing private U.S. telephone companies with NSA-developed technology to bulk-scramble all telephone calls—posed the classic signals intelligence dilemma: "The main problem, from a foreign intelligence perspective, in moving ahead with communications protection is that it may stimulate the Soviets to take even greater protective measures for their own telecommunications and thereby deny us a valuable and possibly irreplaceable source of information."[24]

A plan to jam the Soviets' listening posts was also considered, but rejected for the same reason. Aside from the questionable effectiveness and cost—NSA estimated an initial outlay of $1.8 million just to target the Soviet embassy, the Soviet School in Washington, and the San Francisco consulate—a National Security Council memo noted that "NSA has consistently opposed initiating jamming operations, because they believe the U.S. will be a net loser in a jamming war." The United States might have been more subtle than the Soviets in its use of embassies abroad as surveillance posts, but the fact was that since at least the mid-1960s the U.S. embassy in Moscow was regularly intercepting the conversations of top Soviet officials as they spoke on their car phones while being chauffeured about the Russian capital in their ZIL limousines. That operation, code-named Gamma Guppy, came to an abrupt end in 1971 after investigative reporter Jack Anderson revealed its existence in a column under the headline "CIA Eavesdrops on Kremlin Chiefs." According to a subsequent briefing by U.S. counterintelligence officials, the Soviets changed their car phone communications immediately afterward and a CIA agent who worked as a mechanic on the limousines "was never heard from again and presumed killed."[25]

But Gamma Guppy was just one of many similar embassy-based operations that exploited insecure Soviet communications throughout the latter decades of the Cold War. In a rare admission of a U.S. SIGINT success during this era, Vice Admiral John "Mike" McConnell told a seminar at Harvard's Kennedy School of Government shortly after retiring as NSA director in 1996, "In the mid-1970s, NSA had access to just about everything the Russian leadership said to themselves and about one another. . . . We knew [Soviet leader Leonid] Brezhnev's waist size, his headaches, his wife's problems, his kids' problems, his intentions on the Politburo with regard to positions, his opinion on American leadership, his attitude on negotiating positions." The information more than once allowed Kissinger to outmaneuver Soviet negotiators in the Strategic Arms Limitations Talks. "That's the sort of thing that pays NSA's wages for a year," a senior U.S. official who saw the intercepts told David Kahn.[26]

The reliance on "close access" methods to do what conventional cryptanalysis could not intensified the rivalry between NSA and CIA, both of which laid claim to operate embassy listening posts and more intrusive operations involving taps, bugs, and a wave of new devices that could pick up plaintext directly from electronic equipment such as fax machines, copiers, electric typewriters, and computer terminals.

The two agencies were "competing for targets, locations," said Admiral Bobby Inman, who became NSA director in July 1977; there was also the usual gamesmanship in which neither shared its results with the other until they had already extracted and reported what they could.[27]

CIA undeniably possessed the expertise in surreptitious entry, and its Staff D had a long history of carrying out break-ins in foreign embassies to plant bugs or copy cryptographic material, but Inman saw that not only was there considerable duplication between the two programs but also a pressing need to modernize and miniaturize electronic taps, an area where NSA's technological leadership had not been taken advantage of. CIA for its part saw NSA's moves to take over the program as part of its relentless empire building: "NSA keeps picking, nibbling and lobbying away at CIA SIGINT activities," a CIA memo complained, even though CIA's SIGINT programs "contribute (directly or indirectly) to about 40 percent of NSA's serialized reporting output with an Agency SIGINT budget about one-thirtieth the size of NSA's."[28]

Inman's predecessor, Lew Allen, had worked out a tentative agreement to improve cooperation in these "Special Collection" programs, but in practice every disagreement had to be referred to the CIA director for a decision, and Inman made it a priority upon assuming the directorship to settle the matter once and for all; with the implicit backing of the White House and the House Appropriations Committee, and based on the dictum that, in his words, "if you bring money, people will cooperate," Inman secured NSA control of funding for the entire program, after which a definitive deal was swiftly concluded between the two agencies. The CIA-NSA "peace treaty," as it was called, set up a new joint facility in College Park, Maryland, to oversee and process material from the Special Collection program, with the two agencies rotating the chairmanship every two years. "It led to modernization of that entire collection process," said Inman, "and yielded very high quality return." There were still a few details to iron out, but during the month and a half from mid-February though March 1981 when Inman was acting as both NSA director and CIA deputy director, he resolved the matter by sending memos back and forth to himself approving his solutions. ("Never had the two agencies worked so well together," he observed.)[29]

Up until the very end of the Cold War the U.S.-Soviet battle of the bugs never let up. A KGB officer under diplomatic cover at the Washington embassy who was a member of the academic Washington Operations Research Council noticed that the conference room used for the group's meetings was in the offices of System Planning Corporation, an Arlington, Virginia, company that did contract work for the Pentagon. In September 1980 he succeeded in slipping a bug under the table after one meeting. Over the next year, until the device's battery went dead, the KGB harvested a wealth of classified information from briefings that took place in the room, including reports on U.S. nuclear deployment plans in Europe, military modernization, and SALT II negotiating positions.[30]

In February 1984, tipped off by the discovery of an extremely sophisticated bug in a piece of electronic equipment at another country's Moscow embassy, NSA began quietly swapping out eleven tons of equipment from the U.S. embassy—typewriters, teleprinters, copiers, desktop computers, and cryptographic devices, literally anything that plugged into a wall socket. Back at Fort Meade, the agency's director of communications security, Walter Deeley, offered a $5,000 bonus to the member of his team who could find the first bug in the returned gear. After making detailed X-rays of an IBM Selectric typewriter, a technician spotted an otherwise innocuous metal bar that ran the length of the machine. It turned out to contain an ingenious magnetic detector that measured the movement of the two arms that rotated the typeface "golf ball" at each keystroke. That was sufficient to reveal what letter had been typed; the data was captured in a small electronic memory, then periodically radioed by a miniaturized VHF burst transmitter that sent a signal too brief to be detected by the standard radio spectrum analyzers employed in periodic antibug sweeps by embassy security teams. The bugs were eventually found in sixteen typewriters, and had been in place for as long as eight years. An earlier routine inspection of the typewriters in 1978 had failed to find the well-hidden devices.[31]

This was no doubt the tip of the iceberg of efforts on both sides to exploit the rich possibilities to read messages at their source, without the annoying complications of cryptanalysis, that were offered by the growing use of electronic office equipment everywhere. Although KGB defectors reported that the Soviets obtained considerable information about U.S. military aircraft and other weapons technology projects through their bugging and microwave intercept operations in the United States, particularly those targeting defense contractors on Long Island, and in California, the great irony of electronic surveillance operations as the Cold War entered its final chapter of U.S.-Soviet confrontation in the late 1970s and 1980s was that "the general effect of this intelligence was probably benign—to limit the natural predispositions of [the KGB] to conspiracy theories about American policy," in the words of Mitrokhin and the intelligence historian Christopher Andrew. In 1979, when a Washington political flap arose over accusations by several senators that the Soviets had moved a "combat brigade" to Cuba (in fact it was a unit that had been there ever since the end of the Cuban Missile Crisis in 1962), the KGB's "intercepts of Pentagon telephone discussions and other communications enabled the Washington residency to reassure Moscow that the United States had no plans for military intervention," according to Andrew and Mitrokhin.[32]

Back in 1945, William F. Clarke of GC&CS, contemplating the future of signals intelligence in the postwar world, somewhat idealistically proposed ("this is probably a counsel of perfection," he admitted) that the one step that "would contribute more to a permanent peace than any other" would be the abolishment of all code and cipher communications by international agreement.[33] Three decades later, having fought the code wars to a near draw, the Cold War superpowers were finding in that visionary notion a germ of truth.

At age forty-six the youngest director in NSA's history, Inman brought an energy and determination to take charge of the agency unseen since Canine's tenure as its first chief. Growing up in a small town in East Texas, the son of a gas station owner, Inman had graduated from high school at fifteen and from the University of Texas with a degree in history at nineteen; after joining the Naval Reserve during the Korean War, he had rocketed up through the ranks of naval intelligence despite never having attended the Naval Academy or held a seagoing command, two normally fatal failings when it came to being considered for promotion to admiral. With finely tuned political instincts he sought to raise the agency's profile in Washington, personally working his connections on Capitol Hill and in the White House and giving the first-ever on-the-record newspaper interviews by an NSA director, aided by a dazzling command of whatever subject on which he was holding forth. "Nearly everyone who knows him mentions a piercing intellect, honesty, unusual memory for details and prodigious capacity for work," reported the *Washington Post*'s Bob Woodward. Another observer of the Washington intelligence scene described the indefatigable intellectual energy he brought to the job: "If Inman had a hearing at nine o'clock in the morning, he'd be up at four prepping for it. He'd read the answers to maybe a hundred hypothetical questions. He'd essentially memorize the answers. Then he'd go before the committee and take whatever they threw at him, without referring to a note."[34]

Like Canine, the new director instituted a system of rotation for the agency's top managers. Unlike Canine, his aim was not to pursue some abstract management theory but to break up the clique of civilian czars who had grown accustomed to controlling the organization and to start edging out the World War II generation of cryptologists who still held most of the top positions. As Inman warned one of his successors, left to its own inclinations the NSA staff would always try to make sure the director never actually knew anything or changed anything: "They want to treat you like Pharaoh, to carry you around on a sedan chair and let you have the occasional lunch with a visiting foreign delegation, but to keep you away from anything else that goes on at NSA." He established a review committee to identify the most talented leaders at the GS-14 and GS-15 levels, and promptly assigned all eighty-one of them to new jobs; two years later he shifted them again.[35]

Shaking up NSA's business-as-usual management and cultivating a new generation of civilian leaders was one thing; making progress against the still-impenetrable high-level Soviet ciphers was another. But Inman said that "the most thoughtful analysis" of the problem he read while coming in as director was a study of the state of cryptanalysis done in 1976 by William Perry, a PhD mathematician who headed the Pentagon's R&D programs in the Carter administration and later served as secretary of defense during Bill Clinton's presidency. Earlier outside assessments had affirmed the 1958 Baker Panel's fundamental pessimism about the prospects of breaking any high-level Soviet systems through cryptanalytic means alone. Richard M. Bissell, a former CIA official asked to review NSA's programs in 1965, found that while advances had been made in the diagnosis and recovery of the one Soviet cipher machine that had yielded some results owing to the discovery of bust messages, it was still never more than a small percentage of traffic that could be read at any given time. "A massive intellectual effort was required over a number of years" even to get this far, Bissell noted, adding, "Above all it must be emphasized that the sample of intercepted traffic which turns out to be decipherable is determined by the incidence of [operator] carelessness and of machine malfunctions so that the selection is one over which we have no control." Given the continuing edge codemakers were gaining over codebreakers due to advances in technology, "timely intelligence from high-grade systems on the World War II scale would appear to be out of the question," Bissell concluded.[36]

But Perry's strikingly upbeat assessment eleven years later showed that a fundamental change had occurred in the interim. The key finding of the Perry Committee was that the theoretical-mathematical advances in cryptanalysis that the Baker Panel and others had kept urging NSA to fund were at last on the verge of paying off:

> During World War II, the U.S. and the UK achieved spectacular success in cryptanalysis which had a profound impact on the execution of the war. We stand today on the threshold of a cryptanalytic success of comparable magnitude. . . . No one can guarantee that we will "break" any specific machine of the new generation, but we do not see the problem as being more difficult—relatively speaking—than the one posed . . . thirty-seven years ago by ENIGMA.[37]

NSA had somewhat halfheartedly agreed back in 1959 to set up an outside think tank, as the Baker Panel had recommended, that would bring in top mathematicians from the academic world for whom a regular career within the agency had always been an unattractive proposition. It was a far cry from the Los Alamos–scale project that had been proposed, and for a number of years—just as NSA had stonewalled other outside groups of experts such as SCAMP—the agency refused to send any real problems to them.[38] But the institute, housed at Princeton University under the aegis of an existing military-academic cooperative research organization called the Institute for Defense Analyses (IDA), eventually carried out some particularly groundbreaking work integrating supercomputers into cryptanalysis, having obtained one of the very first all-solid-state supercomputers, the CDC 1604.

In 1976, NSA retired the IBM Harvest system; the kindest thing anyone said about it was that it had done a "journeyman's" job on the standard, tried-and-true massive data runs that were the mainstays of business-as-usual cryptanalysis. Others in the agency more frankly called Harvest a "white elephant" that never lived up to its promise and huge expense. One of the major recommendations of the Perry study was that NSA purchase one of Seymour Cray's revolutionary supercomputers to assist its theoretical, long-term research in cryptanalysis.[39] At a cost of $10 million, the Cray-1 was the fastest computer in the world, and would remain so for the next six years. Forty years later a hundred million iPhone owners would be walking around with the equivalent of a Cray-1 in their pockets, but in 1976 it was ten times faster than any other computer in the world, with an innovative "vector" architecture that represented a huge leap forward in cryptanalytic power in the seesawing race of codemakers versus codebreakers. Though standard histories of Cray Research would persist for decades in stating that the company's first customer was Los Alamos National Laboratory, in fact it was NSA, which, continuing its support of the forefront of computer innovation that began with the founding of ERA in 1946, saw the new Cray-1 as the best hope for achieving its long-sought breakthrough.

In 1979, the confluence of new techniques developed by the mathematicians at the IDA "Communications Research Division" at Princeton and the Cray-1's greatly accelerated computing power produced "the height of American cryptologic Cold War success," in the words of NSA historian Thomas Johnson.[40] The cryptanalytic breakthrough against Soviet ciphers, which involved a fundamental enough mathematical result that NSA still refused to declassify any details more than three and a half decades later—likely because it also had implications for NSA's ability to break digital encryption methods that would assume enduring importance in the Internet era—led to a flow of signals intelligence during the Soviet invasion of Afghanistan, which began on Christmas Day 1979.

It was a triumph; it was also a last hurrah for the golden age of codebreaking that had begun before World War II with William Friedman's founding of the Army Signal Intelligence Service in 1921. The shift to reliance on implanted electronic bugs, direction finding and ELINT, and the interception of microwave and satellite communication links was changing forever the nature of the game. By the 1970s the dwindling importance of conventional communication channels led NSA to shut down its last two major radio intercept posts in the United States, both dating from World War II—the Navy station at Two Rock Ranch in Petaluma, California, and the Army's Vint Hill Station, outside Warrenton, Virginia. Inman pushed to accelerate the development of remotely operated, computer-based intercept receivers to take over the chore of scanning for and collecting the dwindling traffic that still passed on HF manual Morse and radio teleprinter circuits, while upgrading systems for collecting the growing streams of "bauded" signals transmitted by computer modems.[41]

U.S. Navy ELINT satellites launched starting in the fall of 1976 were soon to supplant the need for the huge direction-finding bases; by continuously following the radar signals emitted by Soviet, Chinese, and other warships, they made it possible to track from space, in real time, the movements of individual naval units through their entire deployments, anywhere in the world.[42] The subsequent mass migration of communications to the flow of computer data over uninterceptable fiber-optic cables in the 1980s was only the culmination of a trend long in the making, in which mathematics and the interception and decipherment of encrypted communications en route to their recipients was giving way to electronic and computer engineering and the infiltration of targeted computers and terminals at their source as the mainstay of the entire signals intelligence enterprise.

The Soviets had their last hurrah, too, appropriately involving the spies they had always excelled at recruiting and exploiting right in the very heart of the U.S. and British intelligence agencies. On January 14, 1980, an FBI wiretap on the Soviet embassy in Washington recorded this telephone conversation:

First person: May I know who is calling?
Caller: I would like not to use my name if it's all right for the moment.

First person: Hold on please. Sir?
Caller: Yes, um.
First person: Hold the line, please.
Caller: All right.
Second person: Ah, Vladimir Sorokin speaking. My name is Vladimir.
Caller: Vladimir. Yes. Ah, I have, ah, I don't like to talk on the telephone.
Sorokin: I see.
Caller: Ah, I have something I would like to discuss with you I think that would be very interesting to you.
Sorokin: Uh-huh, uh-huh.
Caller: Is there any way to do so, in, ah, confidence or in privacy?
Sorokin: I see.
Caller: I come from—I, I, I am in, with the United States government.
Sorokin: Ah, huh, United States government. Maybe you can visit.

Making a last-minute change in his arranged meeting time, the unidentified caller just managed to duck an FBI surveillance team outside the embassy the following afternoon. All the FBI got was a photograph of his back.[43]

Five years later a KGB defector named Vitali Yurchenko, who had been the KGB's duty officer at the embassy the day the caller had come in for that first meeting, identified him: he was Ronald Pelton, an NSA cryptanalyst and Russian linguist who had worked on the agency's most sensitive collection projects. In April 1979 he had filed for bankruptcy and three months later resigned from NSA.

Three days after learning of Yurchenko's identification, the FBI tracked Pelton down: he was working as a boat and RV salesman in Annapolis, Maryland, and living with a new girlfriend in an apartment in downtown Washington, where they spent most of their time drinking and taking drugs. Fearing Pelton would flee the country, the FBI at one point had two hundred agents detailed to keeping him under twenty-four-hour surveillance. With no evidence to prove he had engaged in espionage, the FBI agent in charge of the case decided to confront him directly in the hopes of getting a confession. The gambit paid off. After listening to the recording of the telephone call to the Soviet embassy, Pelton offered a full account of his work for the Soviets. In exchange for $35,000 (he had asked for $400,000), he had arranged to meet with KGB officials at the Soviet embassy in Vienna on two occasions, submitting to lengthy interrogations. He told them about A Group's success in breaking Soviet cipher machines, U.S. SIGINT satellites that targeted microwave telephone links throughout the Soviet Union, the U.S. embassy listening post, and an extremely secret Navy-NSA project that had deployed a submarine to install a tap on an undersea cable used by the Soviet Pacific Fleet's headquarters in Vladivostok for its operational communications. The Soviets responded in 1981 by making an across-the-board change in their military encryption systems, bombarding the U.S. embassy with microwave jamming signals, and dispatching a salvage vessel to retrieve the cable tap from the floor of the Sea of Okhotsk.[44]

During Pelton's trial in 1986, President Ronald Reagan personally called the publisher of the *Washington Post,* Katharine Graham, to urge the newspaper not to run a planned story about the cable tapping operation, named Ivy Bells, even threatening prosecution under Section 798 of the Espionage Act. *Post* editor Benjamin Bradlee finally agreed to delete some details about Ivy Bells from the story, even though he concluded that "the Russians already know what we kept out."[45]

Pelton, who did more damage to NSA's work than any spy since William Weisband, was sentenced to three consecutive life sentences. The arrest of another Soviet agent during that "year of the spy," as 1985 came to be known, offered another valedictory chapter to the conventional cryptologic battle in which the United States and the Soviets had been locked for four decades. John Walker's long run as the linchpin of a spy ring that had provided the Soviets the monthly key lists of the KL-47 cipher machine abruptly came to an end when his very drunk but very vengeful ex-wife phoned the FBI and turned him in. For sixteen years, Walker and a fellow Navy cryptotechnician he recruited after his retirement from the Navy in 1976, Jerry Whitworth, made regular copies of the lists just before destroying them in accordance with routine security procedures. Walker would leave them at dead drops around the Washington area for collection by his KGB handlers. The Soviets even supplied him with an ingenious custom-made "rotor reader," as the FBI termed it, a pocket-sized battery-powered device to recover the internal wiring patterns of new rotors for the KL-47: placing the rotor into its circular slot and turning a contact through each sequential position would cause a lamp on a miniature readout board to light showing which contact on the output side corresponded with each on the input side.[46]

The value of Walker's material was abundantly clear both in the considerable sums the Soviets paid—unlike the usual KGB chicken feed, Walker received more than $1 million, Whitworth $400,000—and the extraordinary lengths their handlers took in arranging countersurveillance measures for the drops. Convicted in 1986 of espionage and tax fraud, Whitworth was given a 365-year prison sentence; Walker died in prison in 2014, a year before he would have become eligible for parole, by then an aging relic of a conflict that for the rest of the world had ended twenty-five years earlier in one of the most astonishing, and peaceful, collapses of a totalitarian system in the history of nations.[47]

EPILOGUE

The Collapse of the Wall, and a Verdict

A number of cracks had exposed the fragility of the Soviet hold over its Eastern European empire, not least the courageous challenge to Communist rule by Poland's Solidarity labor movement that began in August 1980. But most of all was the fact that a regime whose authority had always rested upon ideological certainty, the ruthless use of military force, and an all-pervading atmosphere of fear and secrecy was trapped by its own past. When Mikhail Gorbachev was chosen general secretary of the Soviet Communist Party in 1985, becoming the first reform-minded Soviet leader since Khrushchev and the first in decades who was not geriatric, ailing, or (like his immediate predecessor, Konstantin Chernenko) literally senile, his efforts to promote openness and a restructuring of the social and economic system only laid bare how little there was holding up the entire house of cards. In Hungary in 1956 and Czechoslovakia in 1968, Soviet leaders had sent tanks to end popular protests, but once it became clear that Gorbachev would not intervene to suppress the stirring "counterrevolutionary unrest" in Eastern Europe, all the Soviets had left was bluff, and that could not last forever; any small shove could bring the whole edifice down.

It came on November 9, 1989, when, facing weeks of growing antigovernment street protests, the hard-line East German Communist regime desperately tried to save itself with a concession to the protestors, announcing an easing of travel restrictions to the West. Within minutes, crowds of East Berliners surged to the Berlin Wall checkpoints; the guards, caught by surprise and lacking instructions, finally just opened the gates and let them through. In the end it was the spontaneous action of ordinary people, not the calculations of think-tank strategists, intelligence analysts, and Pentagon contingency planners, that brought the ceremonial—and real—end of the Cold War. East and West Berliners surged atop the wall that had symbolized the front line of the conflict, bearing hammers and chisels, and began knocking it to pieces while East German troops stood by and watched.[1] By the end of the year the Marxist-Leninist regimes in Hungary, Bulgaria, Czechoslovakia, and Romania were ousted, joining Poland's earlier escape from the Communist orbit. It was inevitable that the wave Gorbachev had unleashed would not spare the Soviet Union itself, which disintegrated two years later in a dissolution that brought an end to the one-party rule Lenin had so ruthlessly secured for his Bolsheviks seven decades earlier.

Ronald Reagan had given it a decisive shove, too, refusing to accept the Cold War as a permanent condition the world had to live with and deploying his theatrical skills effectively in what was after all a war of global theater and public perception as much as one of military force. But it was in the broadest sense a victory of George Kennan's long policy of containment, the patient belief that if the Soviets could be held in check and if the world did not blow itself up in the meanwhile, the weight of all the contradictions of Soviet power would cause it to fall in on itself. William Odom, NSA's director during the last three years of the Reagan administration, wryly suggested to a reporter that the U.S. intelligence agencies hold a ceremony at CIA headquarters, declare victory, run down the flag from the flagpole out front, and dismantle the entire intelligence system and use the opportunity to start over and do it right this time. It was a sarcastic comment on the crazy jury-rigged American intelligence structure, with its perpetual internal bureaucratic warfare, tangled lines of authority, and wasteful inefficiency and duplication, but it probably also was an admission that even in helping to attain the victory of containment over Soviet Communism the intelligence agencies had often failed spectacularly at crucial moments, and had left in their wake an often sordid trail of transgressions against law, morality, decency, and basic American values.

NSA's most important Cold War achievement was its least visible contribution, but one that undergirded the entire American ability to hold off Soviet military power, and that was its ability to offer the minute-by-minute assurance that no Soviet tank regiment or warship could move and no nuclear-armed bomber or missile could take off without the president knowing about it. Before the deployment in the 1970s and 1980s of the first operational system of space-based infrared sensors to detect missile launches in real time, NSA's global signals intelligence network was the mainstay of tactical warning. The reassurance that the U.S. nuclear force could not be caught in a surprise attack and destroyed on the ground greatly eased the nightmarish need to contemplate pushing the button first in a crisis. The equal reassurance NSA was able to provide the White House during the Suez and Cuban crises that the Soviets were backing away from their saber-rattling threats was of incalculable importance in averting a runaway escalation that was the greatest fear throughout the Cold War confrontation, the danger that even a minor flashpoint could accelerate through miscalculation into an exchange of thermonuclear weapons in which millions would die before it could be brought under control, if it even could once it began.

The other enduring triumph of U.S. signals intelligence in the Cold War was the technical systems NSA worked out, after much bureaucratic delay and confusion, to supply enemy radar tracking data lifted from intercepted signals directly to U.S. fighter aircraft; in the skies over Korea and Vietnam these systems repeatedly proved their worth by extending the warning time American pilots had of approaching MiGs and hugely increasing U.S.-to-enemy kill ratios. This was the genesis of a more comprehensive system of direct support to U.S. troops by NSA's SIGINT operations centers that would reach its maturity in the wars in Iraq and Afghanistan.

NSA's worst failures in the Cold War were the result of its becoming the victim of its own success, and the resulting overreliance on SIGINT by political leaders, which led to two profound miscalculations that framed the American tragedy in Vietnam, the Tonkin Gulf incident and the Tet Offensive. But NSA was not blameless in these fiascoes, nor even in cases such as MacArthur's disastrous decision to ignore the accurate and timely warnings that American cryptologists were able to provide of the impending Chinese intervention in Korea—because all were ultimately rooted in the great dysfunctional flaw that went back to the very start of the U.S. signals intelligence structure in the pre–World War II days and that NSA, both out of bureaucratic inertia and self-interest, never attempted to remedy. This was the fiction that signals intelligence was not intelligence but information, which NSA was not to analyze but merely pass on in its raw form for others to interpret. That, plus the classification rules that always treated SIGINT as more secret than other secrets, led to a tunnel vision in the way intercepts were handled; the absence of definitive analysis by an experienced intelligence officer who could see the whole picture and was trusted by those at the top to do so made it possible for generals and cabinet officials and presidents who were overconfident of their own judgment to do with SIGINT whatever they liked. It led equally to McNamara's seizing on intercepts that confirmed his preconceived beliefs and MacArthur dismissing those that contradicted his. Never having the authority to perform its own analyses, NSA never explicitly developed an ability to do so, yet always did by default with varying degrees of professionalism and success.

It was a system ripe for intellectual corruption for NSA to have so much influence in the highest circles of power without the responsibility for how its wares were used by its eminent customers, and its entrenched bureaucratic managers did not always have the integrity or courage to resist the temptation to place the agency's interests above loyalty to the truth. From the cover-up of its mishandling of the Tonkin Gulf intercepts it was an evolution of degree but not of kind to the more serious political distortions of signals intelligence that occurred in the Reagan administration, when the White House simply edited out portions of intercepted air-to-ground voice communications of a Soviet fighter pilot who shot down a Korean airliner that had strayed into Soviet airspace in 1983, eliminating from the publicly released transcript evidence that the pilot had attempted to signal and warn the aircraft before opening fire; and then to the more egregious manipulation and out-of-context use of signals intelligence by the George W. Bush White House in falsely attempting to make the case that Saddam Hussein's regime possessed weapons of mass destruction prior to the United States' launching its invasion of Iraq in 2003.[2]

More subtly but perhaps of more lasting consequence, the Cold War froze in place expedients adopted to respond to the unprecedented demands of World War II that few imagined would ever become the normal way of business for the government of the United States, with its deeply rooted traditions of moral principle, openness, and the rule of law in international diplomacy, and of personal liberty and the right to privacy at home. What had been acceptable in wartime but anathema in peacetime became the norm for peacetime, too. Eisenhower in the middle of his presidency once tried to reassure himself that in adopting the methods of its adversaries America could still preserve its traditional beliefs in "truth, honor, justice, consideration for others, liberty for all"; he wrote a note to himself in which he suggested the way out was that "we must not confuse these values with mere procedures."[3] But procedures, after decades of repetition, tend to become values, like it or not. NSA's unflagging technologically driven pursuit to "get everything" in the teeth of the Fourth Amendment principle that the burden of proof lies upon the government to establish particularized probable cause for a search; its self-justifying assurance that its business was always too secret to be disclosed to, much less judged by, the people of the democracy that employed it; its reflexive defensiveness that rejected outside criticism as thus inherently uninformed (or, worse, an attack on the honor of its employees)—all showed how far the Cold War calculus had reshaped assumptions.

No one contemplating the crimes of the Communist regimes against their own peoples, the repression of the human spirit that sacrificed generations to an abstract ideological belief, could doubt the worth of the victory gained. In the shadowy four-decade struggle, the Cold War offered little opportunity for moments of glory or exultation that World War II abounded with; the greatest victory was not getting the world blown up along the way so that it was possible for the peaceful end to come when it at last did. The cryptologic struggle that took place in the shadows behind the shadows was as morally ambiguous as everything about the Cold War, and if the breaking of the Russian one-time-pad systems in the 1940s was an echo of the soaring intellectual triumphs of the World War II codebreakers, most of what followed was a far more subdued achievement of quotidian and collective persistence rather than individual inspiration. But the American cryptologists of the Cold War deserve as much credit as anyone for the fact that Americans, Russians, and the rest of the world were never vaporized in a cloud of radioactive ash; without them it is hard to see that containment would have lasted long enough to matter.

Sir Francis Walsingham, the principal secretary to Queen Elizabeth I,

APPENDIX A

Enciphered Codes, Depths, and Book Breaking

Enciphered codes, whether they use one-time pads or books of additive key, are all based on the same simple and robust principle. Security is afforded by two steps that considerably complicate the task of cryptanalysis.

First, the message is encoded using a codebook that provides a (usually) four- or five-digit number for each word. In a one-part code, the words are numbered sequentially in their alphabetical order, which allows for the same codebook to be used for encoding (looking up the word to find its numerical equivalent) and decoding (looking up the number to find its linguistic equivalent). In a two-part, or "hatted," code (because it is as if the words have been drawn out of a hat when assigning their numerical equivalents), the order is random, requiring separate codebooks for each operation: one in numerical order, the other in alphabetical order.

In the second step, which ensures that the same numeral does not stand for the same word in subsequent messages—thereby confounding any straightforward efforts by a codebreaker to guess the meaning of any particular numerical group—a series of four- or five-digit numerals is taken from a book or pad of randomly generated numbers distributed in advance to the sender and recipient, and these are used as additive (commonly, though loosely, also referred to as "key") to obscure the code group values.

So, for example, to encipher the message MEET AT FOUR PM TUESDAY, the code clerk would first look up each word in his codebook and write down the code group numbers that stand for each of the words:

text	MEET	AT	FOUR	PM	TUESDAY
code	0824	8345	0408	6234	2946

He would then choose a one-time-pad page, or a starting point in the additive book, and write out below the code groups the series of additive digits drawn in sequence from the page:

text	MEET	AT	FOUR	PM	TUESDAY
code	0824	8345	0408	6234	2946
key	8243	3825	2391	0158	9813

Finally, he would add together (digit by digit, in modulo 10, noncarrying addition) the code and key to yield the completed, enciphered message to be transmitted:

text	MEET	AT	FOUR	PM	TUESDAY
code	0824	8345	0408	6234	2946
key	8243	3825	2391	0158	9813
cipher	8067	1160	2799	6382	1759

For transmission by telegram, it was the convention to convert the numerals to letters; cable companies charged less for messages containing letters, which were easier to check for accuracy and also avoided the more cumbersome and lengthy Morse code characters for numbers when transmitted manually (for example, U in Morse code is . .—, A is .—, while 2 is . .———, and 8 is ———. .). It was also customary to break the text into five-letter "words." Using the substitution 0 = O, 1 = I, 2 = U, 3 = Z, 4 = T, 5 = R, 6 = E, 7 = W, 8 = A, 9 = P, the cipher text in the example becomes:[1]

AOEWI IEOUW PPEZA UIWRP

The recipient would reverse the process: first subtracting off the additive key, then looking up the meanings of the underlying code groups in the codebook.

The first step in breaking an enciphered code is to locate depths; that is, two or more messages enciphered with the same stretch of additive key. In a true one-time-pad system, there will be no depths to be found: the pad pages are used but a single time, then destroyed, so no two messages are ever enciphered with the same key. But in additive book systems, or one-time-pad systems that contain accidentally duplicated pages, as was the case with the Soviet messages read in the Venona project, depths can sometimes be found through laborious machine-aided searches. The method used in World War II against the many Japanese army and navy enciphered codes and at the start of the Venona project involved IBM punch card runs to look for so-called double hits. The idea was that if the same pair of numerical groups occurred the same number of groups apart in two different messages, this was unlikely to be chance, but could indicate that the two messages contained the same pair of words enciphered with the same run of additive key. (A single hit by contrast did not mean much, as chance alone dictated a one-in-four probability that any two fifty-group messages would have one four-digit numeral in common: in one message 4998 might stand for the code group 1235 plus the additive group 3763, in another it might stand for the code group 7723 plus the additive 7275.)

The still-laborious IBM method used in the 1940s involved punching a card containing the first five or so cipher groups of each of tens of thousands of messages (the opening groups were the most likely to contain stereotyped phrases such as addresses, message numbers, and the like), running the cards through a sorter to place them in numerical order by the first group, then printing out indexes hundreds of pages long which would be scanned by eye to see if any two messages that shared the same first group also had another group in common in another position. The whole process was then repeated with the cards reordered according to the second code group, another index printed, and again scanned by eye.

From two messages in depth, it was possible to calculate the *differences* in values between the underlying code groups, since subtracting two messages enciphered with the same key eliminated the key from the equation altogether. For example, from two messages placed in depth on the basis of a double hit (of the cipher groups 8596 and 1357):

message 1	1432	**8596**	2469	0053	**1357**
message 2	7712	**8596**	9857	4389	**1357**
difference	4720	0000	3612	6774	0000

Then other message pairs placed in depth could be examined to see if any of those same differences occurred in them, too, which would suggest that the same pair of words appeared there as well. Commonly occurring words like STOP, TO, or FROM, and special code groups standing for numerals or indicating "start spell" or "end spell" were the most likely candidates to be identified first. In a one-part code, the book breaker's job was made considerably easier by the fact that the numerical value of a code group relative to other recovered words greatly narrowed the range of alphabetical possibilities of words to consider. But in any case the task required deep familiarity with grammar and usage in the target language. The Jade codebook used with the 1944 and 1945 NKGB one-time-pad messages (also known as Code 2A by Arlington Hall) was a one-part code, and was recovered entirely through Meredith Gardner's book breaking without ever seeing the original. Code 1B, the NKGB codebook that the Russians called Kod Pobeda and which was used from 1939 to November 1943, was a two-part code, and the recovery of a copy of most of the original book by TICOM Team 3 played a significant part in the effort at NSA beginning in the mid-1950s to break most of the 1943 messages.[2]

Appendix B

Russian Teleprinter Ciphers

Captured TICOM documents on the German cryptanalysts' work on the Russian teleprinter cipher (known to the Germans as Bandwurm and to the British as Caviar or the Russian Fish) mentioned that the machine appeared to be similar to the "left portion" of the Germans' SZ40 teleprinter scrambler. It employed a system of five cipher wheels, each of which corresponded to one of the five bits of the Baudot code, to produce a random sequence of marks and spaces as it rotated through each successive position.[1] The output of this bank of key-generating wheels thus corresponded to a single Russian letter in the Baudot code. When that key letter was combined, by noncarrying binary addition, with the plaintext letter, it produced the enciphered letter that was transmitted.

Because the wheels moved to a new position as each letter was sent, the resulting cipher was polyalphabetic: the letter Я might stand for Ш at one position of text, but could stand for И, Н, Л, or any other letter at any another position. The brilliant part of the system was that because noncarrying binary addition is the same as subtraction, exactly the same setting of the machine could be used for both enciphering and deciphering: adding plaintext to key yields cipher text; adding cipher text to key yields plaintext. The basic rules for this noncarrying, modulo 2 addition are the same as the logical exclusive—or:

• + • = •
• + X = X
X + • = X
X + X = •

Thus the bit-by-bit addition of text and key works the same forward and backward; for example:

encipherment

Ш	X • • X •	plain
Г	• X • X •	+ key
Я	X X • • •	= cipher

decipherment

Я	X X • • •	cipher
Г	• X • X •	+ key
Ш	X • • X •	= plain

For the same reason, any number added to itself in modulo 2 addition equals zero. So if two messages are in depth—enciphered with the same sequence of key—then adding the two streams of cipher text together zeroes out the key altogether, leaving a string that is the combination only of the two underlying plaintexts:

$$\begin{aligned} \text{cipher}_1 &= \text{plain}_1 + \text{key} \\ \text{cipher}_2 &= \text{plain}_2 + \text{key} \\ \text{thus, cipher}_1 + \text{cipher}_2 &= \text{plain}_1 + \text{plain}_2 + (\text{key} + \text{key}) \\ &= \text{plain}_1 + \text{plain}_2 + 0 \end{aligned}$$

A report prepared by one of the captured German cryptanalysts for the TICOM investigators provided a table showing the letter produced by adding (or subtracting) any two other letters in the Russian Baudot code; using the table, it is a straightforward matter to find the sum of two streams of cipher text in depth:[2]

cipher text 1	у п ч с д л а г з ш у в н. .
cipher text 2	с н ш ц ю ь я з и й р х б. .
1 + 2	ч а ж к ж ш е к ю д к ы ь. .

	А	Б	В	Г	Д	Е	Ж	З	И	Ю	К	Л	М	Н	О	П	Р	С	Т	У	Ф	Х	Ц	Ч	Ш	Ы	Ь	Я	1	4	5	/
А	/	Ц	Ю	Х	Ф	Я	К	Ь	У	В	Ж	М	Л	П	1	Н	Ы	Т	С	И	Д	Г	Б	5	4	Р	З	Е	О	Ш	Ч	А
Б	Ц	/	М	1	Я	Ф	С	П	4	Л	Т	Ю	В	Ь	Х	З	5	Ж	К	Ш	Е	О	А	Ы	У	Ч	Н	Д	Г	И	Р	Б
В	Ю	М	/	Р	К	С	Ф	У	Ь	А	Д	Ц	Б	4	Ч	Ш	Г	Е	Я	З	Ж	Ы	Л	О	П	Х	И	Т	5	Н	1	В
Г	Х	1	Р	/	У	4	Ь	К	Ф	Ы	З	Ч	5	С	Ц	Т	В	Н	П	Д	И	А	О	Л	Я	Ю	Ж	Ш	Б	Е	М	Г
Д	Ф	Я	К	У	/	Ц	Ю	Р	Х	Ж	В	С	Т	Ч	4	5	З	Л	М	Г	Д	И	Е	Н	1	Ь	Ы	Б	Ш	О	П	Д
Е	Я	Ф	С	4	Ц	/	М	Ч	1	Т	Л	К	Ж	Р	У	Ы	Н	В	Ю	О	Б	Ш	Д	З	Х	П	5	А	И	Г	Ь	Е
Ж	К	С	Ф	Ь	Ю	М	/	Х	Р	Д	А	Я	Е	1	П	О	И	Б	Ц	Ы	В	З	Т	Ш	Ч	У	Г	Л	Н	5	4	Ж
З	Ь	П	У	К	Р	Ч	Х	/	Ю	И	Г	4	Ш	Ц	С	Б	Д	О	1	В	Ы	Ж	Н	Е	М	Ф	А	5	Т	Л	Я	З
И	У	4	Ь	Ф	Х	1	Р	Ю	/	З	Ы	П	Н	М	Я	Л	Ж	5	Ч	А	Г	Д	Ш	Т	Ц	К	В	О	Е	Б	С	И
Ю	В	Л	А	Ы	Ж	Т	Д	И	З	/	Ф	Б	Ц	Ш	5	4	Х	Я	Е	Ь	К	Р	М	1	Н	Г	У	С	Ч	П	О	Ю
К	Ж	Т	Д	З	В	Л	А	Г	Ы	Ф	/	Е	Я	О	Н	1	У	Ц	Б	Р	Ю	Ь	С	4	5	И	Х	М	П	Ч	Ш	К
Л	М	Ю	Ц	Ч	С	К	Я	4	П	Б	Е	/	А	У	Р	И	О	Д	Ф	Н	Т	5	В	Г	Ь	1	Ш	Ж	Ы	З	Х	Л
М	Л	В	Б	5	Т	Ж	Е	Ш	Н	Ц	Я	А	/	И	Ы	У	1	Ф	Д	П	С	Ч	Ю	Х	З	О	4	К	Р	Ь	Г	М
Н	П	Ь	4	С	Ч	Р	1	Ц	М	Ш	О	У	И	/	К	А	Е	Г	Х	Л	5	Т	З	Д	Ю	Я	Б	Ы	Ж	В	Ф	Н
О	1	Х	Ч	Ц	4	У	П	С	Я	5	Н	Р	Ы	К	/	Ж	Л	З	Ь	Е	Ш	Б	Г	В	Ф	М	Т	И	А	Д	Ю	О
П	Н	З	Ш	Т	5	Ы	О	Б	Л	4	1	И	У	А	Ж	/	Я	Х	Г	М	Ч	С	Ь	Ф	В	Е	Ц	Р	К	Ю	Д	П
Р	Ы	5	Г	В	З	Н	И	Д	Ж	Х	У	О	1	Е	Л	Я	/	4	Ш	К	Ь	Ю	Ч	Ц	Т	А	Ф	П	М	С	Б	Р
С	Т	Ж	Е	Н	Л	В	Б	О	5	Я	Ц	Д	Ф	Г	З	Х	4	/	А	Ч	М	П	К	У	Ы	Ш	1	Ю	Ь	Р	И	С
Т	С	К	Я	П	М	Ю	Ц	1	Ч	Е	Б	Ф	Д	Х	Ь	Г	Ш	А	/	5	Л	Н	Ж	И	Р	4	О	В	З	Ы	У	Т
У	И	Ш	З	Д	Г	О	Ы	В	А	Ь	Р	Н	П	Л	Е	М	К	Ч	5	/	Х	Ф	4	С	Б	Ж	Ю	1	Я	Ц	Т	У
Ф	Д	Е	Ж	И	А	Б	В	Ы	Г	К	Ю	Т	С	5	Ш	Ч	Ь	М	Л	Х	/	У	Я	П	О	З	Р	Ц	4	1	Н	Ф
Х	Г	О	Ы	А	И	Ш	З	Ж	Д	Р	Ь	5	Ч	Т	Б	С	Ю	П	Н	Ф	У	/	1	М	Е	В	К	4	Ц	Я	Л	Х
Ц	Б	А	Л	О	Е	Д	Т	Н	Ш	М	С	В	Ю	З	Г	Ь	Ч	К	Ж	4	Я	1	/	Р	И	5	П	Ф	Х	У	Ы	Ц
Ч	5	Ы	О	Л	Н	З	Ш	Е	Т	1	4	Г	Х	Д	В	Ф	Ц	У	И	С	П	М	Р	/	Ж	Б	Я	Ь	Ю	К	А	Ч
Ш	4	У	П	Я	1	Х	Ч	М	Ц	Н	5	Ь	З	Ю	Ф	В	Т	Ы	Р	Б	О	Е	И	Ж	/	С	Л	Г	Д	А	К	Ш
Ы	Р	Ч	Х	Ю	Ь	П	У	Ф	К	Г	И	1	О	Я	М	Е	А	Ш	4	Ж	З	В	5	Б	С	/	Д	Н	Л	Т	Ц	Ы
Ь	З	Н	И	Ж	Ы	5	Г	А	В	У	Х	Ш	4	Б	Т	Ц	Ф	1	О	Ю	Р	К	П	Я	Л	Д	/	Ч	С	М	Е	Ь
Я	Е	Д	Т	Ш	Б	А	Л	5	О	С	М	Ж	К	Ы	И	Р	П	Ю	В	1	Ц	4	Ф	Ь	Г	Н	Ч	/	У	Х	З	Я
1	О	Г	5	Б	Ш	И	Н	Т	Е	Ч	П	Ы	Р	Ж	А	К	М	Ь	З	Я	4	Ц	Х	Ю	Д	Л	С	У	/	Ф	В	1
4	Ш	И	Н	Е	О	Г	5	Л	Б	П	Ч	З	Ь	В	Д	Ю	С	Р	Ы	Ц	1	Я	У	К	А	Т	М	Х	Ф	/	Ж	4
5	Ч	Р	1	М	П	Ь	4	Я	С	О	Ш	Х	Г	Ф	Ю	Д	Б	И	У	Т	Н	Л	Ы	А	К	Ц	Е	З	В	Ж	/	5
/	А	Б	В	Г	Д	Е	Ж	З	И	Ю	К	Л	М	Н	О	П	Р	С	Т	У	Ф	Х	Ц	Ч	Ш	Ы	Ь	Я	1	4	5	/

A cipher square produced by captured German cryptologists, showing the rules for adding (or subtracting) letters in the Russian Baudot teleprinter code.

The next step is to try a piece of likely plaintext—a "crib"—for one message and see if it yields a plausible word (or portion of a word) in Russian in the corresponding message, again using the addition table to combine the plaintext with the stream of summed cipher texts. For example, trying в москву ("to Moscow") at the start of one message:

1 + 2	ч а ж к ж ш е к ю д к ы ь . .
crib 1	в _ м о с к в у _
plaintext 2	о ч е н ь _ с р о

yields очень сро in the second message, which could be the start of the phrase очень сроно ("extremely urgent"); filling out the rest of the letters of the phrase and then working back to the first message:

1 + 2	ч а ж к ж ш е к ю д к ы ь . .
plaintext 2	о ч е н ь _ с р о ч н о _
plaintext 1	в _ м о с к в у _ н о м е

reveals additional letters in the first message—номе, which looks like the start of the word номе, "number." By continuing this seesawing back and forth between the two texts, it is possible with a bit of luck and knowledge of the language to decipher the complete texts of both messages.

Finally, by combing the recovered plaintext of either message with its original cipher text, the actual sequence of key the machine generated to produce the encipherment can be established:

cipher text 1	у п ч с д л а г з ш у в н
plaintext 1	в _ м о с к в у _ н о м е
key	з д х з л е ю д я ю е б р

With enough recovered key, the next step would be to look for repetition cycles in the sequence of marks and spaces in the key stream to begin to reconstruct each cipher wheel and its movement pattern. Once the key-generation system is solved, a general solution of traffic is then possible by using special-purpose analytical machinery or a digital computer to "slide" an intercepted cipher text against all possible key sequences to find the setting that produces likely plaintext or other statistical measures of a correct placement: this was the method the British GC&CS developed for the special-purpose electronic comparator Colossus to implement in its successful breaking of the German Tunny teleprinter cipher during World War II.[3]

APPENDIX C

Cryptanalysis of the Hagelin Machine

Unlike the unpredictable cipher alphabets generated by rotor machines such as the Enigma, the Hagelin machine's encipherment rules followed a simple pattern. The twenty-six different cipher alphabets it employed were generated by simply sliding the letters of an alphabet in reverse Z-to-A order against another alphabet in A-to-Z order.

That meant that if one knew the plaintext equivalent for a cipher letter at a given position of the machine, one automatically knew what every other cipher letter at that same position stood for (which was emphatically not the case with the Enigma). The Hagelin's encipherment formula could be written arithmetically as a simple subtraction rule, where the key is a number from 0 to 25 and the letters of the alphabet are given by A = 0, B = 1, C = 2, D = 3, etc.:

cipher = key − plain (modulo 26)
plain = key − cipher (modulo 26)

(In modulo 26 arithmetic, the resulting number always remains in the range 0 to 25; so, for example, −1 modulo 26 equals 25; −2 equals 24.)

If two messages were in depth—enciphered with the exact same key sequence—then subtracting one from the other eliminated the key from the equation altogether (just as was the case with the example of the Russian teleprinter cipher in appendix B), leaving only the difference between the plaintext values of each:

	A	B	C	D	E	F	G	H	I	J	K	L	M	N	O	P	Q	R	S	T	U	V	W	X	Y	Z
0	A	Z	Y	X	W	V	U	T	S	R	Q	P	O	N	M	L	K	J	I	H	G	F	E	D	C	B
1	B	A	Z	Y	X	W	V	U	T	S	R	Q	P	O	N	M	L	K	J	I	H	G	F	E	D	C
2	C	B	A	Z	Y	X	W	V	U	T	S	R	Q	P	O	N	M	L	K	J	I	H	G	F	E	D
3	D	C	B	A	Z	Y	X	W	V	U	T	S	R	Q	P	O	N	M	L	K	J	I	H	G	F	E
4	E	D	C	B	A	Z	Y	X	W	V	U	T	S	R	Q	P	O	N	M	L	K	J	I	H	G	F
5	F	E	D	C	B	A	Z	Y	X	W	V	U	T	S	R	Q	P	O	N	M	L	K	J	I	H	G
6	G	F	E	D	C	B	A	Z	Y	X	W	V	U	T	S	R	Q	P	O	N	M	L	K	J	I	H
7	H	G	F	E	D	C	B	A	Z	Y	X	W	V	U	T	S	R	Q	P	O	N	M	L	K	J	I
8	I	H	G	F	E	D	C	B	A	Z	Y	X	W	V	U	T	S	R	Q	P	O	N	M	L	K	J
9	J	I	H	G	F	E	D	C	B	A	Z	Y	X	W	V	U	T	S	R	Q	P	O	N	M	L	K
10	K	J	I	H	G	F	E	D	C	B	A	Z	Y	X	W	V	U	T	S	R	Q	P	O	N	M	L
11	L	K	J	I	H	G	F	E	D	C	B	A	Z	Y	X	W	V	U	T	S	R	Q	P	O	N	M
12	M	L	K	J	I	H	G	F	E	D	C	B	A	Z	Y	X	W	V	U	T	S	R	Q	P	O	N
13	N	M	L	K	J	I	H	G	F	E	D	C	B	A	Z	Y	X	W	V	U	T	S	R	Q	P	O
14	O	N	M	L	K	J	I	H	G	F	E	D	C	B	A	Z	Y	X	W	V	U	T	S	R	Q	P
15	P	O	N	M	L	K	J	I	H	G	F	E	D	C	B	A	Z	Y	X	W	V	U	T	S	R	Q
16	Q	P	O	N	M	L	K	J	I	H	G	F	E	D	C	B	A	Z	Y	X	W	V	U	T	S	R
17	R	Q	P	O	N	M	L	K	J	I	H	G	F	E	D	C	B	A	Z	Y	X	W	V	U	T	S
18	S	R	Q	P	O	N	M	L	K	J	I	H	G	F	E	D	C	B	A	Z	Y	X	W	V	U	T
19	T	S	R	Q	P	O	N	M	L	K	J	I	H	G	F	E	D	C	B	A	Z	Y	X	W	V	U
20	U	T	S	R	Q	P	O	N	M	L	K	J	I	H	G	F	E	D	C	B	A	Z	Y	X	W	V
21	V	U	T	S	R	Q	P	O	N	M	L	K	J	I	H	G	F	E	D	C	B	A	Z	Y	X	W
22	W	V	U	T	S	R	Q	P	O	N	M	L	K	J	I	H	G	F	E	D	C	B	A	Z	Y	X
23	X	W	V	U	T	S	R	Q	P	O	N	M	L	K	J	I	H	G	F	E	D	C	B	A	Z	Y
24	Y	X	W	V	U	T	S	R	Q	P	O	N	M	L	K	J	I	H	G	F	E	D	C	B	A	Z
25	Z	Y	X	W	V	U	T	S	R	Q	P	O	N	M	L	K	J	I	H	G	F	E	D	C	B	A

The twenty-six cipher alphabets employed by the Hagelin M-209 machine. The letters inside the grid are the cipher text equivalents of the plaintext letters across the top of the table for each key value from 0 to 25.

$$\text{cipher}_1 = \text{key} - \text{plain}_1$$
$$\text{cipher}_2 = \text{key} - \text{plain}_2$$
$$\text{thus, cipher}_1 - \text{cipher}_2 = \text{plain}_2 - \text{plain}_1$$

Because of the strict alphabetical order followed by the Hagelin's cipher alphabets, this meant that for two messages in depth, the plaintext letters of each had to be the same distance apart in the alphabet as were the corresponding cipher text letters at that same position. (For example, if the cipher letters of two messages in depth were A and D at a given position, the plaintext letters they stood for had to be four letters apart as well, such as B and E, C and F, or Z and C.) The final step in breaking two such paired messages was to “drag” a crib of likely plaintext through every possible position in one message and see if it produced readable plaintext in the matching message. In breaking some Dutch Hagelin traffic in 1944, U.S. Army codebreakers, for example, located two messages in depth whose cipher texts included these sequences:

J E Y M C P A
X H J S T C K

Converting the letters to their numerical values and subtracting, modulo 26, to eliminate the key yielded the following differences between the two strings of plaintext values:

14 3 11 6 17 13 10

Thus, if the first plaintext letter of one message was N, the corresponding letter of the second message would have to come 14 letters earlier in the alphabet, or A. The cryptanalysts found that trying the word LETTERX in this location (“letter” is as good Dutch as it is English, and X was commonly used to represent a space following each word) produced good plaintext in the matching message:

L E T T E R X
14 3 11 6 17 13 10
X B I N N E N

(*Binnen* is Dutch for “within.”) The break was then extended in each direction by guessing additional words that followed or preceded the ones already discovered, revealing additional plaintext in the paired message.[1]

The index of coincidence test (appendix E) offered a trial-and-error method for locating two messages in depth by sliding a set of intercepted messages through every possible relative position. But for a more general solution, it was necessary to reproduce the sequences of key the machine generated so that any message could be broken, not just pairs found to be in depth. The Hagelin used a set of wheels with movable pins to generate the key sequence, and U.S. Army cryptanalysts devised an elaborate procedure to derive the pinwheel settings from the key strings they recovered. (The key strings could be obtained, once the plaintext of a message in depth had been read, simply by adding the recovered plaintext to the cipher text; since cipher = key − plain, therefore key = cipher + plain.) That in turn allowed them to break the indicator system that told the intended recipient of a message what pinwheel settings to use; with the indicator system cracked, the codebreakers could then read every subsequent message directly, without having to rely on finding further depths.[2]

APPENDIX D

Bayesian Probability, Turing, and the Deciban

The tests that cryptanalysts had traditionally relied upon for placing messages in depth or setting them against known key, such as searching for double hits, subtracting hypothetical key and examining the result for high-frequency code groups, or measuring the index of coincidence between two streams of polyalphabetic cipher text (appendix E), were all based on probability: it was not that such coincidences as double hits were impossible to occur by random chance between two strings of unrelated cipher text, just that they were unlikely to. The problem was that once cryptanalysts began to use machine methods to make vast numbers of comparisons, the number of false alarms rose as well: a common problem in early computer-aided cryptanalytic runs was the absence of more sophisticated statistical tests to determine which hits were real and which were just the product of chance.

Alan Turing, among his other contributions to cryptanalysis, developed a method of enduring utility for assessing the odds of whether any given discovered coincidence was "causal" or accidental. The immediate problem Turing was working on involved a method he had developed to reduce the number of tests required to recover the daily key of the naval Enigma. The first step required discovering the relative rotor starting positions of a large number of Enigma messages, which the analysts attempted to do by locating repetitions between two messages where it looked like the same frequently occurring letters or words (often the ubiquitous *eins*) had been enciphered with the same key, allowing them to be placed in proper depth. The method involved sliding punched sheets of messages one against the other and scanning by eye for repetitions of single letters, bigrams, trigrams, tetragrams, and so forth. (The task was "not easy enough to be trivial, but not difficult enough to cause a nervous breakdown," in the words of Turing's colleague Jack Good.)[1]

The chance of any given repeat occurring by chance alone is a basic calculation in probability, akin to the odds of drawing the same two or three or four letters from a hat containing all letters of the alphabet on two successive draws. But Turing noted that this information alone was little help in judging whether a trial alignment of the two messages containing such a repetition was correct or not, given the vastly greater number of "wrong" alignments that are possible for every causal one. What one really wanted to know was if given, say, the discovery of one tetragram, two bigrams, and fifteen single letters lining up between two messages, what was the "weight of evidence" in favor of the hypothesis that this was real versus a false alarm? Each one of these coincidences was, a priori, unlikely to be the product of random chance, but with thousands of pairs of messages and hundreds of places in each message where a repeat could occur, the odds grew that such a random hit would occur. (As Good observed, the phenomenon is akin to the famous "birthday problem" of probability theory. Although the odds of any two people having the same birthday is only 1 in 365, if you have 23 people in a room it is odds-on that two will have the same birthday, because of the much larger number of different pairwise comparisons that are possible with that number of people: 22 + 21 + 20 + 19 + . . . + 3 + 2 + 1 = 253.)

Turing's "factor in favor of a hypothesis" was an application of what is more generally known as Bayesian probability, a method to assess from available evidence which of two competing hypotheses is more likely to be supported. By taking into account the greater number of noncausal comparisons that could be made for every correct one, Turing and Good calculated the odds in favor of any given repetition being causal for each kind of repeat. The total odds were obtained by multiplying together the odds of each individual observed coincidence (a basic law of probability: for example, the odds of two coin tosses coming up heads is the odds of each event multiplied together, $\frac{1}{2} \times \frac{1}{2} = \frac{1}{4}$). To simplify the calculation, Turing expressed the odds as logarithms; adding the logarithms of two numbers together is the same as multiplying the numbers themselves. Turing dubbed his unit for the logarithm of weight of evidence the "ban," an insider-joke reference to the punched paper sheets used in the naval Enigma problem, which were printed in the nearby town of Banbury. One ban equals odds of 10 to 1 in favor; the "deciban," which turned out to be a more practical unit, is equal to a tenth of a ban. A rule of thumb for the Bletchley codebreakers was that odds of 50 to 1 in favor, which equals 1.7 bans or 17 decibans, was a virtual certainty of a comparison being causal.[2]

The power of the method was that it is completely general, applicable to an array of problems in cryptanalysis. Max Newman, the mathematician at Bletchley Park who led the attack on the German teleprinter ciphers, considered Turing's concept "his greatest intellectual contribution during the war"; GCHQ and NSA cryptanalysts still use the ban, an enduring tribute to Turing's legacy to the statistical analysis of cipher problems.[3]

APPENDIX E

The Index of Coincidence

In a polyalphabetic cipher, such as that generated by a rotor machine, the substitution alphabet changes with each successive letter of the message. The resulting stream of cipher text has a flat distribution of letter frequencies: each letter of the alphabet has the same $\frac{1}{26}$ (or 3.8 percent) chance of appearing at any given position of polyalphabetic cipher text. The standard cryptanalyst's trick for solving a simple monoalphabetic substitution cipher—counting the frequency of each letter and assigning the most commonly occurring ones to corresponding high-frequency letters of the underlying language (such as E, T, A, O, I, N, and S in English)—is thus of no help against a string of polyalphabetic cipher.

William Friedman in 1922 developed one of the most fundamental statistical tools to overcome this obstacle and successfully break a polyalphabetic cipher. If any two randomly chosen strings of letters are placed side by side, the chances that any specified letter (say, A) will appear in the same position in both texts is $\frac{1}{26} \times \frac{1}{26}$, or 0.15 percent; the chance that *any* pair of identical letters (A and A, or B and B, or C and C, or so forth) will occur is 26 times that, or 3.8 percent.

Two strings of plaintext, however, will on average show a much greater number of coincidences when placed side by side. This is because the high-frequency letters such as E, T, and A greatly increase the chances of the same letter appearing in the two strings in the same place. For example the plain-language frequency of E, the most common letter in English, is 12 percent; therefore the odds of an E occurring simultaneously at any given point in two strings of plaintext is $\frac{12}{100} \times \frac{12}{100}$, or 1.4 percent, almost ten times the chances of a given coincidence in random letter strings. The chance of two Z's occurring in the same position is obviously much lower (Z's occur 0.3 percent of the time in English plaintext, thus the odds of two of them occurring simultaneously at any given location in two strings of plaintext are $^{0.3}/_{100} \times {}^{0.3}/_{100}$, or 0.0009 percent), but the high-frequency letters so outweigh the low-frequency letters as to elevate the total odds: summing the plaintext coincidence probabilities for all twenty-six letters of the alphabet yields a total probability of about 6.7 percent for a coincidence, nearly double the 3.8 percent of the random situation.

Friedman called this test—the number of such coincidences that occur in two strings of text divided by the total number of letters—the index of coincidence.[1] For example, in these two random sixty-letter strings:

```
DWSWYSRNCDVWRDNFJ[P]PMQITDDTLJXPILDALCD[B]DFDAKEIZDIOSZEYZDOTZUG
KKNOPNSUANGINMKTM[P]MWXSAXRNSEGCEWJZIHO[B]ZHJFRNVLOGBIFTEOGQUSFZ
```

an actual count finds that there are two coincidences, yielding an index of coincidence of $^2/_{60}$, or 3.3 percent; while in these two sixty-letter strings of plaintext:

```
L[E]TUSGOTHENY[O]UANDIWHENTHE[E]VENING[I]SSPREADOUTAGAINSTTHESKYLIKE
W[E]SHALLFIGHT[O]NTHEBEACHESW[E]SHALLF[I]GHTONTHELANDINGGROUNDSWESHA
```

there are four coincidences, for an index value of $^4/_{60}$, or 6.6 percent.

Friedman's striking insight was that if two texts had been enciphered with the same sequence of polyalphabetic key—that is, if they are in depth—this same nonrandom unevenness in the underlying plaintext persists as a ghostly remnant that can be detected by the index of coincidence test. At any given position of two messages that are in depth, each has been enciphered with the same monoalphabetic substitution; thus if, say, E has been enciphered as B at one position, the chance of a B occurring simultaneously in both cipher texts at that particular spot shares the same elevated probability of the E's in the underlying plaintext. To put it another way, any coincidence that occurs in the underlying plaintexts will also occur at the same place in their corresponding polyalphabetic cipher texts if they are in depth: the identities of the letters have been altered, but the coincidences remain.

To locate a depth, two streams of cipher text can thus be "slid" past each other in every possible relative position and the index of coincidence calculated for each trial alignment; the index will suddenly jump from a value closer to the random 3.8 percent to the 6.7 percent of plaintext when they are properly aligned in a true depth. For example, the two plaintext previous sequences above, when each enciphered by an Enigma machine at the same chosen setting, yield the following two cipher texts, which preserve the coincidences of the two underlying plaintexts in the same positions:

```
ADLNNBVLBHGKGTJKLZGMJOYEYWEXBVTXWWLKMFKRUDAWMOMWBBIATCAEFWOD
CDXJLPAOPMEZGYYROOJINLQADWGIZGSQWXWFFCZIDGTZKDQLVGVGXIPLRAIN
```

thus yielding the same index of coincidence of 6.6 percent. But if they are misaligned, for example shifted one letter off:

```
 ADLNNBVLBHGKGTJKLZGMJOYEYWEXBVTXWWLKMFKRUDAWMOMWBBIATCAEFWOD
CDXJLPAOPMEZGYYROOJINLQADWGIZGSQWXWFFCZIDGTZKDQLVGVGXIPLRAIN
```

the number of coincidences falls sharply (in this case to one, for an index value of 1.7 percent).

This simple test was the principle behind many of the electromechanical, and then electronic, cryptanalytic machines developed in the 1940s and later to locate depths, which could then lead to further exploitation of a target cipher system.

NOTES

Full references to the sources cited in shortened form in the notes below are found in the bibliography. Frequently cited works and document collections are identified by the following abbreviations:

AC	Johnson, *American Cryptology During the Cold War*
CCH	NSA Center for Cryptologic History, Web
CIAL	Central Intelligence Agency Library, Web
CW	Gaddis, *Cold War*
DNSArch	Digital National Security Archive, George Washington University, Washington, DC
FRUS	*Foreign Relations of the United States*
HCR	Historic Cryptographic Records, NARA
HV	Benson and Phillips, *History of VENONA*
NARA	Record Group 457, National Archives and Records Administration, College Park, MD
NCM	National Cryptologic Museum Library, Ft. George G. Meade, MD
NSAD	NSA Declassification and Transparency, Web
NSA60	NSA 60th Anniversary, Web
NYT	*New York Times*
OH	oral history
TArch	TICOM Archive, Web
TNA	UK National Archives, Kew, UK
WM	Burke, *Wasn't All Magic*

AUTHOR'S NOTE

1. "Edward Snowden: The Whistleblower Behind the NSA Surveillance Revelations," *Guardian*, June 11, 2013; Fred Kaplan, "Why Snowden Won't (and Shouldn't) Get Clemency," *Slate*, January 3, 2014; "Snowden Persuaded Other NSA Workers to Give Up Passwords," Reuters, November 7, 2013.
2. "Edward Snowden," *Guardian*, June 11, 2013.

3. "A Guide to What We Now Know About the NSA's Dragnet Searches of Your Communications," American Civil Liberties Union, August 9, 2013, Web; "Are They Allowed to Do That? A Breakdown of Government Surveillance Programs," Brennan Center for Justice, New York University School of Law, July 15, 2013, Web; "N.S.A. Said to Search Content of Messages to and from U.S.," *NYT,* August 8, 2013.
4. "SIGINT Enabling Project," Excerpt from 2013 Intelligence Budget Request, "Secret Documents Reveal N.S.A. Campaign Against Encryption," *NYT,* September 5, 2013; "Carnegie Mellon Researcher Warns US Officials of Dangers of Weakening Cybersecurity to Facilitate Government Surveillance," Press Release, Carnegie Mellon University, November 4, 2013, Web.
5. Abelson et al., *Keys Under Doormats,* 2–3.
6. United States District Court for the District of Columbia, Memorandum Opinion, Klayman et al. v. Obama et al., December 16, 2013. In a separate case, a U.S. appellate court in May 2015 found the bulk calling data collection program to be a violation of the law, holding that its "all-encompassing" nature went far beyond what Congress had authorized under the 2001 USA Patriot Act, which allowed the government to seek secret orders from the FISC for production of records related to an "authorized investigation" of foreign espionage or international terrorism: United States Court of Appeals for the Second Circuit, ACLU v. Clapper, May 7, 2015.
7. United States Foreign Intelligence Surveillance Court, Opinion and Order, In Re Orders of this Court Interpreting Section 215 of the Patriot Act, September 13, 2013; "House Votes to End N.S.A.'s Bulk Phone Data Collection," *NYT,* May 13, 2015; "U.S. Surveillance in Place Since 9/11 Is Sharply Limited," *NYT,* June 2, 2015.
8. "Edward Snowden," *Guardian,* June 11, 2013; Greenwald, *No Place to Hide,* 177, 196, 227.
9. Testimony of NSA Deputy Director John "Chris" Inglis before the House Intelligence Committee, June 18, 2013, and his speech to the National Cryptologic Museum Foundation, October 15, 2014; "Ex-CIA Director: Snowden Should Be 'Hanged' If Convicted for Treason," Fox News Politics, December 17, 2013, Web.
10. "Udall, Wyden Call on National Security Agency Director to Clarify Comments on Effectiveness of Phone Data Collection Program," Press Release, Office of Sen. Ron Wyden, June 13, 2013; Bruce Schneier, "How the NSA Threatens National Security," January 6, 2014, Web.
11. McConnell, "Future of SIGINT," 42.

PROLOGUE: "A CATALOGUE OF DISASTERS"

1. Höhne and Zolling, *General Was a Spy,* 160–63; Bower, *Red Web,* 130–31; Hess, "Hans Helmut Klose."
2. Bower, *Red Web,* 143–47.
3. Andrew and Gordievsky, *KGB,* 385; Bower, *Red Web,* 154, 161.
4. Bower, *Red Web,* 145, 152–53, 214.
5. Andrew, *Secret Service,* 343, 438; Macintyre, *Spy Among Friends,* 52.
6. Bower, *Red Web,* 204–5.
7. Ibid., 205, 213; Bower, *Perfect English Spy,* 207.
8. Bower, *Perfect English Spy,* 204–7; Andrew and Gordievsky, *KGB,* 387–89; Aid, "Cold War," 30–31.
9. Macintyre, *Spy Among Friends,* 141; Bethell, *Great Betrayal,* 145, 165, 174–75.
10. Smith, *New Cloak,* 114–15; Aid, "Cold War," 31.
11. *FRUS, Intelligence Community, 1950–55,* 695*n*3.
12. Kistiakowsky quoted in Aid, "Cold War," 31.

1 THE RUSSIAN PROBLEM

1. At the end of World War II, the Army Signal Intelligence Service had 7,848 men and women working at its headquarters in Arlington, Virginia; the Washington complement of the Navy's Communications Intelligence Section, Op-20-G, numbered 5,114: Achievements of the Signal Security Agency in World War II, SRH-349, Studies on Cryptology, NARA, 3; Survey of Op-20-G, Naval Inspector General, July 13, 1945, DNSArch, 16.
2. Budiansky, "Cecil Phillips," 98–99; *HV,* 40*n*16.
3. *HV,* 8, 19.
4. U.S. Army, "Arlington Hall"; Alvarez, *Secret Messages,* 122–23; *HV,* 20*n*34.
5. Achievements of the Signal Security Agency in World War II, SRH-349, Studies on Cryptology, NARA, 3; Treadwell, *Women's Army Corps,* 316; Historical Review of Op-20-G, SRH-152, Studies on Cryptology, NARA.
6. Budiansky, *Battle of Wits,* 225, 262.
7. Standard #530 Bombe, Tentative Brief Descriptions of Cryptanalytic Equipment for Enigma Problems, NR 4645, HCR; Wilcox, *Solving Enigma,* 32, 54.
8. For an explanation of Turing's method and the design of the bombes, see Budiansky, *Battle of Wits,* 127–31.
9. Survey of Op-20-G, Naval Inspector General, July 13, 1945, DNSArch, 5–7.
10. Budiansky, *Battle of Wits,* 243.

11. Ibid.; Achievements of the Signal Security Agency in World War II, SRH-349, Studies on Cryptology, NARA, 3; Benson, *U.S. Communications Intelligence,* 67, 85.
12. History of Cryptanalysis of Japanese Army Codes, NR 3072, HCR, 13.
13. Geoffrey Stevens, September 28, 1942, HW 14/53, TNA.
14. *HV,* 36.
15. *HV,* 17–18, 26, 33; Rowlettt, *Story of MAGIC,* 153; Budiansky, *Battle of Wits,* 163, 218, 321; "Frank W. Lewis, Master of the Cryptic Crossword, Dies at 98," *NYT,* December 3, 2010.
16. Alvarez, *Secret Messages,* 152, 154, 156–57; Alvarez, "No Immunity," 22–23.
17. Callimahos, "Legendary William Friedman," 9, 13; Deavours and Kruh, *Machine Cryptography,* 47.
18. Alvarez, *Secret Messages,* 155.
19. Hinsley et al., *British Intelligence,* I:199*n; HV,* 9.
20. Andrew, *Secret Service,* 268–69, 293–93, 296.
21. Geoffrey Stevens to London, Appendix, August 1, 1942, HW 14/47, TNA.
22. Peterson, "Before BOURBON," 10.
23. Thomas H. Dyer to Chief of Naval Operations, Recommendations Concerning the Post-War Organization of Communications Intelligence Activities, August 1, 1945, DNSArch, 2–3.
24. Benson, *U.S. Communications Intelligence,* 35–36; Alfred McCormack to Generals Bratton and Lee, February 12, 1942, Special Branch, G-2, Military Intelligence Division, NR 3918, HCR.
25. Carter Clarke to Alfred McCormack, May 6, 1942, Special Branch, G-2, Military Intelligence Division, NR 3918, HCR.
26. Benson, *Venona Story,* 12.
27. Joseph R. Redman, Memorandum for OP-02, January 23, 1946, DNSArch, 2.
28. Cook, OH; Rowlett, OH.
29. Frank B. Rowlett, Recollections of Work on Russian, February 11, 1965, Document 6, National Security Archive, *Secret Sentry Declassified,* 4; Peterson, "Before BOURBON," 8–9; *HV,* 9–10. Rowlett recalled that Arlington Hall's shadow teleprinter was copying traffic sent on a landline that the U.S. Army provided the Soviet embassy in Washington to communicate with Ladd Field, Alaska, where aircraft being supplied under Lend-Lease were turned over to the Russians. That was probably a misapprehension on his part; much more likely is that it was copying the Pentagon–Moscow teleprinter link described in Deane, *Strange Alliance,* 64–71.
30. Hatch, *Tordella,* 1–2.
31. *HV,* 20–21; Chronological Record of 3-G-10-Z (The Russian Language

Section), June 1943–March 1947, DNSArch; Cryptographic Codes and Ciphers: Russian: B-01 System, NR 4115, HCR.

32. Chronological Record of 3-G-10-Z (The Russian Language Section), June 1943–March 1947, DNSArch.
33. *HV*, 27–28.
34. Budiansky, *Battle of Wits*, 308–9; SSA General Cryptanalytic Branch—Annual Report FY 1945, NR 4360, HCR, 6; *HV*, 40–41.
35. Budiansky, *Battle of Wits*, 263.
36. SSA General Cryptanalytic Branch—Annual Report FY 1945, NR 4360, HCR, 1; Chronological Record of 3-G-10-Z (The Russian Language Section), June 1943–March 1947, DNSArch; Minutes of a Meeting Held in Captain J. N. Wenger's Office, May 21, 1945, DNSArch.
37. *HV*, 42–43.
38. Budiansky, *Battle of Wits*, 171–79.
39. Agreement Between British Government Code and Cipher School and U.S. War Department Regarding Special Intelligence, NR 2751, HCR; Erskine, "Holden Agreement."
40. Ferris, "British 'Enigma,'" 172–74.
41. Minutes of the Eighth Meeting of the Army-Navy Cryptanalytic Research and Development Committee, February 21, 1945, DNSArch, 3–4, 9; J. N. Wenger, Release of Free French Material to British, May 15, 1945, DNSArch; United Kingdom Base Section London to SSA, November 1, 1944, Clark Files, British Liaison, NR 4566, HCR.
42. Smith, *Station X*, 136–38.
43. Johnson, "Wenger."
44. J. N. Wenger to Edward Travis, January 18, 1946, DNSArch; J. N. Wenger, Collaboration with British on Rattan Project, May 21, 1945, DNSArch.
45. R. S. Edwards, handwritten note on C. W. Cooke, Memorandum for Admiral King, June 4, 1945, DNSArch.
46. Hewlett Thebaud, Memorandum for Admiral King, Rattan Project, Present Status of, June 4, 1945, DNSArch; Memorandum to be Shown (But Not Given) to Colonel H. M. O'Connor, June 12, 1945, DNSArch; Memorandum for Admiral King, Signal Intelligence Activities, and Note for Record by Clayton Bissell, June 6, 1945, DNSArch.
47. *HV*, 29–32; Information from Robert L. Benson.
48. Goldman, *Crucial Decade*, 28; "GIs Protest on Slow Demobilization," *NYT*, January 13, 1946; "Army Seeks a Way Out of Morale Crisis," ibid.; "Caution on Morale Given Eighth Army," *NYT*, January 10, 1946.
49. *CW*, 17, 26–27.

50. *FRUS, Intelligence Establishment, 1945–50,* Introduction.
51. Richard Park, Memorandum for the President, Parts I–III, Rose A. Conway Files, Harry S. Truman Library, Independence, MO; Selected Documents Concerning OSS Operation in Lisbon 5 May–13 Jul 1943, SRH-113, Studies on Cryptology, NARA, 17–18; Recent Messages Dealing with the Compromise of Japanese Codes in Lisbon, NR 2608, HCR; Alvarez, *Spies in Vatican,* 248–52.
52. McCullough, *Truman,* 429.
53. Ibid., 355, 371–73.
54. William D. Leahy, Memorandum for the Secretary of War and the Secretary of State, August 22, 1945, DNSArch.
55. Memorandum, Harry S. Truman, August 28, 1945, NSA60; George C. Marshall and Ernest J. King, Collaboration with the British Foreign Office in the Communication Intelligence Field, Continuation and Extension of, September 1, 1945, DNSArch; Burns, *Origins,* 26.
56. Op-20-G, Memorandum for the Vice Chief of Naval Operations, The Continuation and Development of Communication Intelligence, August 21, 1945, NSA60; James F. Byrnes to Secretary of War, August 17, 1945, DNSArch; Schlesinger, "Cryptanalysis for Peacetime."
57. Chronological Record of 3-G-10-Z (The Russian Language Section), June 1943–March 1947, DNSArch; J. N. Wenger to B. F. Roeder, December 27, 1945, DNSArch; SSA General Cryptanalytic Branch—Annual Report FY 1945, NR 4360, HCR, 3.
58. Alvarez, "Codebreaking in Early Cold War," 869, 875.
59. J. N. Wenger, Army-Navy Collaboration in Communication Intelligence; Reasons for, February 16, 1945, DNSArch.
60. Navy's Interest in Processing of Intercepted Foreign Communications Other Than Military, September 19, 1945, DNSArch; Burns, *Quest for Centralization,* 29; J. N. Wenger to Commander Manson, February 20, 1946, DNSArch.
61. W. F. Clarke, Post-War Organisation of G.C. and C.S., April 1, 1945, HW 3/30, TNA.
62. Plan for Coordination of Army and Navy Communication Intelligence Activities, February 15, 1946, DNSArch; British-U.S. Communication Intelligence Agreement, March 5, 1946, NSA60; *FRUS, Intelligence Establishment, 1945–50,* Introduction; McCullough, *Truman,* 486; Gaddis, *Kennan,* 216–17.

2 UNBREAKABLE CODES

1. Andrew and Gordievsky, *KGB*, 370.
2. Gouzenko, *Iron Curtain*, 264–77.
3. *HV*, 62.
4. Rhodes, *Dark Sun*, 127–28, 150–51.
5. Craig, "Canadian Connection," 216, 223*n*33; Rhodes, *Dark Sun*, 127.
6. *HV*, 63.
7. Ibid., 48.
8. Ibid.; Army Security Agency, Cryptanalytic Branch Annual Report 1945–1946, author's collection, 18.
9. *HV*, 47, 70–73.
10. Levenson, OH, 16; NSA and CIA, *Venona*, Preface; *AC*, I:278.
11. Hatch, *Tordella*, 2; Davis, *Candle in Dark*, 9.
12. *HV*, 22*n*44.
13. Kirby, "Origins of Soviet Problem," 54–55.
14. Rezabek, "Last Great Secret," 514–15; Marshall to Eisenhower, August 7, 1944, General Records, NARA.
15. Levenson, OH, 3, 43–44; Final Report of TICOM Team 1, June 16, 1945, TArch, 1.
16. Rezabek, "Last Great Secret," 516; Final Report of TICOM Team 1, TArch, 24–25; Levenson, OH, 44, 48.
17. Campaigne, OH, 24–25.
18. Final Report of TICOM Team 1, TArch, 39–40.
19. TICOM M-8, Tests on Baudot Equipment Conducted in the UK June 29 to July 8, 1945, TArch; Rezabek, "Russian Fish," 65–67; Rezabek, "Search for OKW/Chi," 147–48.
20. Final Report of TICOM Team 3 on Exploitation of Bergscheidungen, TArch, 1–5; *HV*, 54–59, 73.
21. TICOM I-173, Report by the Karrenberg Party on Russian W/T, December 16, 1945, TArch; TICOM I-168, Report by the Karrenberg Party on Miscellaneous Russian W/T, November 12, 1945, TArch; Final Report of TICOM Team 1, TArch, 40–41. A complete list of Russian code systems studied by the Germans appears in European Axis Signals Intelligence in World War II as Revealed by "TICOM" Investigations, Volume 1, Synopsis, NSAD.
22. TICOM DF-98, Russian Baudot Teletype Scrambler, TArch; TICOM I-169, Report by Uffz. Karrenberg on the Bandwurm, December 2, 1945, TArch.
23. TICOM I-30, Interrogation of Uffz. Karrenberg at Steeple Claydon, July 7, 1945, TArch; TICOM I-169, ibid.

24. TICOM I-169, ibid.
25. Cryptanalysis of "Caviar," ca. 1945, DNSArch.
26. Summary of War Diary G4A April 1946; H. Campaigne, Report on Status of Projects, August 7, 1946, handwritten note with Summary of War Diary G4-A June 1946. I am grateful to Ralph Erskine for copies of these subsequently reclassified Op-20-G4A war diaries.
27. TICOM M-7, Report on Multichannel Intercept Teletype (HMFS), July 22, 1945, TArch; Davis, *Candle in Dark,* 7.
28. The evidence that Longfellow and Caviar are one and the same is circumstantial but convincing. Longfellow is unmistakably referred to in several available documents as a Russian teleprinter encipherment device. Howard Campaigne stated that Longfellow was first used in 1943 (*WM,* 267). All available evidence, from TICOM and elsewhere, points to the Soviets' having only one such teleprinter encipherment system at that time. In the Op-20-G4A war diaries for December 1945 through July 1946, the terms Caviar and Longfellow appear at different times but never together in the same month's report. A list of continuing projects in December 1945 includes "Attack on Caviar"; in the next month's report the list includes "Attack on Longfellow," with no mention of Caviar, even though the project was definitely continuing at that point. In April 1946 the war diary notes that a Lieutenant Clymer has "been borrowed from NY1" group to work on Caviar; in August 1946 the diary reports that "Lt. Clymer has returned to NY1, taking with him the problem of reading depths on Longfellow."
29. J. N. Wenger to Sir Edward W. Travis, ca. February 15, 1947, DNSArch; Tan Analog, Cryptanalytic Machines in NSA, May 30, 1953, Friedman Documents, NSAD; *WM,* 224, 267.
30. Clayton Bissell, Post-War Signal Intelligence Activities, June 16, 1945, DNSArch; U.S. Navy Communication Intelligence Organization Ultimate Post-War Strength, 1945, NSA60.
31. Budiansky, "Cecil Phillips," 98; Alvarez, *Secret Messages,* 121–22; Budiansky, *Battle of Wits,* 231.
32. *The 6813 News,* vol. 1, no. 2, September 11, 1947, NCM.
33. Personnel Policy at Army Security Agency, December 23, 1946, ASA Personnel Strength and Policy Memoranda, 1945–1951, General Records, NARA; Brownell Report, 126.
34. J. N. Wenger, The Continuation and Development of Communication Intelligence, August 21, 1945, NSA60, 4–5; Hatch, *Tordella,* 3; Campaigne, OH, 53; Personnel Requirements for Peacetime Operation of Communication Intelligence Organization, May 9, 1946, DNSArch.

35. Army Security Agency Monthly Status Report for May 1946, DNSArch; Budiansky, *Battle of Wits,* 165, 215.
36. Frank B. Rowlett, Staff Study on Personnel Needs, March 17, 1945, ASA Personnel Strength and Policy Memoranda, 1945–1951, General Records, NARA.
37. Code Word CREAM ASA Interpretation, ca. 1946, DNSArch.
38. Myron C. Cramer, Legality of Signal Intelligence Activities, August 16, 1945, DNSArch.
39. Snider, "Church Committee's Investigation"; U.S. Senate, *Supplementary Staff Reports,* 770–74; Carter W. Clarke, Memorandum for the Honorable James Forrestal, December 13, 1947, DNSArch.
40. Walter L. Pforzheimer, Crypto Security Bill, February 19, 1948, DNSArch.

3 LEARNING TO LIE

1. NSA and CIA, *Venona,* Preface.
2. Benson, *Venona Story,* 15.
3. *HV,* 75, 77.
4. Ibid., 77–80; I. D. Special Analysis Report #1, Covernames in Diplomatic Traffic, August 30, 1947, VENONA Documents, NSAD.
5. Development of the "G"—"HOMER" Case, October 11, 1951, VENONA Documents, NSAD.
6. *HV,* 74, 89; McKnight, "Moscow-Canberra Cables," 164.
7. Arrangements for Contact with "DAN" in Great Britain, September 15, 1945, VENONA Documents, NSAD.
8. Haynes, Klehr, and Vassiliev, *Spies,* 400.
9. Pipes, *Russian Revolution,* 109–19.
10. Gaddis, *Kennan,* 97, 195, 220.
11. Ibid., 220.
12. Ibid., 189–90, 226; *CW,* 29.
13. Gaddis, *Kennan,* 221–22, 229–30.
14. *CW,* 11; Montefiore, *Stalin,* 5; McCullough, *Truman,* 418–19.
15. Montefiore, *Stalin,* 6–7.
16. Pipes, *Russian Revolution,* 789–805; Pipes, *Bolshevik Regime,* 121.
17. Montefiore, *Stalin,* 228–29, 232–33, 243, 244; *CW,* 99.
18. Gaddis, *Kennan,* 105.
19. Montefiore, *Stalin,* 505–7.
20. Pipes, *Russian Revolution,* 801.
21. McCullough, *Truman,* 533, 561; *CW,* 30.
22. McCullough, *Truman,* 541, 545, 561.

23. Ibid., 565; Goldman, *Crucial Decade,* 77.
24. National Security Council Directive on Office of Special Projects, NSC 10/2, June 18, 1948, *FRUS, Intelligence Establishment, 1945–1950,* 714.
25. *CW,* 163–64.
26. "Public Trust in Government: 1958–2014," Pew Research Center, November 13, 2014, Web.
27. Gaddis, *Kennan,* 221.
28. United States Objectives and Programs for National Security, NSC-28, April 14, 1950, *FRUS 1950, National Security,* 234–92.
29. *CW,* 165–66.
30. NSA and CIA, *Venona,* Preface; Haynes and Klehr, *Venona,* 157–60.
31. McCullough, *Truman,* 537–39, 860.
32. Ibid., 551–53, 652, 768; NSA and CIA, *Venona,* Preface.
33. D. M. Ladd to H. B. Fletcher, [Redacted] Material, October 18, 1949, DNSArch.
34. *WM,* 224.
35. William F. Friedman, A Progenitor of [Pagoda]: Or, Can Cryptologic History Repeat Itself, July 21, 1948, Friedman Documents, NSAD.
36. Summary of War Diary G4A April 1946, copy from Ralph Erskine.
37. Comdr. P. H. Currier to Captain Wenger, April 8, 1947, DNSArch; Denham, "Hugh Alexander," 31; Stork, Piccolo I–IV, Cryptanalytic Machines in NSA, May 30, 1954, Friedman Documents, NSAD.
38. Analysis of the Hagelin Cryptograph B-211, Technical Paper, Signal Intelligence Service, 1939, NR 3833, HCR, 1–7.
39. Hagelin, "Story of Cryptos," 216; Summary of War Diary G4A March 1946, copy from Ralph Erskine; Alvarez, "Behind Venona," 180, 185*n*4; Davis, *Candle in Dark,* 16.
40. C to Prime Minister, June 24, 1941, HW 1/6, TNA.
41. Andrew and Gordievsky, *KGB,* 216–17, 304–5, 400; Benson, *Venona Story,* 50; Information from Baron, April 3, 1941, VENONA Documents, NSAD.
42. Kahn, "Soviet COMINT," 14.
43. Burns, *Origins,* 7–9.
44. Budiansky, *Battle of Wits,* 228–29; Commander Humphrey Sandwith, May 20, 1942, HW 14/46, TNA.
45. Burns, *Origins,* 37–39; *HV,* 87.
46. Burns, *Origins,* 53–55; Kirby, OH, 16.
47. F. L. Lucs, quoted in Smith, *Station X,* 35.
48. Bennett, "Hut 3," 34–35; Hinsley et al., *British Intelligence,* III:402.
49. Peterson, "Beyond BOURBON," 2–3; Kirby, OH, 18.

4 DIGITAL DAWN

1. Goldstine, *Computer,* 167.
2. Ibid., 182.
3. Ibid., 191, 214.
4. Tomash and Cohen, "Birth of ERA," 84.
5. Campaigne, OH, 54; Pendergrass, OH, 13.
6. Pendergrass, OH, 10–11.
7. J. T. Pendergrass, Cryptanalytic Use of High-Speed Digital Computing Machines, 1946, NCM, 1.
8. Tomash and Cohen, "Birth of ERA," 84–85; Development Contract with Northwestern Aeronautical Corporation—Summary of Background Information, Bureau of Ships, August 16, 1946, DNSArch.
9. Parker, OH, 15.
10. Tomash and Cohen, "Birth of ERA," 85–88; Parker, OH, 23, 27.
11. Tomash and Cohen, "Birth of ERA," 90–91; Gray, "Atlas Computer"; Eachus et al., "Computers at NSA," 7.
12. Pendergrass, OH, 15–16.
13. Tomash and Cohen, "Birth of ERA," 90; Snyder, *NSA Computers,* 8.
14. History of Machine Branch, NR 3247, HCR, 40–41; History of Cryptanalysis of Japanese Army Codes, NR 3072, HCR, 36–38.
15. History of Machine Branch, NR 3247, HCR, Appendix II; IBM Slide Runs, Use of High-Speed Cryptanalytic Equipment, September 14, 1945, NR 2807, HCR.
16. Brief Descriptions of RAM Equipment, October 1947, NR 1494, HCR, 14–19; *WM,* 171–73.
17. William F. Friedman, Development of RAM, June 14, 1945, NR 2808, HCR; Good, "Enigma and Fish," 163; Brief Descriptions of RAM Equipment, October 1947, NR 1494, HCR.
18. *WM,* 178–79, 199.
19. Snyder, *NSA Computers,* 8–9.
20. Ibid., 17–19.
21. Ibid., 21–22.
22. Dumey, OH, 7.
23. Snyder, *NSA Computers,* 22.
24. *WM,* 221–24; Goldberg, Cryptanalytic Machines in NSA, May 30, 1953, Friedman Documents, NSAD; Demon I–III, Brief Description of Analytic Machine Fourth Installment, September 20, 1954, ibid.; Gray, "Atlas Computer."

25. Clay, *Decision in Germany,* 366.
26. *CW,* 33, 71.
27. McCullough, *Truman,* 630–31, 647–48; Rhodes, *Dark Sun,* 329–30.
28. Lilienthal, *Journals,* 391.
29. Developments in Soviet Cypher and Signals Security, 1946–1948, ca. December 21, 1948, DNSArch; *WM,* 267; Alvarez, "Behind Venona," 185*n*4.
30. TI Item #137, NT-1 Traffic Intelligence, November 2, 1948, DNSArch.
31. Information from Robert L. Benson.
32. *On Watch,* 19.
33. Peterson, "Beyond BOURBON," 17; *WM,* 274–75.
34. Information from Robert L. Benson.
35. Peterson, "Beyond BOURBON," 24–25.
36. Ibid., 11–12.
37. Davis, *Candle in Dark,* 16.
38. Ibid., 16–17.
39. Peterson, "Beyond BOURBON," 8, 33; Davis, *Candle in Dark,* 16–19. Although the entire text of the initial 1946 agreement has been declassified, NSA as of 2015 has refused to release any of the thirty pages of Appendix K. (UKUSA COMINT Agreement and Appendices Thereto, 1951–1953, NSAD.)
40. Davis, *Candle in Dark,* 18, 25.
41. Ibid., 29, 53.
42. History of Cryptanalysis of Japanese Army Codes, NR 3072, HCR; Davis, *Candle in Dark,* 10–11, 15–17.
43. Davis, *Candle in Dark,* 6–7.
44. Williams, *Invisible Cryptologists,* 8–9.
45. Ibid., 10–12, 24–25.
46. Brownell Report, 86; Williams, *Invisible Cryptologists,* 19–23.
47. Williams, *Invisible Cryptologists,* 21.
48. Brownell Report, 86; Peterson, "Beyond BOURBON," 33; Williams, *Invisible Cryptologists,* 20–21.
49. Williams, *Invisible Cryptologists,* 27–29, 35, 39.

5 SHOOTING WARS

1. East, "European Theater and VQ-2."
2. Schindler, *Dangerous Business,* 3.
3. East, "Pacific and VQ-1."
4. McCullough, *Truman,* 748–49.
5. Ibid., 756–58; Rhodes, *Dark Sun,* 405.

6. Rosenberg, "Origins of Overkill," 16; Rhodes, *Dark Sun,* 347.
7. Budiansky, *Air Power,* 348–50, 361–63, 365.
8. Rosenberg, "Origins of Overkill," 17.
9. Ibid., 21*n*58.
10. East, "European Theater and VQ-2."
11. Peterson, "Soviet Shootdowns," 5; *AC,* I:141; Schindler, *Dangerous Business,* 1.
12. "Partial History of ELINT," 55–56.
13. Rosenberg, "Origins of Overkill," 47.
14. Robertson Committee, Appendix III:1–2.
15. Ibid., Appendix VI:1–4.
16. Budiansky, *Battle of Wits,* 48.
17. J. H. Plumb, Notes on the Future of Signals Analysis, October 18, 1944, HW 3/30, TNA, 1; Callimahos, "Traffic Analysis."
18. Robertson Committee, 24, Appendix IX:3
19. Frahm, *SIGINT and Pusan,* 3–5; Brownell Report, 66.
20. Burns, *Quest for Centralization,* 70.
21. Aid, "US Humint and Comint," 25–27.
22. Ibid.
23. *CW,* 41–43.
24. Hatch and Benson, *Korean War,* 4–5, 8.
25. Frahm, *SIGINT and Pusan,* 6–7, 11–13; Aid, "US Humint and Comint," 46–47; Aid, *Secret Sentry,* 27.
26. *CW,* 45.
27. Vanderpool, "PRC Intervention," 9–14.
28. Aid, *Secret Sentry,* 32; Halberstam, *Coldest Winter,* 373–77.
29. Drea, *MacArthur's ULTRA,* 28–30; Aid, "US Humint and Comint," 45, 52–53.
30. Halberstam, *Coldest Winter,* 371; *FRUS 1950, Korea,* 953.
31. Aid, "US Humint and Comint," 49; Aid, *Secret Sentry,* 31–32.
32. Vanderpool, "PRC Intervention," 17–18.
33. Halberstam, *Coldest Winter,* 476–77.
34. *AC,* I:55.
35. *AC,* I:40. For the history of Y units, see Howe, *Signal Intelligence.*
36. Aid, "US Humint and Comint," 42–43.
37. Futtrell, "Intelligence in Korean War," 280–81; *AC,* I:41–42; Aid, "US Humint and Comint," 46.
38. *AC,* I:48–49; Aid, "Comint in Korean War, Part II," 38–41.
39. *AC,* I:49–51.

40. Wiley D. Ganey to Curtis LeMay, June 20, 1952, DNSArch; *CW,* 60.
41. Hatch and Benson, *Korean War SIGINT,* 15.
42. Brownell Report, 105.
43. Burns, *Quest for Centralization,* 73–74, 77–78; *AC,* I:27, 30.
44. Johnson, "Cryptology During Korean War"; J. N. Wenger, Memorandum for Mr. Brownell, March 13, 1952, NSA60; Aid, "US Humint and Comint," 49–51; Brownell Report, 79; *AC,* I:29–30.
45. Brownell Report, 89, 119–20; Burns, *Quest for Centralization,* 79, 87–88; Kirby, "Origins of Soviet Problem," 57.
46. Burns, *Quest for Centralization,* 54.
47. Ibid., 90–92; Truman Memorandum, Communications Intelligence Activities, October 24, 1952, NSA60.
48. Macintyre, *Spy Among Friends,* 143, 148.
49. Ibid., 28, 36.
50. Ibid., 19, 46, 70–71, 132–35.
51. Haynes and Klehr, *Venona,* 52–53.
52. Lamphere, *FBI-KGB,* 133–35; Benson, *Venona Story,* 19–20.
53. Rhodes, *Dark Sun,* 55, 153–55, 422.
54. Benson, *Venona Story,* 13–14; Lamphere, *FBI-KGB,* 178–86; J. Edgar Hoover to Rear Admiral Robert L. Dennison, July 18, 1950, Document 32, NSA and CIA, *Venona,* Part I.
55. Andrew, *History of MI5,* 376–78.
56. NSA and CIA, *Venona,* Preface, *n*59; Haynes and Klehr, *Venona,* 52.
57. Macintyre, *Spy Among Friends,* 144, 147, 149; New York to Moscow, June 28, 1944, VENONA Documents, NSAD; Development of the "G"—"HOMER" Case, October 11, 1951, ibid.
58. Macintyre, *Spy Among Friends,* 146–49.
59. Ibid., 150–51, 154–55.
60. Ibid., 159, 163–64, 167, 195, 258.
61. William Wolf Weisband, Washington Field Office, FBI, November 27, 1953, Document 34, NSA and CIA, *Venona,* Part I.
62. Haynes, Klehr, and Vassiliev, *Spies,* 402–4; Haynes and Klehr, "Vassiliev's Notebooks," 17–24.
63. Haynes, Klehr, and Vassiliev, *Spies,* 405; *AC,* I:277–78.

6 "AN OLD MULE SKINNER"

1. Appointment Books, 1951–57, Canine, Papers.
2. "Glimpses of a Man," 31; Gurin, "Ralph J. Canine," 8, 10; Canine, OH, I:2, III:20.

3. Canine, OH, I:2.
4. Charles P. Collins, quoted in *AC,* I:62.
5. "From Chaos Born," 28.
6. Gurin, "Ralph J. Canine," 7; Canine, OH, I:3.
7. Gurin, "Ralph J. Canine," 8.
8. Retirement Book, Canine, Papers; Gurin, "Ralph J. Canine," 9–10; *AC,* I:64–67.
9. Canine, OH, II:7.
10. Ibid., II:17–18.
11. Ibid., II:5; Gurin, "Ralph J. Canine," 8.
12. *AC,* I:64.
13. Johnson, "Human Side."
14. Information from Lewis Lord.
15. Johnson, "Human Side."
16. NSA Personnel Newsletters, 1953–1960, NCM.
17. Canine, OH, III:21.
18. Clark, *Man Who Broke Purple,* 199.
19. Johnson, "War's Aftermath"; Boone, "Management Challenges."
20. Public Law 81-513, codified as 18 USC §798 of the Espionage Act; Croner, "Espionage Act Constitutionality."
21. Recommendations Relative to the Film "Venetian Incident," January 12, 1953, Physical Security, Security Policy and Direction Files, NARA.
22. Clark, *Man Who Broke Purple,* 194; Coffey, OH; Sheldon, *Friedman Collection,* 6–7; Classification of Material Received from Mr. Friedman, February 6, 1959, Friedman Documents, NSAD.
23. Minutes of the Second Meeting of the USCIB Committee on Personnel Security Standards and Practices, October 11, 1955, DNSArch, 14.
24. J. Edgar Hoover quoted in U.S. Congress, *Security Practices,* 14–15; National Research Council, *Polygraph and Lie Detection.*
25. U.S. Congress, *Security Practices,* 11; Report and Recommendations on Procedures Implementing USCIB Directive #5, September 19, 1955, DNSArch, 2.
26. *AC,* I:73–74; Duncan Sinclair, Memorandum for the Record, August 7, 1956, Modified Interview Program—1956, General Records, NARA.
27. *AC,* I:74; Sheldon, *Friedman Collection,* 12, 71; MacDonald, "Lie-Detector Era," 23–25, 29; Howard Campaigne, Memorandum for the Chief of Staff, Subject: The Polygraph, May 11, 1953, Friedman Documents, NSAD.
28. The Petersen Case, Memorandum for the Members of USCIB, May 5, 1955, DNSArch, 2.

29. Wiebes, "Petersen Spy Case," 490–93.
30. Ibid., 494, 497–99.
31. *AC*, I:279; Mowry, "Petersen."
32. "Dutch Say Petersen Gave Data, but They Thought He Had Right," *NYT*, October 20, 1954; *AC*, I:280.
33. The Petersen Case, Memorandum for the Members of USCIB, May 5, 1955, DNSArch, 1; Wiebes, "Petersen Spy Case," 500–501, 505–7, 510, 528.
34. William P. Rogers to Allen W. Dulles, November 18, 1954, DNSArch; Wiebes, "Petersen Spy Case," 520–21, 523.
35. The Petersen Case, Memorandum for the Members of USCIB, May 5, 1955, DNSArch, 1–2.
36. Ibid., 2; Wiebes, "Petersen Spy Case," 504; Frank Lewis, letter to author, March 20, 1999.
37. Aldrich, "GCHQ in Cold War," 91.
38. *WM*, 277; Alleviation of Overcrowding upon Initial Occupancy of NSA Operations Building at Fort Meade, June 1, 1955, NSA60.
39. *AC*, I:206–9; Task Force Report on Intelligence Activities, Appendix I, May 1955 [Clark Report], DNSArch, 1:11.
40. *AC*, I:50–51, 112, 118; Aid, "Comint in Korean War, Part II," 38.
41. Smith, *New Cloak*, 180–82; Aldrich, "GCHQ in Cold War," 74–78.
42. UKUSA COMINT Agreement and Appendices Thereto, March 5, 1946, Parts 6 and 8, NSAD; Aid, *Secret Sentry*, 12–13; Aldrich, "GCHQ in Cold War," 76–77.
43. Aldrich, "GCHQ in Cold War," 81, 84.
44. Bower, *Perfect English Spy*, 159–61; Macintyre, *Spy Among Friends*, 198–202.
45. Taubman, *Khrushchev*, 355–58; "CIA Files Reveal Cold War Leader's Anger," BBC News, November 6, 2009, Web; Macintyre, *Spy Among Friends*, 203.
46. "British to Debate Frogman Episode," *NYT*, May 11, 1956; Macintyre, *Spy Among Friends*, 202–5.
47. Aldrich, "GCHQ in Cold War," 83–85.
48. Easter, "GCHQ in 1960s," 683; Aldrich, "GCHQ in Cold War," 67.
49. Hodges, *Turing*, 496–97.
50. Canine, OH, I:8.
51. Ibid., III:14–15; Management Development at the Executive Level: A Planned Program, July 1, 1952, Canine, Papers.
52. Gurin, "Cryptologic Mission"; Exinterne, "Intern Program, Part Three," 16.
53. Lewis, "May Rub Off," 2.
54. "Finding a Home for AFSA," 1; *AC*, I:66.

55. "Finding a Home for AFSA," 2; Preliminary Field Survey, Louisville and Fort Knox Area, Administrative and Morale Section, 8 to 18 August, 1951, NSA60, 3, 5, 17.
56. Johnson, "The Move," 95–96.
57. Ibid., 97–99; Canine, OH, II:9; Collins, *History of SIGINT,* III:40.

7 BRAINS VERSUS BUGS

1. Brownell Report, 83, 137–38; Robertson Committee, 2; Task Force Report on Intelligence Activities [Clark Report], May 1955, DNSArch, 1:48, 57a.
2. "Before Super-Computers"; Lavington, "Description of Oedipus"; Baker Panel, 7.
3. Summary of Early Operations on the Robin Machinery, May 18, 1951, document released to author under NSA Mandatory Declassification Review, May 13, 2015; *WM,* 274–75; Meeting of Special Communications Advisory Group (SCAG), March 12, 1952, Friedman Documents, NSAD, 12–13, 47.
4. Stamp and Chan, "SIGABA Keyspace," 208.
5. Antal and Zajac, "Key Space of Fialka"; "M-125 (Fialka)," Crypto Museum, Web.
6. Baker Panel, 1, 3, Technical Adjunct III:3.
7. Gaddis, *Kennan,* 89, 106; Kahn, "Soviet COMINT," 16–17.
8. Andrew and Mitrokhin, *Sword and Shield,* 338.
9. Andrew and Gordievsky, *KGB,* 226–28, 451–53.
10. Gaddis, *Kennan,* 464–65; Brooker and Gomez, "Termen's Bug," 4; Andrew and Mitrokhin, *Sword and Shield,* 338.
11. Nikitin, "Theremin," 254–55; Brooker and Gomez, "Termen's Bug," 5–8.
12. Nikitin, "Theremin," 252–53; Haynes, Klehr, and Vassiliev, *Spies,* 362–66.
13. Hove, *Diplomatic Security,* 175–78; Andrew and Gordievsky, *KGB,* 452–53.
14. Andrew and Gordievsky, *KGB,* 454; Kahn, "Soviet COMINT," 7, 9–10.
15. *AC,* I:93; Sinkov, OH, 119; Buffham, OH, 17.
16. Staff D Comments on Part I of Clark Report, DNSArch; Aid, *Secret Sentry,* 46.
17. Murphy, Kondrashev, and Bailey, *Battleground Berlin,* 208; "Engineering the Berlin Tunnel."
18. Steury, ed., *Intelligence War in Berlin,* Part V; Murphy, Kondrashev, and Bailey, *Battleground Berlin,* 206; *Operation REGAL,* 2–3.
19. Thomas, *Very Best Men,* 130–32; "America's James Bond: The New Biography of William King Harvey," *Tribune-Star* (Terre Haute, IN), April 28, 2007.
20. CIA, *Berlin Tunnel Operation,* 3–4.

21. Murphy, Kondrashev, and Bailey, *Battleground Berlin,* 213, 219.
22. "Engineering the Berlin Tunnel"; CIA, *Berlin Tunnel Operation,* 17, 21, 25–26.
23. *Operation REGAL,* 3, 10; Hatch, "Berlin Tunnel, Part II"; CIA, *Berlin Tunnel Operation,* Appendix B: Recapitulation of the Intelligence Derived.
24. Murphy, Kondrashev, and Bailey, *Battleground Berlin,* 216; Macintyre, *Spy Among Friends,* 233–39.
25. "Revealed: Grim Fate of the MI6 Agents Betrayed by George Blake," *Telegraph,* March 14, 2015.
26. Steury, ed., *Intelligence War in Berlin,* Part V; Gurin, "Ralph J. Canine," 10; Murphy, Kondrashev, and Bailey, *Battleground Berlin,* 227.
27. CIA, *Berlin Tunnel Operation,* 21, 27, Appendix D: Round-Up of East German Press Reaction; *Operation REGAL,* 19; Murphy, Kondrashev, and Bailey, *Battleground Berlin,* 232.
28. CIA, *Berlin Tunnel Operation,* Appendix C: Typical American Press Comment.
29. *Operation REGAL,* 1–2; CIA, *Berlin Tunnel Operation,* 2; "TEMPEST: Signal Problem," 26–27.
30. "TEMPEST: Signal Problem," 27–28.
31. Baker Panel, 4, 9.
32. Hannah, "Frank B. Rowlett," 10.
33. *CW,* 107–8.
34. Ibid.; Murphy, Kondrashev, and Bailey, *Battleground Berlin,* 159–61.
35. *CW,* 108–9.
36. Hungary, October 25, 1956, Document 0000119733, CIAL; Colley, "Shadow Warriors," 30; Military Activity Connected with the Hungarian Crisis, October 27, 1956, Document 0000119739, CIAL; Current Intelligence Weekly Review, November 8, 1956, Document 0000119763, CIAL, Part I, 8.
37. *Suez Crisis,* 19–20; Aid, *Secret Sentry,* 48.
38. *Suez Crisis,* 31–32.
39. *AC,* I:235.
40. *Suez Crisis,* 30–31; *AC,* I:239.
41. Baker Panel, 56–57, Digest: 3.
42. Ibid., Synopsis:7.
43. *WM,* 308–9.
44. Brown, *Scientific Advisory Board,* 8; Special Cryptologic Advisory Group (SCAG), Agenda for First Conference of SCAG, 4–5 June 1951, Friedman Documents, NSAD; Meeting of Special Communications Advisory Group

(SCAG), March 12, 1952, ibid., 37; John A. Samford to W. O. Baker, September 24, 1957, Baker Panel 1957–1963, Baker, Papers.

45. Comments on SCAMP, October 22, 1952, Friedman Documents, NSAD; C. B. Tompkins, Summary of a Meeting at Which NSA and SCAMP Problems Were Discussed, April 17, 1955, ibid., 5.
46. C. B. Tompkins, ibid., 7, 9, 16.
47. Snyder, "Influence on Computer Industry," 81; HARVEST in Perspective, HARVEST General Purpose Computer & 7950 Data Processing System, 1957–1959, Research Files Relating to HARVEST, NARA, 5.
48. Meyer, "Wailing Wall," 73, 78.
49. Report of Special Study Group on FARMER-NOMAD, November 15, 1954, Friedman Documents, NSAD, 15–23; *WM*, 314.
50. Recommendations for a Full-Scale Attack on the Russian High-Level Systems, June 8, 1956, document released to author under NSA Mandatory Declassification Review, May 13, 2015.
51. *WM*, 295–98; Estimated EDPM Rental Budget, HARVEST General Purpose Computer & 7950 Data Processing System, 1957–1959, Research Files Relating to HARVEST, NARA; Snyder, *NSA Computers*, 40, 60–61, 93–94.
52. *WM*, 298, 312; HARVEST Purchase Description, Research Files Relating to HARVEST, NARA; Snyder, *NSA Computers*, 22, 42, 55.
53. Snyder, "Influence on Computer Industry," 80; *WM*, 301–7; Brown, *Scientific Advisory Board*, 28.
54. *WM*, 313–16, 324*n*150; Baker Panel, Appendix I:1; Bissell, *NSA Cryptanalytic Efforts*, 12–14.
55. Baker Panel, 24, 62, Appendix II:16–17.
56. Baker Panel, 28, 56–57; *WM*, 301.
57. Lewis, "May Rub Off," 5.
58. Baker Panel, Technical Adjunct III:5; Shannon, "Entropy of English," 50.
59. Baker Panel, Appendix II:1–14, Technical Adjunct I:1.
60. Buffham, OH, 18; Brown, *Scientific Advisory Board*, 44.
61. Baker Panel, Technical Adjunct III:4.

8 DAYS OF CRISIS

1. Audio Recording of Press Conference, William H. Martin and Bernon F. Mitchell, Moscow, September 6, 1960, NSA60; "Two Code Clerks Defect to Soviet, Score U.S. 'Spying,'" *NYT*, September 7, 1960.
2. *CW*, 167–68.
3. Taubman, *Khrushchev*, 460, 462–63, 467, 468; *CW*, 168.

4. The Security Program of AFSA and NSA, 1949–1962, NSA Historian, Office of Central Reference, October 1963, 207–27; Summary of Office of Security Services investigation, July 26, 1960, to February 1, 1961. I am grateful to Rick Anderson for supplying copies of these declassified documents. Additional details on NSA's investigation mentioned here are found in Missing NSA Personnel, Memorandum for Deputy Secretary Douglas, August 1, 1960, NSA60.
5. Rick Anderson, "The Worst Internal Scandal in NSA History Was Blamed on Cold War Defectors Homosexuality," *Seattle Weekly News,* July 17, 2007; Predeparture Statement, "Text of Statements Read in Moscow by Former U.S. Security Agency Workers," *NYT,* September 7, 1960.
6. U.S. Congress, *Security Practices,* 14–15; Security Program of AFSA and NSA, 226–27.
7. Anderson, "Worst Internal Scandal."
8. Sergeant Jack E. Dunlap, U.S. Army, November 22, 1963, DNSArch; Andrew and Gordievsky, *KGB,* 459–60.
9. Ibid.; Kahn, *Codebreakers,* 696–97; *AC,* II:470.
10. U.S. Congress, *Security Practices,* 10–12; Kahn, *Codebreakers,* 695–96.
11. Gurin, "Ralph J. Canine," 10.
12. Paul S. Willard, Memo for the Record, October 5, 1960, Martin and Mitchell Defection, General Records, NARA.
13. Burke, ed., "Last Bombe Run."
14. Budiansky, *Battle of Wits,* 293.
15. Clark, *Man Who Broke Purple,* 162.
16. Notes and Draft of Historical Record of Boris Hagelin, Friedman Documents, NSAD; "The Hagelin Cryptographer, Type CX-52, Instructions for Operation," Crypto Museum, Web.
17. Personal for Friedman from Canine, July 22, 1954, Personal Messages Concerning Hagelin Machines, Friedman Documents, NSAD; Ralph J. Canine, Memorandum for the Members of USCIB, December 17, 1954, ibid.
18. Currier from Shinn, November 15, 1955, Personal Messages Concerning Hagelin Machines, ibid.; Friedman Letter to Hagelin, Oral Report from Barlow on Visit to Hagelin, November 29, 1956, ibid.
19. Report of Visit to Crypto A.G. (Hagelin) by William F. Friedman, 21–28 February 1955, ibid., 20–21; Report by William F. Friedman to the Director, Armed Forces Security Agency on the Aktiebolaget Cryptoteknik, Stockholm, Sweden (The Hagelin Cryptograph Company), May 22, 1951, Negotiations in Regard to Hagelin Machines, ibid., 1, 5; Personal for Fried-

man from Canine, July 22, 1954, Personal Messages Concerning Hagelin Machines, ibid.

20. Friedman Memorandum to General Canine Passing on Extract of Letter from Hagelin Commenting on Siemens Random Tape Device, August 7, 1956, ibid.
21. Memorandum for the Record, Subject: Hagelin Negotiations, December 18, 1957, ibid., 14, 17–18. The more secure versions of the machine were most likely supplied with extra rotors that could be swapped in and out, considerably increasing the number of permutations that a codebreaker would need to try.
22. Report of Visit to Crypto A.G. (Hagelin) by William F. Friedman, 21–28 February 1955, ibid., 21; Memorandum for Colonel Davis, Subject: 16 June Comments of Mr. Friedman, June 17, 1955, ibid.
23. Bamford, *Puzzle Palace,* 409.
24. Clark, *Man Who Broke Purple,* ix–x, 188–89. Clark stated that NSA officials told him such a disclosure "might deprive the NSA of the daily information enabling it to read the secret messages of other NATO countries." Boris Hagelin's August 7, 1956, letter to Friedman, however, alluded to a different concern: that any devices made by Hagelin for NATO not be sold to other countries.
25. Scott Shane and Tom Bowman, "Rigging the Game: Spy Sting," *Baltimore Sun,* December 10, 1995.
26. Clark, *Man Who Broke Purple,* 186–87, 199–200.
27. Ibid., 201.
28. "Piggy-Back Satellites Hailed as Big Space Gain for U.S.," *Washington Post,* June 23, 1960.
29. Manchester, *Glory and Dream,* 788–89, 792–93.
30. McDonald and Moreno, *Grab and Poppy,* 2–3; "GRAB: World's First Reconnaissance Satellite," Naval Research Laboratory, Web.
31. McDonald and Moreno, *Grab and Poppy,* 9.
32. Samos Special Satellite Reconnaissance System, Document 628, WS117L, SAMOS, and SENTRY Records, National Reconnaissance Office Declassified Records, Web.
33. Ibid.; *AC,* II:404; Bernard, *ELINT at NSA,* 2–7.
34. *AC,* II:403–5.
35. Ibid.; Day, "Ferrets Above," 451–52.
36. *AC,* II:338–41; Kirby, OH, 61; Campaigne, OH, 125–26; Day, "Ferrets Above," 452.

37. O'Hara, "Telemetry Signals."
38. *AC,* II:343–44, 409–10.
39. Wagoner, *Space Surveillance,* 3–6, 19, 32–33; Matthew Aid, "Listening to the Soviets in Space," March 10, 2013, Web.
40. *CW,* 76.
41. Taubman, *Khrushchev,* 537; "U.S. and Russian Strategic Nuclear Forces," Natural Resources Defense Council, Web; Blight, Allyn, and Welch, *Cuba on the Brink,* 130, 135–36.
42. Taubman, *Khrushchev,* 537, 541.
43. Blight, Allyn, and Welch, *Cuba on the Brink,* 84–86; Blight and Welch, *On the Brink,* 291.
44. Hatch, "Juanita Moody"; Budiansky, *Battle of Wits,* 311–12.
45. Johnson and Hatch, *NSA and Cuban Crisis,* 3; Spanish-Speaking Pilots Training at Trencin Airfield, Czechoslovakia, June 19, 1961, Cuban Missile Crisis Documents, NSAD.
46. *AC,* II:318–19; Wigglesworth, "Cuban Crisis SIGINT," 79, 83.
47. Operators' Chatter Reveals Cuban Air Force Personnel to Learn Russian, May 23, 1961, Cuban Missile Crisis Documents, NSAD.
48. *AC,* II:315–16, 320–22; Wigglesworth, "Cuban Crisis SIGINT," 79.
49. First ELINT Evidence of Scan Odd Radar in Cuban Area, June 6, 1962, and Unusual Number of Soviet Passenger Ships En Route to Cuba, July 24, 1962, Cuban Missile Crisis Documents, NSAD.
50. Johnson and Hatch, *NSA and Cuban Crisis,* 4–5; New Radar Deployment in Cuba, September 19, 1962, Cuban Missile Crisis Documents, NSAD.
51. *AC,* II:327; Johnson and Hatch, *NSA and Cuban Crisis,* 6–9.
52. *AC,* II:325–26.
53. Ibid., II:329.
54. Ibid.
55. Blight, Allyn, and Welch, *Cuba on the Brink,* 163; Taubman, *Khrushchev,* 567–70; Manchester, *Glory and Dream,* 971.
56. Taubman, *Khrushchev,* 573–75.
57. Alvarez, "Cuban Crisis," 176*n*10; *AC,* II:317.
58. *AC,* I:253–56, II:366, 374; Klein, *TSEC/KW-26,* 10–11.
59. Gallo, "NSA Signal Collection," 54–55; *AC,* II:350, 362, 368–73.
60. *AC,* II:347–49; Johnson, "War's Aftermath"; Comments on the National SIGINT Program, May 20, 1965, President's Foreign Intelligence Advisory Board, 1958–1965, Baker, Papers, 1.
61. *AC,* II:332, 352–53; Interview with Admiral Bobby Inman, March 12, 2015.

9 REINVENTING THE WHEEL

1. *AC,* II:353, 356; Desmond Morton to C, September 27, 1940, HW 1/1, TNA.
2. Manchester, *Glory and Dream,* 1012.
3. Ibid., 1040; Memorandum from the Secretary of Defense to the President, March 16, 1964, Document 84, *FRUS, Vietnam, 1964; CW,* 168–69.
4. Sheehan, *Bright Shining Lie,* 136–37; Halberstam, *Best and Brightest,* 149–50.
5. Sheehan, *Bright Shining Lie,* 122–24, 289–90.
6. Hanyok, "Skunks, Bogies," 4–7.
7. Ibid., 12; Moïse, *Tonkin Gulf,* 60; Zaslow, OH, 33–34.
8. 2/G11/VHN/R01-64, August 1, 1964, From Phu Bai, SIGINT Reports and Translations, Gulf of Tonkin Release, NSAD.
9. 2/Q/VHN/R26-64, August 1, 1965, From San Miguel, Philippines, SIGINT Reports and Translations, Gulf of Tonkin Release, NSAD.
10. Hanyok, "Skunks, Bogies," 13–17.
11. Ibid., 15, 18; Aid, *Secret Sentry,* 88–89.
12. *On Watch,* 46–49; Hanyok, "Skunks, Bogies," 23–24.
13. Hanyok, "Skunks, Bogies," 22, 25; *AC,* II:523; *On Watch,* 50.
14. Moïse, *Tonkin Gulf,* 197.
15. Hanyok, "Skunks, Bogies," 26–27, 38.
16. Ibid., 26–36.
17. Ibid., 2–4, 40–44.
18. Scott Shane, "Vietnam Study, Casting Doubts, Remains Secret," *NYT,* October 31, 2005.
19. *AC,* II:503.
20. Hanyok, *Spartans in Darkness,* 125–29; *AC,* II:504.
21. Hanyok, *Spartans in Darkness,* 129–31.
22. Ibid., 134–36.
23. *AC,* II:543, 547.
24. Ibid., II:502, 506, 534.
25. Ibid., II:511; Hanyok, *Spartans in Darkness,* 146–49.
26. *AC,* II:551, 553; *PURPLE DRAGON,* 10.
27. Maneki, "Lessons of Vietnam," 19, 31; *AC,* II:551–55.
28. *AC,* II:380, 551–55; Maneki, "Lessons of Vietnam," 34–35.
29. Budiansky, *Air Power,* 384; *AC,* II:538.
30. *AC,* II:545–55; Hanyok, *Spartans in Darkness,* 235–37.
31. *AC,* II:549, 580–81; "TEABALL," 93–95.

32. Hanyok, *Spartans in Darkness,* 149–50, 307; *AC,* II:561.
33. *AC,* II:561–62; Hanyok, *Spartans in Darkness,* 327.
34. Hanyok, *Spartans in Darkness,* 344–45.
35. *AC,* II:563–64.
36. Hanyok, *Spartans in Darkness,* 344–45.
37. Ibid., 325, 337–39.
38. Ibid., 309, 339, 348; Manchester, *Glory and Dream,* 1053, 1125–26.
39. Hanyok, *Spartans in Darkness,* 444–45; *AC,* III:10, 12.
40. Newton, *Pueblo and SIGINT,* 9–11, 42, 46–47.
41. Ibid., 18–20.
42. Ibid., 21, 31, 49.
43. Ibid., 26–27, 29–31.
44. Ibid., 48–52.
45. Ibid., 54–61.
46. Ibid., 55, 61.
47. Ibid., 63–65, 155.
48. Ibid., 114, 126, 132–34.
49. Ibid., 160–63; Memorandum for USIB Principals, Assessment of the Loss of the USS PUEBLO, May 13, 1968, USS Pueblo Release, NSAD.
50. Louis Tordella, Report on the Assessment of Cryptographic Damage Resulting from the Loss of the USS PUEBLO (AGER-2), July 28, 1969, USS Pueblo Release, NSAD; Heath, *Security Weaknesses,* 54–58.
51. Newton, *Pueblo and SIGINT,* 11; W. O. Baker, Memorandum to Honorable Clark Clifford, January 26, 1968, President's Foreign Intelligence Advisory Board, 1959–1970, Baker, Papers, 3.
52. Newton, *Pueblo and SIGINT,* 165, 174–75.
53. Howe, "Civilians in Field Operations," 5–6; *AC,* II:393.
54. "Civilianization of Harrogate," 10–11.
55. Ibid., 12–13.
56. *AC,* II:297, III:22, 209.
57. Ibid., II:382–88, 391–93.

10 BRUTE FORCE AND LEGERDEMAIN

1. Aid, *Secret Sentry,* 128; *AC,* II:293, III:21; Aid, "Time of Troubles," 7.
2. "2,000-Year-Old Transcriber," 17–18. A complete (though highly redacted) series of *Cryptologs* is at NSAD.
3. Exinterne, "Intern Program," 16, 18; Exinterne, "Intern Program, Part Three," 17.

4. Buffham, OH, 20; interview with Thomas R. Johnson, June 13, 2015; interview with Robert L. Benson, April 12, 2015.
5. Interview with Thomas R. Johnson, June 13, 2015.
6. Comments on the National SIGINT Program, May 20, 1965, President's Foreign Intelligence Advisory Board, 1958–1965, Baker, Papers, 8; *AC*, III:192.
7. *AC*, II:478; Comments on the National SIGINT Program, May 20, 1965, President's Foreign Intelligence Advisory Board, 1958–1965, Baker, Papers, 4–5.
8. Garthoff, *Directors of Central Intelligence*, 35–36.
9. Memorandum for Assistant Comptroller, Requirements and Evaluation, The CIA/NSA Relationship, August 20, 1976, DNSArch.
10. Turner, *Secrecy and Democracy*, 235.
11. *AC*, II:357, 485; Hersh, *Price of Power*, 207.
12. *AC*, III:87–88; U.S. Senate, *Huston Plan*, 6–7; U.S. Senate, *National Security Agency*, 26–27.
13. Hersh, *Price of Power*, 207–8.
14. *AC*, II:487.
15. *CW*, 177.
16. Ibid., 157, 177–78; *AC*, III:91; Seymour M. Hersh, "Huge CIA Operation Reported in U.S. Against Antiwar Forces, Other Dissidents in Nixon Years," *NYT*, December 22, 1974.
17. Snider, "Church Committee's Investigation."
18. Ibid.; *AC*, III:99, 106.
19. *AC*, III:84–85; U.S. Senate, *Supplementary Staff Reports*, 779–80.
20. *AC*, III:85; Snider, "Church Committee's Investigation"; Summary of Task Force Report on Inquiry into CIA-Related Electronic Surveillance Activities Disclosed in Rockefeller Commission Report, Draft, March 4, 1977, posted in James Bamford, "The NSA and Me," *Intercept*, October 2, 2014, Web.
21. Foreign Intelligence Surveillance Act of 1978, Public Law 95-111.
22. Memorandum from the President's Assistant for National Security Affairs (Kissinger) to President Ford, Document 175, *FRUS*, *National Security Policy, 1973–1976;* National Security Decision Memorandum 266, August 15, 1974, Document 176, ibid., *n*2.
23. Memorandum from the President's Assistant for National Security Affairs (Kissinger) to President Ford, Document 175, ibid.; Ball, *Soviet SIGINT*, 45–51; Andrew and Mitrokhin, *Sword and Shield*, 348.
24. Memorandum from the President's Assistant for National Security Affairs

(Scowcroft) and the President's Assistant for Domestic Affairs and Director of the Domestic Council (Cannon) to President Ford, January 6, 1977, Document 181, *FRUS, National Security Policy, 1973–1976.*

25. Memorandum for Brent Scowcroft, Soviet Microwave Interception Program, National Security Council, April 5, 1976, NSA60; Office of the National Counterintelligence Executive, Unauthorized Disclosure of Classified Information, September 11, 2011, National Security Archive, *CIA and Signals Intelligence.*
26. McConnell quoted in Aid, *Secret Sentry,* 152; Kahn, "Big Ear."
27. Interview with Admiral Bobby Inman, March 12, 2015.
28. Memorandum for Assistant Comptroller, Requirements and Evaluation, The CIA/NSA Relationship, August 20, 1976, DNSArch.
29. *AC,* III:224–31; interview with Admiral Bobby Inman, March 12, 2015.
30. Andrew and Mitrokhin, *Sword and Shield,* 350–51; "KGB Defector Tells of Soviet Bugging Operations in U.S.," *Washington Post,* September 14, 1999.
31. Maneki, *GUNMAN Project,* 8–9, 13–14, 17–18, 23.
32. Andrew and Mitrokhin, *Sword and Shield,* 348.
33. W. F. Clarke, Post-War Organisation of G.C. and C.S., April 1, 1945, HW 3/30, TNA.
34. *AC,* III:189–90.
35. Ibid., III:191; Hayden, "Interview"; Interview with Admiral Bobby Inman, March 12, 2015.
36. Interview with Admiral Bobby Inman, March 12, 2015; Bissell, *NSA Cryptanalytic Efforts,* 13–16.
37. *AC,* III:218.
38. *WM,* 309, 322–23*n*125.
39. Campaigne, OH, 69; *AC,* III:218.
40. Johnson, "Human Side."
41. *AC,* III:192, 208–10, 221–23.
42. Aid, *Secret Sentry,* 155–56.
43. *AC,* IV:409.
44. Ibid., IV:410–13; Aid, *Secret Sentry,* 184–85.
45. *AC,* IV:415–16.
46. Ibid., IV:417–22; Hunter, *Spy Hunter,* 85; Melton, *Spy Book,* 54.
47. Hunter, *Spy Hunter,* 184, 195–97.

EPILOGUE: THE COLLAPSE OF THE WALL, AND A VERDICT

1. *CW*, 245–46.
2. Aid, *Secret Sentry*, 175–77, 236–41.
3. *CW*, 165.

APPENDIX A: ENCIPHERED CODES, DEPTHS, AND BOOK BREAKING

1. Haynes and Klehr, *Venona*, 27, 397*n*7.
2. *HV*, 73.

APPENDIX B: RUSSIAN TELEPRINTER CIPHERS

1. TICOM DF-98, Russian Baudot Teletype Scrambler, September 20, 1943, TArch.
2. TICOM I-169, Report by Uffz. Karrenberg on the Bandwurm, December 2, 1945, TArch.
3. For a technical description of GC&CS's cryptanalysis of the German teleprinter ciphers, see Budiansky, "Colossus, Codebreaking."

APPENDIX C: CRYPTANALYSIS OF THE HAGELIN MACHINE

1. Solution of a Three Deep Hagelin, June 7, 1944, NR 2340, HCR.
2. An Insecure Use of the Hagelin Cryptograph Leading to the Discovery of Messages in Depth and the Reconstruction of Base Settings, November 20, 1944, NR 2818, HCR. Churchhouse, *Codes and Ciphers*, 133–52, has a good explanation of the process for recovering the pinwheel settings from a sequence of known key.

APPENDIX D: BAYESIAN PROBABILITY, TURING, AND THE DECIBAN

1. Good, "Enigma and Fish," 157.
2. A. M. Turing, Paper on Statistics of Repetitions, ca. 1941, HW 25/38, TNA; Good, "Turing's Statistical Work"; Copeland, "Turingery," 380.
3. Copeland, "Turingery," 380; Lobban, "Turing's Legacy."

APPENDIX E: THE INDEX OF COINCIDENCE

1. Friedman, *Index of Coincidence*.

Bibliography

ARCHIVAL SOURCES AND DOCUMENT COLLECTIONS

Central Intelligence Agency Library. Web. (CIAL)

Digital National Security Archive. *The National Security Agency, Organization and Operations, 1945–2009.* George Washington University, Washington, DC. (DNSArch)

Foreign Relations of the United States. Washington, DC: Department of State. (*FRUS*)

———. *Emergence of the Intelligence Establishment, 1945–1950.*

———. *1950. Volume I, National Security Affairs; Foreign Economic Policy.*

———. *1950. Volume VII, Korea.*

———. *The Intelligence Community, 1950–1955.*

———. *1964–1968. Volume I, Vietnam, 1964.*

———. *1969–1976. Volume XXXV, National Security Policy, 1973–1976.*

National Archives and Records Administration. Records of the National Security Agency, Record Group 457. College Park, MD. (NARA)

———. General Records, 1948–1969. Entry P5.

———. Historic Cryptographic Records, 1949–1981. Entry A1 9032. (HCR)

———. Research Project Case Files Relating to the HARVEST System, 1957–1962. Entry A1 9037.

———. Security Policy and Direction Files, 1944–1969. Entry P14.

———. Studies on Cryptology, ca. 1952–ca. 1994, SRH series. Entry A1 9002.

National Cryptologic Museum Library. Ft. George G. Meade, MD. (NCM)

National Reconnaissance Office. Declassified Records. Web.

National Security Agency. Center for Cryptologic History. Web. (CCH)

———. Declassification and Transparency. Web. (NSAD)

———. NSA 60th Anniversary. Web. (NSA60)

TICOM Archive. Randy Rezabek. Web. (TArch)

UK National Archives. Government Code and Cypher School, HW series. Kew, UK. (TNA)

———. Signals Intelligence Passed to the Prime Minister, Messages and Correspondence. HW 1.

———. Personal Papers, Unofficial Histories. HW 3.
———. Directorate, Second World War Policy Papers. HW 14.
———. Cryptographic Studies. HW 25.

BOOKS, ARTICLES, AND OTHER PUBLICATIONS

Abelson, Harold, et al. *Keys Under Doormats: Mandating Insecurity by Requiring Government Access to All Data and Communications.* Computer Science and Artificial Intelligence Laboratory Technical Report, MIT-CSAIL-TR-2015-026. Cambridge, MA: Massachusetts Institute of Technology, 2015.

Aid, Matthew M. "US Humint and Comint in the Korean War: From the Approach of War to the Chinese Intervention." *Intelligence and National Security* 14, no. 4 (1999): 17–63.

———. "American Comint in the Korean War (Part II): From the Chinese Intervention to the Armistice." *Intelligence and National Security* 15, no. 1 (2000): 14–49.

———. "The Time of Troubles: The US National Security Agency in the Twenty-First Century." *Intelligence and National Security* 15, no. 3 (2000): 1–32.

———. "The National Security Agency in the Cold War." *Intelligence and National Security* 16, no. 1 (2001): 27–66.

———. *The Secret Sentry: The Untold Story of the National Security Agency.* New York: Bloomsbury, 2009.

Aldrich, Richard J. "GCHQ and Sigint in the Early Cold War, 1945–70." *Intelligence and National Security* 16, no. 1 (2001): 67–96.

Alvarez, David. "No Immunity: Signals Intelligence and European Neutrals, 1939–45." *Intelligence and National Security* 12, no. 2 (1997): 22–43.

———. "Behind Venona: American Signals Intelligence in the Early Cold War." *Intelligence and National Security* 14, no. 2 (1999): 179–86.

———. "American Signals Intelligence and the Cuban Missile Crisis." *Intelligence and National Security* 15, no. 1 (2000): 169–76.

———. *Secret Messages: Codebreaking and American Diplomacy, 1930–1945.* Lawrence: University Press of Kansas, 2000.

———. *Spies in the Vatican: Espionage and Intrigue from Napoleon to the Holocaust.* Lawrence: University Press of Kansas, 2002.

———. "Trying to Make the MAGIC Last: American Diplomatic Codebreaking in the Early Cold War." *Diplomatic History* 31, no. 5 (2007): 865–82.

Andrew, Christopher. *Her Majesty's Secret Service: The Making of the British Intelligence Community.* 1985. Reprint, New York: Penguin, 1987.

———. *The Defence of the Realm: The Authorised History of MI5.* London: Allen Lane, 2009.

Andrew, Christopher, and Oleg Gordievsky. *KGB: The Inside Story of Its Foreign Operations from Lenin to Gorbachev.* New York: HarperCollins, 1990.

Andrew, Christopher, and Vasili Mitrokhin. *The Sword and the Shield: The Mitrokhin Archive and the Secret History of the KGB.* New York: Basic Books, 1999.

Antal, Eugen, and Pavol Zajac. "Key Space and Period of Fialka M-125 Cipher Machine." *Cryptologia* 39, no. 2 (2015): 126–44.

Baker, William O. Papers. Mudd Manuscript Library, Princeton University, Princeton, NJ.

Baker Panel. *Scientific Judgments on Foreign Communications Intelligence.* Special Intelligence Panel of the Science Advisory Committee to the President, January 23, 1958. DNSArch.

Ball, Desmond. *Soviet Signals Intelligence (SIGINT).* Canberra Papers on Strategy and Defence No. 47. Canberra, Australia: Australia National University, 1989.

Bamford, James. *The Puzzle Palace: A Report on America's Most Secret Agency.* Boston: Houghton Mifflin, 1982.

"Before Super-Computers: NSA and Computer Development." *Cryptologic Almanac,* May–June 2002. NSAD.

Bennett, Ralph. "The Duty Officer, Hut 3." In *Codebreakers: The Inside Story of Bletchley Park,* edited by F. H. Hinsley and Alan Stripp. 1993. Reprint, Oxford: Oxford University Press, 1994.

Benson, Robert Louis. *The Venona Story.* Ft. George G. Meade, MD: Center for Cryptologic History, National Security Agency. n.d. CCH.

———. *A History of U.S. Communications Intelligence During World War II: Policy and Administration.* Ft. George G. Meade, MD: Center for Cryptologic History, National Security Agency, 1997. NSAD.

Benson, Robert Louis, and Cecil J. Phillips. *History of VENONA.* Ft. George G. Meade, MD: Center for Cryptologic History, National Security Agency, 1995. NCM.

Bernard, Richard L. *Electronic Intelligence (ELINT) at NSA.* Ft. George G. Meade, MD: Center for Cryptologic History, National Security Agency, 2009. CCH.

Bethell, Nicholas. *The Great Betrayal: The Untold Story of Kim Philby's Greatest Coup.* London: Hodder & Stoughton, 1984.

Bissell, Richard M. *Review of Selected NSA Cryptanalytic Efforts.* February 18, 1965. Document 4, *The National Security Agency Declassified,* National Security Archive Electronic Briefing Book No. 24. George Washington University, Washington, DC. Web.

Blight, James G., and David A. Welch. *On the Brink: Americans and Soviets Reexamine the Cuban Missile Crisis.* 1989. 2nd ed., New York: Noonday Press, 1990.

Blight, James G., Bruce J. Allyn, and David A. Welch. *Cuba on the Brink: Castro, the Missile Crisis, and the Soviet Collapse.* 1993. Revised ed., Lanham, MD: Rowman & Littlefield, 2002.

Boone, James V. "Applying Advanced Technology to Cryptologic Systems: Some Special Management Challenges." Paper presented at the 2013 Cryptologic History Symposium, Johns Hopkins Applied Physics Laboratory, Laurel, MD, October 17–18, 2013.

Bower, Tom. *The Red Web: MI6 and the KGB Master Coup.* 1989. Reprint, London: Mandarin, 1993.

———. *The Perfect English Spy: Sir Dick White and the Secret War, 1935–90.* 1995. Reprint, London: Mandarin, 1996.

Brooker, Graham, and Jairo Gomez. "Lev Termen's Great Seal Bug Analyzed." *IEEE A&E Systems Magazine,* November 2013, 4–11.

Brown, Anne S. *The National Security Agency Scientific Advisory Board, 1952–1963.* Ft. George G. Meade, MD: NSA Historian, Office of Central Reference, 1965. Government Attic, Web.

Brownell Committee Report. June 13, 1952. SRH-123, Studies on Cryptology, RG 457, NARA.

Budiansky, Stephen. "A Tribute to Cecil Phillips—and Arlington Hall's 'Meritocracy.' " *Cryptologia* 23, no. 2 (1999): 97–107.

———. *Battle of Wits: The Complete Story of Codebreaking in World War II.* New York: Free Press, 2000.

———. *Air Power: The Men, Machines, and Ideas that Revolutionized War, from Kitty Hawk to Gulf War II.* New York: Viking, 2004.

———. "Colossus, Codebreaking, and the Digital Age." In *Colossus: The Secrets of Bletchley Park's Codebreaking Computers,* edited by B. Jack Copeland. Oxford: Oxford University Press, 2006.

Buffham, Benson. Oral History. NSA-OH-51-99. National Security Agency, 1979. NSAD.

Burke, Colin B. *It Wasn't All Magic: The Early Struggle to Automate Cryptanalysis, 1930s–1980s.* Ft. George G. Meade, MD: Center for Cryptologic History, National Security Agency, 2002. NSAD.

———, ed. "From the Archives: The Last Bombe Run, 1955." *Cryptologia* 32, no. 3 (2008): 277–78.

Burns, Thomas L. *The Origins of the National Security Agency, 1940–1952.* Ft.

George G. Meade, MD: Center for Cryptologic History, National Security Agency, 1990. NSAD.
———. *The Quest for Centralization and the Establishment of NSA: 1940–1952.* Ft. George G. Meade, MD: Center for Cryptologic History, National Security Agency, 2005. NSAD.
Callimahos, Lambros D. "Introduction to Traffic Analysis." *NSA Technical Journal,* April 1958, 1–11. NSAD.
———. "The Legendary William F. Friedman." *Cryptologic Spectrum,* winter 1974, 6–17. NSAD.
Campaigne, Howard. Oral History. NSA-OH-20-83. National Security Agency, 1983. NSAD.
Canine, Ralph J. Oral History. Part I, OH-2012-80. Part II, OH-2012-81. Part III, OH-2012-82. National Security Agency, ca. 1965. NSAD.
———. Papers. National Cryptologic Museum Library, Ft. George G. Meade, MD.
Central Intelligence Agency. *The Berlin Tunnel Operation, 1952–1956.* Clandestine Services History. Washington, DC: 1967. CIAL.
Churchhouse, Robert. *Codes and Ciphers: Julius Caesar, the Enigma, and the Internet.* Cambridge: Cambridge University Press, 2002.
"The Civilianization of Harrogate." *Cryptologic Spectrum,* summer 1970, 8–16. NSAD.
Clark, Ronald W. *The Man Who Broke Purple: The Life of the World's Greatest Cryptologist, William F. Friedman.* London: Weidenfeld & Nicolson, 1977.
Clay, Lucius D. *Decision in Germany.* Garden City, NY: Doubleday, 1950.
Coffey, Donald F. Oral History. NSA-OH-23-82. National Security Agency, 1982. NSAD.
Colley, David. "Shadow Warriors: Intelligence Operatives Waged Clandestine Cold War." *VFW Magazine,* September 1997, 24–30.
Collins, Charles P. *The History of SIGINT in the Central Intelligence Agency, 1947–1970.* Washington, DC: CIA History Office, 1971. CIAL.
"COMINT Satellites—A Space Problem." *NSA Technical Journal,* October 1959, 39–48. NSAD.
Cook, Earle F. Oral History. National Security Agency, 1982. NCM.
Copeland, B. Jack. "Turingery." In *Colossus: The Secrets of Bletchley Park's Codebreaking Computers,* edited by B. Jack Copeland. Oxford: Oxford University Press, 2006.
Craig, Bruce. "A Matter of Espionage: Alger Hiss, Harry Dexter White, and Igor Gouzenko—The Canadian Connection Reassessed." *Intelligence and National Security* 15, no. 2 (2000): 211–24.

Croner, Andrew. "Snake in the Grass?: Section 798 of the Espionage Act and Its Constitutionality as Applied to the Press." *George Washington Law Review* 77, no. 3 (2009): 766–98.

Davis, Carol B. *Candle in the Dark: COMINT and Soviet Industrial Secrets, 1946–1956.* Ft. George G. Meade, MD: Center for Cryptologic History, National Security Agency, 2008. NCM.

Day, Dwayne A. "Ferrets Above: American Signals Intelligence Satellites During the 1960s." *International Journal of Intelligence and Counterintelligence* 17, no. 3 (2004): 449–67.

Deane, John R. *The Strange Alliance: The Story of Our Efforts at Wartime Cooperation with Russia.* 1946. Reprint, Bloomington: Indiana University Press, 1973.

Deavours, Cipher A., and Louis Kruh. *Machine Cryptography and Modern Cryptanalysis.* Norwood, MA: Artech, 1985.

Denham, Hugh. "In Memoriam: Conel Hugh O'Donel Alexander." *Cryptologic Spectrum,* summer 1974, 30–32. NSAD.

Drea, Edward J. *MacArthur's ULTRA: Codebreaking and the War Against Japan, 1942–1945.* Lawrence: University Press of Kansas, 1992.

Dumey, Arnold. Oral History. Charles Babbage Institute, Center for the History of Information Processing, University of Minnesota, 1984.

Eachus, Joseph, et al. "Growing Up with Computers at NSA: A Panel Discussion." *NSA Technical Journal,* Special Issue, 1972, 3–14. NSAD.

East, Don C. "A History of U.S. Navy Fleet Reconnaissance: Part I, The Pacific and VQ-1." *The Hook: Journal of Carrier Aviation,* spring 1987.

———. "A History of U.S. Navy Fleet Reconnaissance: Part II, The European Theater and VQ-2." *The Hook: Journal of Carrier Aviation,* spring 1987.

Easter, David. "GCHQ and British External Policy in the 1960s." *Intelligence and National Security* 23, no. 5 (2008): 681–706.

"Engineering the Berlin Tunnel: Turning a Cold War Scheme into Reality." *Studies in Intelligence* 48, no. 2 (2004). CIAL.

Erskine, Ralph. "The Holden Agreement on Naval Sigint: The First BRUSA?" *Intelligence and National Security* 14, no. 2 (1999): 187–97.

Exinterne, Anne [pseud.]. "A Long Hard Look at the Intern Program." *Cryptolog,* September 1974, 16–18. NSAD.

———. "A Long Hard Look at the Intern Program, Part Three: Motivation and Morale." *Cryptolog,* November 1974, 15–17. NSAD.

Ferris, John. "The British 'Enigma': Britain, Signals Security, and Cipher Machines, 1906–1953." In *Intelligence and Strategy: Selected Essays.* New York: Routledge, 2005.

"Finding a Home for AFSA 1949–1952; or, Why We Don't Have Bluegrass in Front of the Building and Mint Juleps in the Dining Parlour." *Cryptolog,* April 1985, 1–3. NSAD.

Frahm, Jill. *So Power Can Be Brought into Play: SIGINT and the Pusan Perimeter.* Ft. George G. Meade, MD: Center for Cryptologic History, National Security Agency, 2000. CCH.

Friedman, William F. *The Index of Coincidence and Its Applications in Cryptology.* Department of Ciphers, Riverbank Laboratories, Publication 22. Geneva, IL, 1922.

"From Chaos Born: General Canine's First Charge to the NSA Workforce." *Cryptologic Quarterly,* summer 1987, 23–29. NSAD.

Futtrell, Robert F. "A Case Study: USAF Intelligence in the Korean War." In *The Intelligence Revolution: A Historical Perspective,* edited by Walter T. Hitchcock. Proceedings of the Thirteenth Military History Symposium, U.S. Air Force Academy, Colorado Springs, CO, October 12–14, 1988. Washington, DC: Office of Air Force History, United States Air Force, 1991.

Gaddis, John Lewis. *The Cold War: A New History.* New York: Penguin, 2005.

———. *George F. Kennan: An American Life.* New York: Penguin, 2011.

Gallo, Jules. "NSA Signal Collection Equipment and Systems: The Early Years—Magnetic Tape Recorders." *Cryptologic Quarterly,* spring 1996, 45–62. NSAD.

Garthoff, Douglas F. *Directors of Central Intelligence as Leaders of the U.S. Intelligence Community, 1946–2005.* Washington, DC: Potomac Books, 2007.

"Glimpses of a Man: The Life of Ralph J. Canine." *Cryptologic Quarterly,* summer 1987, 31–38. NSAD.

Goldman, Eric F. *The Crucial Decade—and After: America, 1945–1960.* New York: Vintage, 1960.

Goldstine, Herman H. *The Computer from Pascal to von Neumann.* Princeton, NJ: Princeton University Press, 1972.

Good, I. J. "Studies in the History of Probability and Statistics. XXXVII. A. M. Turing's Statistical Work in World War II." *Biometrika* 66, no. 2 (1979): 393–96.

———. "Enigma and Fish." In *Codebreakers: The Inside Story of Bletchley Park,* edited by F. H. Hinsley and Alan Stripp. 1993. Reprint, Oxford: Oxford University Press, 1994.

Gouzenko, Igor. *The Iron Curtain.* New York: Dutton, 1948.

Gray, George. "Engineering Research Associates and the Atlas Computer (UNIVAC 1101)." *Unisys History Newsletter,* June 1999.

Greenwald, Glenn. *No Place to Hide: Edward Snowden, the NSA, and the U.S. Surveillance State.* New York: Metropolitan, 2014.

Gurin, Jacob. "Ralph J. Canine." *Cryptologic Spectrum,* fall 1969, 7–10. NSAD.

———. "Let's Not Forget Our Cryptologic Mission." *Cryptolog,* February 1979, 16–17. NSAD.

Hagelin, Boris C. W. "The Story of Hagelin Cryptos." *Cryptologia* 18, no. 3 (1994): 204–42.

Halberstam, David. *The Best and the Brightest.* New York: Random House, 1972.

———. *The Coldest Winter: America and the Korean War.* 2007. Reprint, New York: Hachette, 2008.

Hannah, Theodore M. "Frank B. Rowlett." *Cryptologic Spectrum,* spring 1981, 4–21. NSAD.

Hanyok, Robert J. "Skunks, Bogies, Silent Hounds, and the Flying Fish: The Gulf of Tonkin Mystery, 2–4 August, 1964." *Cryptologic Quarterly,* winter 2000–spring 2001, 1–55. NSAD.

———. *Spartans in Darkness: American SIGINT and the Indochina War, 1945–1975.* Ft. George G. Meade, MD: Center for Cryptologic History, National Security Agency, 2002. NSAD.

Hatch, David A. *In Memoriam Dr. Louis Tordella.* Ft. George G. Meade, MD: Center for Cryptologic History, National Security Agency. NSAD.

———. "The Berlin Tunnel, Part II: The Rivals." *Cryptologic Almanac,* March–April 2002. NSAD.

———. "Juanita Moody." *Cryptologic Almanac,* March–April 2002. NSAD.

Hatch, David A., and Robert Louis Benson. *The Korean War: The SIGINT Background.* Ft. George G. Meade, MD: Center for Cryptologic History, National Security Agency, 2000. CCH.

Hayden, Michael V. "Overseeing an Era of Change: An Interview with NSA Director Lt. Gen. Michael V. Hayden." *Studies in Intelligence* 44, no. 1 (2000). CIAL.

Haynes, John Earl, and Harvey Klehr. *Venona: Decoding Soviet Espionage in America.* New Haven, CT: Yale University Press, 1999.

———. "Alexander Vassiliev's Notebooks and the Documentation of Soviet Intelligence Activities in the United States During the Stalin Era." *Journal of Cold War Studies* 11, no. 3 (2009): 6–25.

Haynes, John Earl, Harvey Klehr, and Alexander Vassiliev. *Spies: The Rise and Fall of the KGB in America.* New Haven, CT: Yale University Press, 2009.

Heath, Laura J. *An Analysis of the Systematic Security Weaknesses of the U.S. Navy Fleet Broadcasting System, 1967–1974, as Exploited by CWO John Walker.* Master's Thesis, U.S. Army Command and General Staff College, Ft. Leavenworth, KS, 2005.

Hersh, Seymour M. *The Price of Power: Kissinger in the Nixon White House.* New York: Summit Books, 1983.

Hess, Sigurd. "The Clandestine Operations of Hans Helmut Klose and the British Baltic Fishery Protection Service (BBFPS), 1949–1956." *Journal of Intelligence History* 1, no. 2 (2001): 169–78.

Hinsley, F. H., et al. *British Intelligence in the Second World War.* 5 vols. New York: Cambridge University Press, 1979–90.

Hodges, Andrew. *Alan Turing: The Enigma.* New York: Simon & Schuster, 1983.

Höhne, Heinz, and Hermann Zolling. *The General Was a Spy: The Truth About General Gehlen and His Spy Ring.* New York: Coward, McCann & Geoghegan, 1972.

Hove, Mark T. *History of the Bureau of Diplomatic Security of the United States Department of State.* Washington, DC: Department of State, 2011.

Howe, George F. "A History of US Civilians in Field Comint Operations, 1953–1960. Part I." *Cryptologic Spectrum,* spring 1973, 5–9. NSAD.

———. *American Signal Intelligence in Northwest Africa and Western Europe.* Ft. George G. Meade, MD: Center for Cryptologic History, National Security Agency, 2010. CCH.

Hunter, Robert W. *Spy Hunter: Inside the FBI Investigation of the Walker Espionage Case.* Annapolis, MD: Naval Institute Press, 1999.

Johnson, Thomas R. "The Move, or How NSA Came to Fort Meade." *Cryptologic Quarterly,* summer 1995, 93–99. NSAD.

———. *American Cryptology During the Cold War, 1945–1989.* 4 vols. Ft. George G. Meade, MD: Center for Cryptologic History, National Security Agency, 1995–99. NSAD.

———. "American Cryptology During the Korean War." *Studies in Intelligence* 45, no. 3 (2001). CIAL.

———. "Joseph N. Wenger." *Cryptologic Almanac,* March–April 2002. NSAD.

———. "A Cryptologist Encounters the Human Side of Intelligence." *Studies in Intelligence,* 2007. CIAL.

———. "In War's Aftermath: Cryptology and the Early Cold War." Paper presented at the 2013 Cryptologic History Symposium, Johns Hopkins Applied Physics Laboratory, Laurel, MD, October 17–18, 2013.

Johnson, Thomas R., and David A. Hatch. *NSA and the Cuban Missile Crisis.* Ft. George G. Meade, MD: Center for Cryptologic History, National Security Agency, 1998. CCH.

Kahn, David. *The Codebreakers: The Story of Secret Writing.* 1967. Revised ed., New York: Scribner, 1996.

———. "Big Ear or Big Brother?" *New York Times Magazine,* May 16, 1976.

———. "Soviet COMINT in the Cold War." *Cryptologia* 22, no. 1 (1998): 1–24.

Kirby, Oliver R. "The Origins of the Soviet Problem: A Personal View." *Cryptologic Quarterly,* winter 1992, 51–58. NSAD.

———. Oral History. NSA-OH-20-93. National Security Agency, 1993. NSAD.

Klein, Melville. *Securing Record Communications: The TSEC/KW-26.* Ft. George G. Meade, MD: Center for Cryptologic History, National Security Agency, 2003. CCH.

Lamphere, Robert J. *The FBI-KGB War: A Special Agent's Story.* New York: Random House, 1986.

Lavington, Simon. "In the Footsteps of Colossus: A Description of Oedipus." *IEEE Annals of the History of Computing,* April–June 2006, 44–55.

Levenson, Arthur J. Oral History. NSA-OH-40-80. National Security Agency, 1980. NSAD.

Lewis, Frank W. "Something May Rub Off!" *NSA Technical Journal,* winter 1965, 1–5. NSAD.

Lilienthal, David E. *The Journals of David E. Lilienthal.* Vol. 2, *The Atomic Energy Years, 1945–1950.* New York: Harper & Row, 1964.

Lobban, Iain. "GCHQ and Turing's Legacy." Speech delivered at Leeds University, October 4, 2012. GCHQ. Web.

McConnell, J. M. "The Future of SIGINT: Opportunities and Challenges in the Information Age." *Defense Intelligence Journal* 9, no. 2 (2000): 39–49.

McCullough, David. *Truman.* New York: Simon & Schuster, 1992.

MacDonald, Dwight. "The Lie-Detector Era II. 'It's a Lot Easier, and It Don't Leave Marks.'" *Reporter,* June 22, 1954, 22–29.

McDonald, Robert A., and Sharon K. Moreno. *Grab and Poppy: America's Early ELINT Satellites.* Chantilly, VA: Center for the Study of National Reconnaissance, National Reconnaissance Office, 2005.

Macintyre, Ben. *A Spy Among Friends: Kim Philby and the Great Betrayal.* New York: Crown, 2014.

McKnight, David. "The Moscow–Canberra Cables: How Soviet Intelligence Obtained British Secrets Through the Back Door." *Intelligence and National Security* 13, no. 2 (1998): 159–70.

Manchester, William. *The Glory and the Dream: A Narrative History of America, 1932–1972.* 1974. Reprint, New York: Bantam, 1990.

Maneki, Sharon A. "Remembering the Lessons of the Vietnam War." *Cryptologic Quarterly,* spring–summer 2004, 19–36. NSAD.

———. *Learning from the Enemy: The GUNMAN Project.* Ft. George G. Meade, MD: Center for Cryptologic History, National Security Agency, 2012. NSAD.

Melton, H. Keith. *The Ultimate Spy Book*. New York: DK Publishing, 1996.

Meyer, J. A. "Computers—The Wailing Wall." *NSA Technical Journal,* October 1956, 69–90. NSAD.

Moïse, Edwin E. *Tonkin Gulf and the Escalation of the Vietnam War*. Chapel Hill: University of North Carolina Press, 1996.

Montefiore, Simon Sebag. *Stalin: The Court of the Red Tsar.* New York: Random House, 2003.

Mowry, David P. "Betrayers of the Trust: Joseph Sidney Petersen." *Cryptologic Almanac,* March–April 2002. NSAD.

Murphy, David E., Sergei A. Kondrashev, and George Bailey. *Battleground Berlin: CIA vs. KGB in the Cold War.* New Haven, CT: Yale University Press, 1997.

National Research Council. *The Polygraph and Lie Detection*. Washington, DC: National Academies Press, 2003. Web.

National Security Agency and Central Intelligence Agency. *Venona: Soviet Espionage and the American Response, 1939–1957.* Washington, DC, 1996. CIAL.

National Security Archive. *The Secret Sentry Declassified.* Electronic Briefing Book No. 278. George Washington University, Washington, DC. Web.

———. *The CIA and Signals Intelligence*. Electronic Briefing Book No. 506. Web.

Newton, Robert E. *The Capture of the USS* Pueblo *and Its Effect on SIGINT Operations*. Ft. George G. Meade, MD: Center for Cryptologic History, National Security Agency, 1992. NSAD.

Nikitin, Pavel. "Leon Theremin (Lev Termen)." *IEEE Antennas and Propagation Magazine* 54, no. 5 (2012): 252–57.

O'Hara, John. "Analog Recording of Telemetry Signals Prior to the Advent of Digital Processing." Paper presented at the 2013 Cryptologic History Symposium, Johns Hopkins Applied Physics Laboratory, Laurel, MD, October 17–18, 2013.

On Watch: Profiles from the National Security Agency's Past 40 Years. Ft. George G. Meade, MD: National Cryptologic School, National Security Agency. NSAD.

Operation REGAL: The Berlin Tunnel. Ft. George G. Meade, MD: Office of Cryptologic Archives and History, National Security Agency, 1988. NSAD.

Parker, John E. Oral History. Charles Babbage Institute, Center for the History of Information Processing, University of Minnesota, 1986. Web.

"A Partial History of ELINT at NSA." *Cryptologic Quarterly,* spring–summer 2004, 55–64. NSAD.

Pendergrass, James T. Oral History. Charles Babbage Institute, Center for the History of Information Processing, University of Minnesota, 1985. Web.

Peterson, Michael L. "Maybe You Had to Be There: SIGINT on Thirteen Soviet Shootdowns of U.S. Reconnaissance Aircraft." *Cryptologic Quarterly,* summer 1993, 1–44. NSAD.

———. "Before BOURBON: American and British COMINT Efforts Against Russia and the Soviet Union Before 1945." *Cryptologic Quarterly,* fall–winter 1993, 1–20. NSAD.

———. "Beyond BOURBON—1948: The Fourth Year of Allied Collaborative COMINT Effort Against the Soviet Union." *Cryptologic Quarterly,* spring 1995, 1–59. Interagency Security Classification Appeals Panel Released Decisions. Web.

Pipes, Richard. *The Russian Revolution.* New York: Knopf, 1990.

———. *Russia Under the Bolshevik Regime.* New York: Knopf, 1994.

PURPLE DRAGON: The Origin and Development of the United States OPSEC Program. Ft. George G. Meade, MD: Center for Cryptologic History, National Security Agency, 1993. NSAD.

Rezabek, Randy. "TICOM: The Last Great Secret of World War II." *Intelligence and National Security* 27, no. 4 (2012): 513–30.

———. "TICOM and the Search for OKW/Chi." *Cryptologia* 37, no. 2 (2013): 139–53.

———. "The Russian Fish with Caviar." *Cryptologia* 38, no. 1 (2014): 61–76.

Rhodes, Richard. *Dark Sun: The Making of the Hydrogen Bomb.* 1995. Reprint, New York: Simon & Schuster, 2005.

Robertson Committee. *The Potentialities of Comint for Strategic Warning.* Special Study Group of the NSA Scientific Advisory Board, October 20, 1953. Government Attic, Web. (Another partially released version of this report, with different passages redacted, is in Friedman Documents, NSAD.)

Rosenberg, David Alan. "The Origins of Overkill: Nuclear Weapons and American Strategy, 1945–1960." *International Security* 7, no. 4 (1983): 3–71.

Rowlett, Frank B. Oral History. National Security Agency, 1974, 1976. NCM.

———. *The Story of MAGIC: Memoirs of an American Cryptologic Pioneer.* Laguna Hills, CA: Aegean Park Press, 1998.

Schindler, John R. *A Dangerous Business: The U.S. Navy and National Reconnaissance During the Cold War.* Ft. George G. Meade, MD: Center for Cryptologic History, National Security Agency. CCH.

Schlesinger, Stephen. "Cryptanalysis for Peacetime: Codebreaking and the Birth and Structure of the United Nations." *Cryptologia* 19, no. 3 (1995): 217–35.

Shannon, C. E. "Prediction and Entropy of Printed English." *Bell System Technical Journal* 30, no. 1 (1951): 50–64.

Sheehan, Neil. *A Bright Shining Lie: John Paul Vann and America in Vietnam.* New York: Random House, 1998.

Sheldon, Rose Mary. *The Friedman Collection: An Analytic Guide.* George C. Marshall Research Library, Lexington, VA.

Sinkov, Abraham. Oral History. NSA-OH-02-79 through 04-79. National Security Agency, 1979. NSAD.

Smith, Michael. *New Cloak, Old Dagger: How Britain's Spies Came In from the Cold.* London: Victor Gollancz, 1996.

———. *Station X: The Codebreakers of Bletchley Park.* London: Channel 4 Books, 1998.

Snider, Britt L. "Unlucky SHAMROCK: Recollections from the Church Committee's Investigation of NSA." *Studies in Intelligence,* winter 1999–2000. CIAL.

Snyder, Samuel S. *History of NSA General-Purpose Electronic Digital Computers.* Washington, DC: Department of Defense, 1964.

———. "Influence of US Cryptologic Organizations on the Digital Computer Industry." *Cryptologic Quarterly,* fall 1987–spring 1988, 65–82. NSAD.

Stamp, Mark, and Wing On Chan. "SIGABA: Cryptanalysis of the Full Keyspace." *Cryptologia* 31, no. 3 (2007): 201–22.

Steury, Donald P., ed. *On the Front Lines of the Cold War: Documents on the Intelligence War in Berlin, 1946 to 1961.* Washington, DC: Center for the Study of Intelligence, Central Intelligence Agency, 1999. CIAL.

The Suez Crisis: A Brief Comint History. Ft. George G. Meade, MD: Office of Cryptologic Archives and History, National Security Agency, 1988. NSAD.

Taubman, William. *Khrushchev: The Man and His Era.* New York: Norton, 2003.

"TEABALL: Some Personal Observations of SIGINT at War." *Cryptologic Quarterly,* winter 1991, 91–97. NSAD.

"TEMPEST: A Signal Problem." *Cryptologic Spectrum,* summer 1972, 26–30. NSAD.

Thomas, Evan. *The Very Best Men: Four Who Dared; The Early Years of the CIA.* New York: Simon & Schuster, 1996.

Tomash, Erwin, and Arnold A. Cohen. "The Birth of an ERA: Engineering Research Associates, Inc., 1946–1955." *Annals of the History of Computing* 1, no. 2 (1979): 83–97.

Treadwell, Mattie E. *The Women's Army Corps.* Washington, DC: Office of the Chief of Military History, Department of the Army, 1954.

Turner, Stansfield. *Secrecy and Democracy: The CIA in Transition.* Boston: Houghton Mifflin, 1985.

"The 2,000-Year-Old Transcriber." *Cryptolog,* October 1979, 16–19. NSAD.

U.S. Army. "Forty One and Strong: Arlington Hall Station." *Cryptologia* 9, no. 4 (1985): 306–10.

U.S. Congress. House Un-American Activities Committee. *Security Practices in the National Security Agency (Defection of Bernon F. Mitchell and William H. Martin).* 87th Cong., 2nd Sess. Washington, DC: GPO, 1962.

U.S. Senate. *Huston Plan. Hearings Before the Senate Committee to Study Governmental Operations with Respect to Intelligence Activities, Volume 2.* 94th Cong., 1st Sess. Washington, DC: GPO, 1976.

———. *The National Security Agency and Fourth Amendment Rights. Hearings Before the Senate Committee to Study Governmental Operations with Respect to Intelligence Activities, Volume 5.* 94th Cong., 1st Sess. Washington, DC: GPO, 1976.

———. *Supplementary Detailed Staff Reports on Intelligence Activities and the Rights of Americans. Final Report of the Select Committee to Study Governmental Operations with Respect to Intelligence Activities, Book III.* Report 94-755. 94th Cong., 2nd Sess. Washington, DC: GPO, 1976.

Vanderpool, Guy R. "COMINT and the PRC Intervention in the Korean War." *Cryptologic Quarterly,* summer 1996, 1–26. NSAD.

Wagoner, H. D. *Space Surveillance Sigint Program.* Ft. George G. Meade, Md.: History and Publications Staff, National Security Agency, 1980. NSAD.

Wiebes, Cees. "Operation 'Piet': The Joseph Sidney Petersen Jr. Spy Case, a Dutch 'Mole' Inside the National Security Agency." *Intelligence and National Security* 23, no. 4 (2008): 488–535.

Wigglesworth, Donald C. "The Cuban Missile Crisis: A SIGINT Perspective." *Cryptologic Quarterly,* spring 1994, 77–97. NSAD.

Wilcox, Jennifer. *Solving the Enigma: History of the Cryptanalytic Bombe.* Revised ed. Ft. George G. Meade, MD: Center for Cryptologic History, National Security Agency, 2004.

Williams, Jeannette. *The Invisible Cryptologists: African-Americans, WWII to 1956.* Ft. George G. Meade, MD: Center for Cryptologic History, National Security Agency, 2001. NSAD.

Zaslow, Milton. Oral History. September 14, 1993. Gulf of Tonkin Documents, National Security Agency. NSAD.

INDEX

Page numbers in *italic* refer to illustrations

Illustration Credits

All photographic images are courtesy of NSA with the exception of those listed below.

Arlington Hall: Courtesy of Arlington Public Library
B-211 cipher machine: Crypto AG
C-47s at Tempelhof Airport: U.S. Air Force
Truman at Wake Island: Department of Defense
U.S. Marines at Chosin: U.S. Navy
Russian Fialka cipher machine: Mark J. Blair
Cavity bug in the Great Seal: U.S. State Department
Khrushchev: CIA
Checkpoint Charlie: U.S. Information Agency
USS *Pueblo:* U.S. Navy
Khe Sanh: U.S. Air Force
"Rotor reader": FBI
Lech Wałęsa: Government of Poland
Berlin Wall: Department of Defense

A Note About the Author

Stephen Budiansky was the national security correspondent and foreign editor of *U.S. News & World Report,* Washington editor of *Nature,* and editor of *World War II* magazine. He is the author of six books of military and intelligence history, including *Blackett's War,* a *Washington Post* Notable Book. He has served as a Congressional Fellow and frequently lectures on intelligence and military history, and his articles have appeared in *The New York Times, The Washington Post, The Wall Street Journal, The Atlantic, The Economist,* and other publications. He is a member of the editorial board of *Cryptologia,* the leading academic journal of codes, codebreaking, and cryptologic history.

A Note on the Type

This book was set in Minion, a typeface produced by the Adobe Corporation specifically for the Macintosh personal computer and released in 1990. Designed by Robert Slimbach, Minion combines the classic characteristics of old-style faces with the full complement of weights required for modern typesetting.

Typeset by North Market Street Graphics,
Lancaster, Pennsylvania

Printed and bound by Berryville Graphics, Berryville, Virginia

Designed by Betty Lew